P9-AQP-882

Methods in Enzymology

Volume 386
IMAGING IN BIOLOGICAL RESEARCH
Part B

METHODS IN ENZYMOLOGY

EDITORS-IN-CHIEF

John N. Abelson   Melvin I. Simon

DIVISION OF BIOLOGY
CALIFORNIA INSTITUTE OF TECHNOLOGY
PASADENA, CALIFORNIA

FOUNDING EDITORS

Sidney P. Colowick and Nathan O. Kaplan

QP601
M49
V. 386

Methods in Enzymology

Volume 386

Imaging in Biological Research

Part B

EDITED BY

P. Michael Conn

OREGON NATIONAL PRIMATE RESEARCH CENTER
OREGON HEALTH AND SCIENCE UNIVERSITY
BEAVERTON, OREGON

ELSEVIER
ACADEMIC
PRESS

AMSTERDAM • BOSTON • HEIDELBERG • LONDON
NEW YORK • OXFORD • PARIS • SAN DIEGO
SAN FRANCISCO • SINGAPORE • SYDNEY • TOKYO
Academic Press is an imprint of Elsevier

NO LONGER THE PROPERTY
OF THE
UNIVERSITY OF R. I. LIBRARY

Elsevier Academic Press
525 B Street, Suite 1900, San Diego, California 92101-4495, USA
84 Theobald's Road, London WC1X 8RR, UK

This book is printed on acid-free paper. ∞

Copyright © 2004, Elsevier Inc. All Rights Reserved.

No part of this publication may be reproduced or transmitted in any form or by any
means, electronic or mechanical, including photocopy, recording, or any information
storage and retrieval system, without permission in writing from the Publisher.

The appearance of the code at the bottom of the first page of a chapter in this book
indicates the Publisher's consent that copies of the chapter may be made for
personal or internal use of specific clients. This consent is given on the condition,
however, that the copier pay the stated per copy fee through the Copyright Clearance
Center, Inc. (www.copyright.com), for copying beyond that permitted by
Sections 107 or 108 of the U.S. Copyright Law. This consent does not extend to
other kinds of copying, such as copying for general distribution, for advertising
or promotional purposes, for creating new collective works, or for resale.
Copy fees for pre-2004 chapters are as shown on the title pages. If no fee code
appears on the title page, the copy fee is the same as for current chapters.
0076-6879/2004 $35.00

Permissions may be sought directly from Elsevier's Science & Technology Right
Department in Oxford, UK: phone: (+44) 1865 843830, fax: (+44) 1865 853333,
E-mail: permissions@elsevier.com.uk. You may also complete your request on-line
via the Elsevier homepage (http://elsevier.com), by selecting
"Customer Support" and then "Obtaining Permissions."

For all information on all Academic Press publications
visit our Web site at www.academicpress.com

ISBN: 0-12-182791-7

PRINTED IN THE UNITED STATES OF AMERICA
04 05 06 07 08 9 8 7 6 5 4 3 2 1

Table of Contents

Section I. Disease Models

Section II. Preparation of Materials

Section III. General Methods

Contributors to Volume 386

Article numbers are in parentheses and following the names of contributors.
Affiliations listed are current.

ELLEN ACKERSTAFF (1), *Department of Radiology, The Johns Hopkins University School of Medicine, Baltimore, Maryland 21205*

SHIVANI AGARWAL (2), *Department of Neurological Surgery, Columbia University, New York, New York 10032*

NOAM ALPERIN (16), *Physiologic Imaging and Modeling Lab, Department of Radiology, The University of Illinois at Chicago, Chicago, Illinois 60612*

CAROLYN J. ANDERSON (11), *Mallinckrodt Institute of Radiology, Washington University School of Medicine, St. Louis, Missouri 63110*

ALI S. ARBAB (13), *Laboratory of Diagnostic Radiology Research, Experimental Neuroimaging Section, National Institutes of Health, Bethesda, Maryland 20892*

DMITRI ARTEMOV (1), *Department of Radiology, The Johns Hopkins University School of Medicine, Baltimore, Maryland 21205*

MARK D. BEDNARSKI (10), *Lucas MRI Research Center, Department of Radiology, Stanford University, Stanford, California, 94305-5488*

JIMMY D. BELL (14), *Molecular Imaging Group, MRC Clinical Sciences Centre, Imperial College London, Hammersmith Hospital, London W12 0HS, United Kingdom*

KISHORE K. BHAKOO (14), *Stem Cell Imaging Group, MRC Clinical Sciences Centre, Faculty of Medicine, Imperial College London, Hammersmith Hospital, London W12 0HS, United Kingdom*

ZAVER M. BHUJWALLA (1), *Department of Radiology, The Johns Hopkins University School of Medicine, Baltimore, Maryland 21205*

DAVID A. BLUEMKE (5), *Department of Radiology, Johns Hopkins Hospital, The Johns Hopkins University School of Medicine, Baltimore, Maryland 21287*

JEFF W. M. BULTE (13), *Department of Radiology and Institute for Cell Engineering, The Johns Hopkins University School of Medicine, Baltimore, Maryland 21205-2195*

E. SANDER CONNOLLY, JR. (2), *Department of Neurological Surgery, Columbia University, New York, New York 10032*

I. JANE COX (14), *Imaging Sciences Department, MRC Clinical Sciences Centre, Imperial College London, Hammersmith Hospital, London W12 0HS, United Kingdom*

ANTHONY L. D'AMBROSIO (2), *Department of Neurological Surgery, Columbia University, New York, New York 10032-2699*

MILIND Y. DESAI (5), *Division of Cardiology, Department of Internal Medicine, The Johns Hopkins University School of Medicine, Baltimore, Maryland 21287*

T. Douglas (13), *Department of Chemistry and Biochemistry, Montana State University, Bozeman, Montana 59717*

J. A. Frank (13), *Laboratory of Diagnostic Radiology Research, Experimental Neuroimaging Section, National Institutes of Health, Bethesda, Maryland 20892*

Barjor Gimi (1), *Department of Radiology, The Johns Hopkins University School of Medicine, Baltimore, Maryland 21205*

Kristine Glunde (1), *Department of Radiology, The Johns Hopkins University School of Medicine, Baltimore, Maryland 21205*

Yueqing Gu (17), *Biomedical Engineering Program, The University of Texas at Arlington, Arlington, Texas 76019*

Samira Guccione (10), *Lucas MRI Research Center, Department of Radiology, Stanford University School of Medicine, Stanford, California 94305-5488*

Chien Ho (3), *Pittsburgh NMR Center for Biomedical Research, Department of Biological Sciences, Carnegie Mellon University, Pittsburgh, Pennsylvania 15213*

Lan Jiang (18), *The University of Texas Southwestern Medical Center at Dallas, Dallas, Texas 75390*

Jae G. Kim (17), *Biomedical Engineering Program, The University of Texas at Arlington, Arlington, Texas 76019*

Ryan G. King (2), *Department of Neurological Surgery, Columbia University, New York, New York 10032*

Kinc C. P. Li (10), *Department of Radiology, Stanford University, Stanford, CA 94305-5488*

João A. Lima (5), *Division of Cardiology, Department of Internal Medicine, The Johns Hopkins University School of Medicine, Baltimore, Maryland 21287*

Pengnian Charles Lin (4), *Department of Radiation Oncology, Vanderbilt University Medical Center, Nashville, Tennessee 37232*

Hanli Liu (17), *Biomedical Engineering Program, The University of Texas at Arlington, Arlington, Texas 76019*

Jia Lu (8), *Defence Medical Research Institute, Singapore 117510*

Mark F. Lythgoe (6), *RCS Unit of Biophysics, Institute of Child Health, University College London, London WC1N 1EH, United Kingdom*

William J. Mack (2), *Department of Neurological Surgery, Columbia University, New York, New York 10032*

Pasquina Marzola (7), *Anatomy and Histology Section, Department of Morphological and Biomedical Sciences, University of Verona, Verona 37134 Italy*

Ralph P. Mason (17, 18), *Department of Radiology, The University of Texas Southwestern Medical Center at Dallas, Dallas, Texas 75390-9058*

J. Mocco (2), *Department of Neurological Surgery, Columbia University, New York, New York 10032*

Shabbir Moochhala (8), *Defence Medical Research Institute, Singapore 117510*

Arvind P. Pathak (1), *Department of Radiology, The Johns Hopkins University School of Medicine, Baltimore, Maryland 21205*

Richard L. Roberts (4), *Department of Pathology, Vanderbilt-Ingram Cancer Center, Vanderbilt University Medical Center, Nashville, Tennessee 37232*

KAZUYA SATO (3), *Pittsburgh NMR Center for Biomedical Research, Department of Biological Sciences, Carnegie Mellon University, Pittsburgh, Pennsylvania 15213*

ANDREA SBARBATI (7), *Anatomy and Histology Section, Department of Morphological and Biomedical Sciences, University of Verona, Verona, Italy 37134*

DANIEL M. SFORZA (15), *Department of Molecular and Medical Pharmacology, UCLA School of Medicine, Los Angeles, California 90095-1735*

DESMOND J. SMITH (15), *Department of Molecular and Medical Pharmacology, UCLA School of Medicine, Los Angeles, California 90095-1735*

PETER M. SMITH-JONES (12), *Nuclear Medicine Service, Department of Radiology, Memorial Sloan-Kettering Cancer Center, New York, New York 10021*

DAVID B. SOLIT (12), *Department of Medicine, Memorial Sloan-Kettering Cancer Center, New York, New York 10021*

MICHAEL E. SUGHRUE (2), *Department of Neurological Surgery, Columbia University, New York, New York 10032*

XIANKAI SUN (11), *Mallinckrodt Institute of Radiology, Washington University School of Medicine, St. Louis, Missouri 63110*

SIMON D. TAYLOR-ROBINSON (14), *Department of Medicine, Imperial College London, Hammersmith Hospital, London W12 0HS, United Kingdom*

DAVID L. THOMAS (6), *Wellcome Trust High Field MR Research Laboratory, Department of Medical Physics and Bioengineering, University College London, London WC1E6JA, United Kingdom*

JOHN S. THORNTON (6), *Lysholm Department of Neuroradiology, National Hospital for Neurology and Neurosurgery, UCLH NHS Trust, London WC1N 3BG, United Kingdom*

LOUISE VAN DER WEERD (6), *RCS Unit of Biophysics, Institute of Child Health, University College London, London WC1N 1EH, United Kingdom*

HUA WU (9), *Department of Nuclear Medicine, Tongji Hospital, Tongji Medical College, Huazhong University of Science and Technology, Wuhan 430030, China*

YI-JEN L. WU (3), *Pittsburgh NMR Center for Biomedical Research, Department of Biological Sciences, Carnegie Mellon University, Pittsburgh, Pennsylvania 15213*

QING YE (3), *Pittsburgh NMR Center for Biomedical Research, Department of Biological Sciences, Carnegie Mellon University, Pittsburgh, Pennsylvania 15213*

XUEMEI ZHANG (9), *Department of Nuclear Medicine, Tongji Hospital, Tongji Medical College, Huazhong University of Science and Technology, Wuhan 430030, China*

ZHENWEI ZHANG (9), *Department of Nuclear Medicine, Tongji Hospital, Tongji Medical College, Huazhong University of Science and Technology, Wuhan 430030, China*

DAWEN ZHAO (18), *The University of Texas Southwestern Medical Center at Dallas, Dallas, Texas 75390*

MING ZHAO (9), *Department of Nuclear Medicine, Tongji Hospital, Tongji Medical College, Huazhong University of Science and Technology, Wuhan 430030, China*

Preface

As these volumes were being completed, American Paul C. Lauterbur and Briton Sir Peter Mansfield won the 2003 Nobel Prize for medicine for discoveries leading to the development of MRI.

The *Washington Post* story on October 6, 2003 announced the accolade, noting: "Magnetic resonance imaging, or MRI, has become a routine method for medical diagnosis and treatment. It is used to examine almost all organs without need for surgery, but is especially valuable for detailed examination of the brain and spinal cord." Unfortunately, the article overlooked the growing usefulness of this technique in basic research.

MRI, along with other imaging methods, has made it possible to glance inside the living system. For patients, this may obviate the need for surgery; for researchers, it becomes a noninvasive method that enables the model systems to continue "doing what they do" without being disturbed. The value and potential of these techniques is enormous, and that is why these once clinical methods are finding their way to the laboratory.

Authors have been selected based on research contributions in the area about which they have written and based on their ability to describe their methodological contributions in a clear and reproducible way. They have been encouraged to make use of graphics and comparisons to other methods, and to provide tricks and approaches that make it possible to adapt methods to other systems.

The editor wants to express appreciation to the contributors for providing their contributions in a timely fashion, to the senior editors for guidance, and to the staff at Academic Press for helpful input.

P. MICHAEL CONN

METHODS IN ENZYMOLOGY

VOLUME 226. Metallobiochemistry (Part C: Spectroscopic and
Physical Methods for Probing Metal Ion Environments in Metalloenzymes
and Metalloproteins)
Edited by JAMES F. RIORDAN AND BERT L. VALLEE

VOLUME 227. Metallobiochemistry (Part D: Physical and Spectroscopic
Methods for Probing Metal Ion Environments in Metalloproteins)
Edited by JAMES F. RIORDAN AND BERT L. VALLEE

VOLUME 228. Aqueous Two-Phase Systems
Edited by HARRY WALTER AND GÖTE JOHANSSON

VOLUME 229. Cumulative Subject Index Volumes 195–198, 200–227

VOLUME 230. Guide to Techniques in Glycobiology
Edited by WILLIAM J. LENNARZ AND GERALD W. HART

VOLUME 231. Hemoglobins (Part B: Biochemical and Analytical Methods)
Edited by JOHANNES EVERSE, KIM D. VANDEGRIFF, AND ROBERT M. WINSLOW

VOLUME 232. Hemoglobins (Part C: Biophysical Methods)
Edited by JOHANNES EVERSE, KIM D. VANDEGRIFF, AND ROBERT M. WINSLOW

VOLUME 233. Oxygen Radicals in Biological Systems (Part C)
Edited by LESTER PACKER

VOLUME 234. Oxygen Radicals in Biological Systems (Part D)
Edited by LESTER PACKER

VOLUME 235. Bacterial Pathogenesis (Part A: Identification and Regulation of
Virulence Factors)
Edited by VIRGINIA L. CLARK AND PATRIK M. BAVOIL

VOLUME 236. Bacterial Pathogenesis (Part B: Integration of Pathogenic
Bacteria with Host Cells)
Edited by VIRGINIA L. CLARK AND PATRIK M. BAVOIL

VOLUME 237. Heterotrimeric G Proteins
Edited by RAVI IYENGAR

VOLUME 238. Heterotrimeric G-Protein Effectors
Edited by RAVI IYENGAR

VOLUME 239. Nuclear Magnetic Resonance (Part C)
Edited by THOMAS L. JAMES AND NORMAN J. OPPENHEIMER

VOLUME 240. Numerical Computer Methods (Part B)
Edited by MICHAEL L. JOHNSON AND LUDWIG BRAND

VOLUME 241. Retroviral Proteases
Edited by LAWRENCE C. KUO AND JULES A. SHAFER

VOLUME 242. Neoglycoconjugates (Part A)
Edited by Y. C. LEE AND REIKO T. LEE

VOLUME 243. Inorganic Microbial Sulfur Metabolism
Edited by HARRY D. PECK, JR., AND JEAN LEGALL

Section I

Disease Models

[1]   Molecular and Functional Imaging of Cancer: Advances in MRI and MRS

By Arvind P. Pathak, Barjor Gimi, Kristine Glunde, Ellen Ackerstaff, Dmitri Artemov, and Zaver M. Bhujwalla

Introduction

Cancer is a disease that exhibits a degree of multiplicity and redundancy of pathways almost protean in nature. To understand and exploit molecular pathways in cancer for therapeutic strategies, it is essential not only to detect and image the expression of these pathways, but also to determine the impact of this expression on function at the cellular level, as well as within the complex system, which is a tumor. Multiparametric molecular and functional imaging techniques have several key roles to play in cancer treatment, such as revealing key targets for therapy, visualizing delivery of the therapy, and assessing the outcome of treatment. As a technique, magnetic resonance (MR) has a formidable array of capabilities to characterize function. Noninvasive multinuclear magnetic resonance imaging (MRI) and MR spectroscopic imaging (MRSI) provide a wealth of spatial and temporal information on tumor vasculature, metabolism, and physiology. MR is therefore particularly applicable to investigating a complex disease such as cancer. Several of the MRI techniques are also translatable into the clinic, and are therefore compatible with "bench to bedside" applications.

Tumor vasculature plays an important role in growth, treatment, and metastatic dissemination. The first section in this chapter therefore describes the use of MRI techniques and the underlying assumptions and mechanisms in characterizing tumor vasculature. Recent advances in the development of targeted contrast agents have significantly increased the versatility of MR for molecular imaging. Although MR techniques provide a wealth of structural and functional information, MR suffers from poor sensitivity. The second section discusses the use of targeted contrast agents and amplification strategies to increase the sensitivity of detection of molecular targets in MR molecular imaging of cancer. Technical strategies to improve the signal to noise ratio (SNR) for applications of MR microscopy in cancer are also included in this section. Because MR spectroscopy (MRS) and MRSI provide information on metabolism and pH, MRS applications in cancer are reviewed in the third section. One of the most exciting aspects of MR is the ability to perform multiparametric imaging.

Copyright 2004, Elsevier Inc.
All rights reserved.
0076-6879/04 $35.00

In the fourth section, we present two examples of the use of multiparametric imaging in understanding cancer cell invasion and in characterizing the relationship between tumor vasculature and metabolism.

Vascular Imaging of Tumors with MRI

MRI techniques can be used to characterize several aspects of tumor vasculature. Tumor vasculature is typified by structural and functional anomalies that include alterations in hemodynamics, blood rheology, permeability, and drainage, and plays a critical role in cancer growth, treatment, and metastasis. Vascular MR methods are therefore useful in cancer treatment and management. An overview of the endogenous and exogenous MR contrast mechanisms utilized in characterizing tumor vasculature is presented in this section.

MR Relaxation Mechanisms and the Basis of Contrast

Every contrast mechanism for probing the tumor vasculature, including the use of exogenous MR contrast agents, is in some way a result of changes in the MR signal intensity brought about by changes in tissue relaxation times (T_1, T_2, or T_2^*). Briefly, T_1, the spin–lattice or longitudinal relaxation time, is the time constant that characterizes the exponential process by which the magnetization returns or "relaxes" to its equilibrium position. It does so by exchanging energy with its surroundings, or lattice, at the Larmor frequency. T_1 relaxation occurs at the molecular level through several pathways, including interactions between protons in tissue water and those on macromolecules or proteins, and by interactions with paramagnetic substances (i.e., substances with unpaired electrons in their outermost shells). T_1-based MR contrast results from differences in T_1 dominating the MR signal intensity. For example, tissues with short T_1s (such as fat) appear bright in T_1-weighted MRI, since the transverse magnetization recovers to equilibrium rapidly compared with tissues with long T_1s (such as cerebrospinal fluid).

Microscopic magnetic field heterogeneities in the main field, as well as variations in local magnetic susceptibility due to the physiologic microenvironment, cause spins contributing to the transverse magnetization to lose phase coherence. The process through which this occurs is known as T_2^* relaxation. The loss in transverse coherence attributable to static magnetic field heterogeneities can be recovered using a spin–echo sequence or a refocusing pulse. However, as protons diffuse through the microscopic field inhomogeneities, they also lose phase coherence due to their Brownian random walks through the magnetic field gradients, which result in phase dispersion that cannot be reversed by the application of a refocusing pulse. This process is known as T_2 relaxation. In T_2-weighted MR images, tissues with short T_2s, such as the liver, appear dark due to the rapid decay of transverse magnetization compared with those with long T_2s, such as fat. Similarly, in T_2^*-weighted images, regions with large susceptibility gradients, such as air–tissue interfaces of the inner ear or orbits of the eye, or large veins carrying deoxygenated blood, appear hypointense.

In general, the addition of a paramagnetic solute causes an increase in the $1/T_1$ and $1/T_2$ of solvent nuclei. The diamagnetic and paramagnetic contributions to the relaxation rates of such solutions are additive and are expressed as[1]:

$$\left(\frac{1}{Ti}\right)_{obs} = \left(\frac{1}{Ti}\right)_d + \left(\frac{1}{Ti}\right)_p \quad i = 1, 2 \tag{1}$$

where $(1/Ti)_{obs}$ is the observed solvent relaxation rate in the presence of a paramagnetic species (e.g., contrast agent), $(1/Ti)_d$ is the (diamagnetic) solvent relaxation rate in the absence of a paramagnetic species, and $(1/Ti)_p$ represents the additional paramagnetic contribution. In the absence of any solute–solute interactions, the solvent relaxation rates (in solution) are linearly dependent on the concentration of the paramagnetic species $[M]$, and if $(1/Ti)$ or the relaxivity R_i, is defined as the slope of this dependence in $mM^{-1}\ s^{-1}$, we may write Eq. (1) as:

$$\left(\frac{1}{Ti}\right)_{obs} = \left(\frac{1}{Ti}\right)_d + R_i[M] \quad i = 1, 2 \tag{2}$$

All molecules, large and small, are in a constant state of motion, tumbling and colliding with other molecules. Intramolecular motion, as well as interaction with nearby molecules, produces fluctuations in the local magnetic field experienced by a proton. It turns out that these magnetic interactions can promote both T_1 and T_2 relaxation, but whether they do so depends on the rate at which their magnetic fields fluctuate. For example, a small molecule such as water moves quickly, so that it produces rapid magnetic fluctuations. A large molecule such as a protein moves more slowly and produces magnetic fluctuations at a correspondingly lower rate. From the relaxation theory described by Solomon–Bloembergen,[2,3] three primary factors that regulate the dipole–dipole interactions responsible for both T_1 and T_2 relaxation are: (1) the strength of the magnetic

[1] R. B. Lauffer, *Chem. Rev.* **87**, 901 (1987).
[2] I. Solomon, *Phys. Rev.* **99**, 559 (1955).
[3] N. Bloembergen, *J. Chem. Phys.* **27**, 572 (1957).

moment, (2) the separation between the two dipoles, and (3) the relative motion of the two dipoles.

Intrinsic or Endogenous Contrast

Probing tumor vasculature using intrinsic contrast produced by deoxyhemoglobin in tumor microvessels is based on the blood oxygenation level dependent (BOLD) contrast mechanism first proposed by Ogawa.[4] The concentration of endogenous paramagnetic deoxyhemoglobin is one of the primary determinants of the eventual image contrast observed. The presence of deoxyhemoglobin in a blood vessel causes a susceptibility difference between the vessel and its surrounding tissue, inducing microscopic magnetic field gradients that cause dephasing of the MR proton signal, leading to a reduction in the value of T_2^* (Fig. 1). Because oxyhemoglobin is diamagnetic and does not produce the same dephasing, changes in oxygenation of the blood can be observed as signal changes in T_2^*-weighted images. The functional dependence of T_2^* on oxygenation in a tissue is expressed as:

$$\frac{1}{T_2^*} \propto (1 - Y)b \tag{3}$$

where Y is the fraction of oxygenated blood and b the fractional blood volume. In hypoxic tumors where $0 < Y < 0.2$, the contrast produced by the method is primarily dependent on b. This method works best in poorly oxygenated tumors such as subcutaneous models, and in human xenografts with random orientation of sprouting capillaries, and it provides a fast and noninvasive measurement of tumor fractional blood volume because exogenous contrast is not required. However, the method cannot provide quantitative measurements of tumor vascular volume, vascular permeability, or blood flow. Nonetheless, this technique has been used to detect changes in tumor oxygenation and vascularization following induction of angiogenesis by external angiogenic agents,[5] as well as to obtain maps of the "functional" vasculature in genetically modified HIF-1 (+/+ and −/−) animal models.[6] BOLD contrast is not solely related to the oxygenation status of blood, but is also affected by factors such as oxygen saturation, the hematocrit, blood flow, blood volume, vessel orientation, and

[4] S. Ogawa, *Magn. Reson. Med.* **14,** 68 (1990).
[5] R. Abramovitch, H. Dafni, E. Smouha, L. Benjamin, and M. Neeman, *Cancer Res.* **59,** 5012 (1999).
[6] P. Carmeliet, Y. Dor, J.-M. Herbert, D. Fukumura, K. Brusselmans, M. Dewerchin, M. Neeman, F. Bono, R. Abramovitch, P. Maxwell, C. J. Koch, P. Ratcliffe, L. Moons, R. K. Jain, D. Collen, and E. Keshet, *Nature* **394,** 485 (1998).

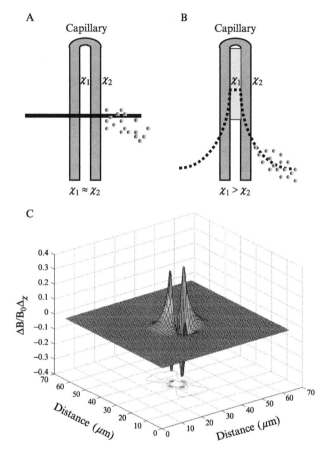

Fig. 1. Schematic illustrating the premise of the BOLD effect. (A) In the absence of a susceptibility difference between (oxygenated) blood (χ_1) and the surrounding tissue (χ_2), no microscopic field gradient is set up and diffusing water protons "see" the same local magnetic field. (B) When there is a susceptibility difference between (deoxygenated) blood (χ_1) and the surrounding tissue (χ_2), a microscopic field gradient (—) is set up and diffusing water protons "see" different local magnetic fields, leading to loss of phase coherence, reduction in T_2^*, and MR signal attenuation. (C) Surface plot illustrating the three-dimensional aspects of mathematically simulated microscopic field gradients induced around a microvessel. (See color insert.)

geometry,[7] which should be considered when interpreting BOLD maps. In a recent study, Silva et al.[8] demonstrated the feasibility of imaging blood flow in a rodent brain tumor model at a high magnetic field, using another endogenous contrast MR technique known as arterial spin labeling (ASL). In this approach, arterial blood water is used as the perfusion tracer, and it is magnetically tagged proximal to the tissue of interest, using spatially selective inversion pulses. The effect of arterial tagging on downstream images can be quantified in terms of tissue blood flow, since changes in signal intensity depend on the regional blood flow and degree of T_1 relaxation. Tissue blood flow (F) images can be computed from magnetically tagged and control images according to the expression:

$$F = \frac{\lambda}{T_1} \frac{S_{control} - S_{label}}{2\alpha S_{control}} \tag{4}$$

where $\lambda = 0.9$, is the tissue–blood partition coefficient for water, $S_{control}$ is the control image signal intensity, S_{label} is the tagged image signal intensity, T_1 is the precalculated T_1-map, and $\alpha = 0.8$, is the tagging efficiency.[9] Although ASL exhibits sufficient sensitivity at high magnetic field for mapping heterogeneities in tumor blood flow, it may not do as well when the blood flow (F) is very low, since tagged arterial spins may not reach the tissue in time, relative to their T_1s (i.e., the spins will fully relax by the time they enter the imaging slice).

The advantages of vascular characterization using endogenous contrast are that the administration of an external agent is not required, making it entirely noninvasive, and repeated measurements are only limited by the constraints of anesthesia, providing dynamic data with high temporal resolution. Endogenous contrast methods, however, cannot quantify tumor vascular volume or vascular permeability, for which exogenous contrast MRI techniques are required.

Extrinsic or Exogenous Contrast

MR contrast agents (CA), unlike dyes or agents used with nuclear medicine or X-ray techniques, are not visualized directly in the MR image, but indirectly from the changes they induce in water proton relaxation behavior. The most commonly used MR CA are paramagnetic gadolinium chelates. These agents are tightly bound complexes of the rare earth element gadolinium (Gd) and various chelating agents. The seven unpaired electrons of Gd produce a large magnetic moment that results in shortening of both T_1 and T_2 of tissue water. Because tissue T_2 values are intrinsically shorter than the corresponding T_1 values, the T_1 effect of the contrast agent predominates and tissues that take up the agent are brightened in T_1-weighted images.

---

[7] A. P. Pathak, S. D. Rand, and K. M. Schmainda, *J. Magn. Reson. Imaging* **18,** 397 (2003).
[8] A. C. Silva, S.-G. Kim, and M. Garwood, *Magn. Reson. Med.* **44,** 169 (2000).
[9] A. C. Silva, W. Zhang, D. S. Williams, and A. P. Koretsky, *Magn. Reson. Med.* **33,** 209 (1995).

Susceptibility effects of Gd-based CA resulting in the shortening of T_2 and T_2^* relaxation times are also used to measure tumor vascular volume and flow. Tissues that take up the paramagnetic agent are darkened in T_2- and T_2^*-weighted images. Vascular parameters can be calculated from tracer kinetics or mass balance principles, using the tissue concentration of Gd-based agents.

Several Gd complexes are either under development or in use and may be broadly classified as either low molecular weight (≈ 0.57 kD) agents, for example, the gadolinium diethylenetriamine pentaacetic acid (GdDTPA) compounds used clinically for contrast enhancement of various lesions, including malignant tumors, or macromolecular agents (≈ 90 kD) such as albumin–GdDTPA, which remain in the intravascular space for up to several hours.

Based upon the physical properties of the CA used, brief descriptions of current MR methods used to characterize tumor vascularization are presented here.

Low Molecular Weight Contrast Agents. These are the only class of paramagnetic agents approved for routine clinical use, and there are several reports describing applications of these agents to image a variety of tumors, including breast,[10] brain,[11] and uterine tumors.[12] Most T_1 methods involve the analyses of relaxivity changes induced by the contrast agent to determine influx and outflux transfer constants, as well as the extracellular extravascular volume fraction based on one of several compartmental models.[13] Although these agents are not freely diffusible and remain in the extracellular compartment, three standard kinetic parameters that can be derived from dynamic contrast-enhanced T_1-weighted MRI of a diffusible tracer are (1) K^{trans} (min^{-1}), which is the volume transfer constant between the blood plasma and the extravascular extracellular space (EES); (2) k_{ep} (min^{-1}), which is the rate constant between the EES and blood plasmas; and (3) v_e (%), which is the volume of the EES per unit volume of tissue (i.e., the volume fraction of the EES).[14] These three parameters are related by:

[10] E. Furman-Haran, R. Margalit, D. Grobgeld, and H. Degani, *Proc. Natl. Acad. Sci. USA* **93,** 6247 (1996).
[11] H. J. Aronen, M. S. Cohen, J. W. Belliveau, J. A. Fordham, and B. R. Rosen, *Top. Magn. Reson. Imaging* **5,** 14 (1993).
[12] H. Hawighorst, P. G. Knapstein, W. Weikel, M. V. Knopp, I. Zuna, A. Knof, G. Brix, U. Schaeffer, C. Wilkens, S. O. Schoenberg, M. Essig, P. Vaupel, and G. van Kaick, *Cancer Res.* **57,** 4777 (1997).
[13] P. S. Tofts, *J. Magn. Reson. Imaging* **7,** 91 (1997).
[14] P. S. Tofts, G. Brix, D. L. Buckley, J. L. Evelhoch, E. Henderson, M. V. Knopp, H. B. Larsson, T. Y. Lee, N. A. Mayr, G. J. Parker, R. E. Port, J. Taylor, and R. M. Weisskoff, *J. Magn. Reson. Imaging* **10,** 223 (1999).

$$k_{ep} = \frac{K^{trans}}{v_e} \tag{5}$$

k_{ep} can be derived from the shape of the tracer concentration–time curve, but the determination of K^{trans} requires absolute values of the tracer concentration. K^{trans} has several different connotations depending on the balance between blood flow and capillary permeability in the tissue of interest. Ignoring the contribution of intravascular tracer to the total tissue concentration, as well as the possibility of further compartmentalization within the voxel, simple two-compartment (blood space and the EES) models can be broadly classified into four main types as previously described by Tofts et al.[14]:

1. *Flow-limited (high permeability) or Kety Model:* This model assumes that arterial and venous blood pools have distinct CA concentrations, and because permeability (P) is high and the transendothelial flux flow (F) is limited (i.e., permeability surface area product $PS \gg F$), it assumes that the venous blood exits the tissue with a concentration that is in equilibrium with the tissue. For an extracellular tracer, the differential equation relating the tissue concentration (C_t) to the plasma concentration (C_p) is:

$$\frac{dC_t}{dt} = F\rho(1 - Hct)\left(C_p - \frac{C_t}{v_e}\right) \tag{6}$$

Here, $K^{trans} = F\rho(1-Hct)$, where F (ml g^{-1} min^{-1}) is the flow of whole blood per unit mass of tissue, ρ is the tissue density (g ml^{-1}), and Hct is the hematocrit, making $(1-Hct)$ the plasma fraction.

2. *PS-limited Model:* If the flow (F) is high, the arterial and venous concentrations can be considered equal and the rate of tissue uptake is then limited by the permeability surface area product (PS) of the vessel wall and the concentration gradient between the plasma and EES compartments. Thus when $PS \ll F$, the differential equation relating the tissue concentration (C_t) to the plasma concentration (C_p) is given by:

$$\frac{dC_t}{dt} = PS\rho\left(C_p - \frac{C_t}{v_e}\right) \tag{7}$$

Here, $K^{trans} = PS\rho(1-Hct)$, P (cm min^{-1}) is the permeability of the vessel wall, S (cm^2 g^{-1}) is the surface area per unit mass of tissue, ρ is the tissue density (g ml^{-1}), and $(1-Hct)$ is the plasma fraction.

3. *Mixed Flow and PS-limited Model:* In the instance where tracer uptake might be limited by both flow and permeability, one can consider an additional parameter, the initial extraction ratio (E), given by:

$$E = \frac{(C_q - C_v)}{C_a} \tag{8}$$

that describes the reduction in the blood concentration as it transits the tissue bed. The differential equation relating the tissue concentration (C_t) to the plasma concentration (C_p) is given by:

$$\frac{dC_t}{dt} = EF\rho(1 - Hct)\left(C_p - \frac{C_t}{v_e}\right) \tag{9}$$

In the flow-limited case $(PS \gg F)$, ignoring tracer backflow if initial extraction is complete (i.e., $E = 1$), Eq. (9) reduces to the Kety Eq. (6). In the PS-limited case $(PS \ll F)$, $E = PS/F(1-Hct)$, and Eq. (9) reduces to Eq. (7).

4. *Clearance Model:* Finally, if we define clearance (CL) as a constant relating the rate of tracer elimination from the tissue to the current tracer concentration, we can relate the tissue concentration (C_t) to the plasma concentration (C_p) as:

$$\frac{dC_t}{dt} = \frac{CL}{V_t}\left(C_p - \frac{C_t}{v_e}\right) \tag{10}$$

where CL (ml min^{-1}) is the clearance and V_t (ml) is the total tissue volume. Because all of the previous differential equations have the same general form, we may formulate a generalized kinetic model as[14]:

$$\frac{dC_t}{dt} = K^{trans}\left(C_p - \frac{C_t}{v_e}\right) = K^{trans}C_p - k_{ep}C_t \tag{11}$$

This model reduces to the previous forms under the appropriate boundary conditions. An excellent review of the previous models, terminology, and definitions can be found elsewhere.[14] The arterial input function that these models require can be measured separately or defined in real time using voxels localized within large blood vessels. The analytic solution of the model can be derived by approximating the arterial input function [$C_a = (1-Hct)C_p$] to a multiexponential decay as first described by Ohno *et al.*[15] In all of the previous models, the concentration of GdDTPA is measured from changes in the T_1 relaxation rate, assuming that water is in fast exchange between the vascular and extracellular compartments. It has recently been demonstrated that the accuracy of the tissue vascular volume measurement critically depends on the validity of this assumption.[16]

[15] K. Ohno, K. D. Pettigrew, and S. I. Rapoport, *Stroke* **10**, 62 (1979).
[16] Y. R. Kim, K. J. Rebro, and K. M. Schmainda, *Magn. Reson. Med.* **47**, 1110 (2002).

As mentioned earlier, the low molecular weight Gd chelates employed in MRI produce both T_1 and T_2 relaxation effects. However, when high doses of these agents are employed, the induced bulk susceptibility differences between the intravascular and extravascular spaces dominate the classical dipolar effects. There are two related mechanisms by which MRI contrast can be engendered from local magnetic field heterogeneities. (1) *Through diffusion*—as protons diffuse through the microscopic field inhomogeneities, they lose phase coherence due to their random Brownian motion through magnetic field gradients that are present. (2) *Through intervoxel dephasing*—even without the diffusive movement of water, there exists a heterogeneity of resonant frequencies due to the presence of microscopic field inhomogeneities within an imaging voxel, which in turn affects the MR signal intensity (in gradient–echo images) by causing intravoxel dephasing. The effect of magnetic field inhomogeneities on transverse relaxation can be characterized as[17,18]:

$$\frac{1}{T_2{}^*} = \frac{1}{T_2} + \frac{1}{T_2'}$$

(12)

The relaxation rate $1/T_2{}^*$ ($R_2{}^*$) is the rate of free induction decay or the rate at which the gradient–echo amplitude decays. The relaxation rate, $1/T_2'$ (R_2'), is the water resonance linewidth, which is a measure of the frequency distribution within a voxel. In the presence of a magnetic field perturber (such as a tumor vessel), the relative R_2 and $R_2{}^*$ relaxation rates depend on the diffusion coefficient (D) of spins in the vicinity of the induced field inhomogeneities, the radius (R) of the field perturber (i.e., tumor vessel caliber), and the variation of the Larmor frequency at the surface of the perturber.[17–20] The two physical characteristics (R and D) can be collapsed into one term, and the proton correlation time τ_D can be described as:

$$\tau_D = \frac{R^2}{D}$$

(13)

and the variation in the Larmor frequency ($d\omega$), at the surface of the perturber, is given by:

$$d\omega = \gamma(\Delta_\chi)B_0$$

(14)

[17] C. R. Fisel, J. L. Ackerman, R. B. Buxton, L. Garrido, J. W. Belliveau, R. B. Rosen, and T. J. Brady, *Magn. Reson. Med.* **17,** 336 (1991).
[18] R. P. Kennan, *Magn. Reson. Med.* **31,** 9 (1994).
[19] R. M. Weisskoff, C. S. Zuo, J. L. Boxerman, and B. R. Rosen, *Magn. Reson. Med.* **31,** 601 (1994).
[20] D. A. Yablonskiy, *Magn. Reson. Med.* **32,** 749 (1994).

where γ is the proton gyromagnetic ratio, Δ_χ is the susceptibility difference between the perturber and its background, and B_0 is the strength of the applied magnetic field. Depending on the relative magnitudes of these variables, the magnitude of the susceptibility-induced relaxation effects are commonly described by three regimes.[17–21]

1. *Fast Exchange Regime:* In this regime, the rate of diffusion $(1/\tau_D)$ is substantially greater than the frequency variation $(d\omega)$ (i.e., $\tau_D \, d\omega << 1$). The high diffusion rate causes all the spins to experience a similar range of field inhomogeneities within an echo time, causing minimal loss of phase coherence, as well as similar loss of phase coherence between gradient– and spin–echo sequences. This is also known as the "motional-averaged" or "motional-narrowed" regime because the susceptibility-induced local magnetic field gradients are averaged out.[17,18,22]

2. *Slow Exchange Regime:* In this regime, the rate of diffusion $(1/\tau_D)$ is substantially smaller than the frequency variation $(d\omega)$ (i.e., $\tau_D \, d\omega >> 1$). Thus the phase that a proton accumulates as it passes one perturber is large (i.e., the effect is the same as it would be for the case of static field inhomogeneities). Due to the absence of motion averaging, the gradient–echo relaxation rate tends to be greater than the spin–echo relaxation rate. Also, there will be no signal attenuation on a T_2-weighted scan because the $180°$ pulse during the spin–echo sequence refocuses static magnetic field inhomogeneities, whereas intravoxel dephasing still occurs in a gradient–echo sequence (due to the absence of a similar refocusing RF pulse) (Fig. 2).

3. *Intermediate Exchange Regime:* In this regime, $\tau_D \, d\omega \sim 1$ (i.e., water diffusion is neither fast enough to be fully motionally narrowed, nor slow enough to be approximated as linear gradients, making the description of the susceptibility-induced contrast more complex). In this regime, spin–echo relaxation is maximum and the gradient–echo relaxation is not very different from what it would be in the slow exchange regime (Fig. 2). In this regime, analytic solutions to estimate signal loss in the presence of diffusion become complicated due to the large spatial heterogeneity of the induced field gradients and numerical simulations are required.[18,19,22]

From the preceding description, it is apparent that spin–echo (SE) and gradient–echo (GE) sequences have greatly differing sensitivities to the size and scale of the field inhomogeneities, resulting in a differential

[21] A. Villringer, B. R. Rosen, J. W. Belliveau, J. L. Ackerman, R. B. Lauffer, R. B. Buxton, Y. S. Chao, V. J. Wedeen, and T. J. Brady, *Magn. Reson. Med.* **6,** 164 (1988).
[22] J. L. Boxerman, L. M. Hamberg, B. R. Rosen, and R. M. Weisskoff, *Magn. Reson. Med.* **34,** 555 (1995).

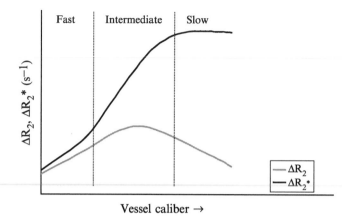

Fig. 2. Schematic to illustrate the three regimes of susceptibility-induced relaxation effects and the differential sensitivity of gradient–echo (ΔR_2^*) and spin–echo (ΔR_2) relaxation rates to vessel caliber. This sensitivity to vessel size constitutes the basis for imaging macrovascular and microvascular blood volume, as well as imaging vessel size.

sensitivity to vessel caliber. The SE relaxation rate change (ΔR_2) increases, reaches a maximum for capillary-sized vessels (5–10 μm), and then decreases inversely with vessel radius. The GE relaxation rate change (ΔR_2^*) increases and then plateaus to remain independent of vessel radius beyond capillary-sized vessels (Fig. 2). A consequence of this result is that the SE relaxation rate changes are maximally sensitive to the microvascular blood volume, whereas the GE changes are more sensitive to the total blood volume. Based on this observation, SE sequences have been used in many tumor studies with the assumption that tumor angiogenesis is primarily characterized by an increase in the microvasculature.[23] However, given the large (>20 μm) tortuous vessels usually found in tumors,[24,25] whether either SE or GE methods are most appropriate remains to be determined. Several investigators have acquired relative cerebral blood volume (rCBV) maps from first-pass dynamic susceptibility contrast (DSC) studies, with good spatio-temporal resolution.[26,27] With this technique, preliminary results indicate that MRI-derived rCBV may better

[23] H. J. Aronen, I. E. Gazit, D. N. Louis, B. R. Buchbinder, F. S. Pardo, R. M. Weisskoff, G. R. Harsh, G. R. Cosgrove, E. F. Halpern, F. H. Hochberg, and B. R. Rosen, *Radiology* **191,** 41 (1994).

[24] B. R. Deane and P. L. Lantos, *J. Neurological Science* **49,** 55 (1981).

[25] A. P. Pathak, K. M. Schmainda, B. D. Ward, J. R. Linderman, K. J. Rebro, and A. S. Greene, *Magn. Reson. Med.* **46,** 735 (2001).

differentiate histologic tumor types than conventional MRI[23] and provide information to predict tumor grade.[27] For example, to quantitatively measure relative cerebral blood volume (rCBV) or cerebral blood flow (CBF), regional changes in signal intensity versus time need to be converted into concentration versus time curves. As mentioned earlier, both empiric data and modeling indicate that for a given echo time (TE), the T_2^* rate change $(\Delta R_2^* = \Delta(1/T_2^*) = (1/T_2^*_{postcontrast} - 1/T_2^*_{precontrast})$ is proportional to the brain tissue concentration:

$$\Delta R_2^* = k[conc.] \tag{15}$$

where k is a tissue-specific MR pulse sequence and field strength–dependent calibration factor. Assuming monoexponential signal decay, signal intensity change following Gd injection is:

$$S(t) = S_0 e^{-TE[\Delta R_2^*(t)]} \tag{16}$$

yielding:

$$\frac{-1}{TE} \ln\left[\frac{S(t)}{S_0}\right] = kC(t) \tag{17}$$

where S_0 is the baseline signal intensity before contrast administration, $S(t)$ is the tissue signal with contrast, TE is the echo time, and $C(t)$ is the concentration–time curve. The area under the concentration–time curve is proportional to the rCBV.

$$\frac{-1}{TE} \int_0^\infty \ln \frac{S(t)}{S_0} dt \; \alpha \; \text{rCBV} \tag{18}$$

These steps are summarized in Fig. 3. These curves often include contributions due to recirculation that must be eliminated before tracer-kinetic principles may be used to extract volume and flow information. This is usually accomplished by exponential extrapolation or fitting to a gamma-variate function with recirculation cut-off.[28] For an instantaneous bolus injection, the central volume principle states that CBF = CBV/MTT, where MTT is the mean transit time of contrast agent through the vascular network.[29] However, most injections are of finite duration, and the observed

[26] B. R. Rosen, J. W. Belliveau, B. R. Buchbinder, R. C. McKinstry, L. M. Porkka, D. N. Kennedy, M. S. Neuder, C. R. Fisel, H. J. Aronen, K. K. Kwong, R. M. Weisskoff, M. S. Cohen, and T. J. Brady, *Magn. Reson. Med.* **19**, 285 (1991).

[27] M. Maeda, S. Itoh, H. Kimura, T. Iwasaki, N. Hayashi, K. Yamamoto, Y. Ishii, and T. Kubota, *Radiology* **189**, 233 (1993).

[28] H. K. J. Thompson, C. F. Starmer, R. E. Whalen, and H. D. McIntosh, *Circulation Res.* **14**, 502 (1964).

[29] K. L. Zierler, *Circulation Res.* **10**, 393 (1962).

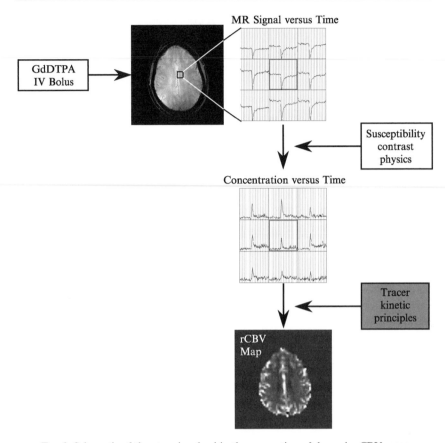

Fɪɢ. 3. Schematic of the steps involved in the generation of dynamic rCBV maps.

concentration–time curve is the convolution of the ideal tissue–transit curve with the arterial input function. Thus measurement of the blood flow requires knowledge of the arterial input curve to deconvolve the observed concentration–time curve.[30]

A potential complication with using first-pass rCBV techniques with low molecular weight Gd contrast agents is that with elevated permeability, as is often observed in tumor vasculature, or with significant blood–brain barrier (BBB) disruption, as is often the case with brain tumors, contrast agent leaks out of the vasculature into the brain or tumor tissue, resulting

[30] L. Axel, *Radiology* **137,** 679 (1980).

in enhanced T_1 relaxation effects. Signal increases due to T_1 effects may then mask signal decreases due to T_2 or T_2^* effects, leading to an underestimation of rCBV. To address this issue, a method of analysis has been devised that corrects for these leakage effects when the leakage is not extreme.[31] Donahue et al.[31] recently showed that although GErCBV (total tumor blood volume) correlated strongly with tumor grade, when the GErCBV data were not corrected for leakage effects, the correlation with tumor grade was no longer significant. Another obstacle to the application of the central volume principle for the calculation of blood flow is the direct measurement of the mean transit time. Weisskoff et al.[32] have demonstrated that MTT, which relates tissue blood volume to blood flow from the central volume principle, is not the first moment of the concentration–time curve for MR of intravascular tracers, and although first-moment methods cannot be used by themselves to determine absolute flow, they do provide a useful relative measure of flow.

More recently, the differential sensitivities of GE and SE methods to vessel radius have been further exploited to provide a measure of the averaged vessel diameter by measuring the ratio of GE and SE relaxation rates ($\Delta R_2^*/\Delta R_2$). From Fig. 3 it can be seen that as the perturber (i.e., vessel) size increases, so does the ratio $\Delta R_2^*/\Delta R_2$. Dennie et al.[33] have shown that using an intravascular superparamagnetic iron oxide nanoparticle (MION) contrast agent, this ratio compared favorably to a predicted ratio using histologically determined vessel sizes and the theoretical Monte Carlo modeling results. More recently, Donahue et al.[31] have demonstrated that clinically, the ratio $\Delta R_2^*/\Delta R_2$ correlated strongly with tumor grade and was a promising marker for the evaluation of tumor angiogenesis in patients.

Finally, all dynamic susceptibility-based contrast measurements are made assuming that the calibration factor "k" [see Eq. (13)] is the same for all tissue types and independent of tissue condition. However, a recent study has shown that k is the same for brain gray and white matter but *not* the same for normal brain and tumor tissue.[34] This difference may be attributed to the grossly different vascular morphology of tumors, due to tumor angiogenesis, compared with normal brain and/or possibly differing

[31] K. M. Donahue, H. G. J. Krouwer, S. D. Rand, A. P. Pathak, C. S. Marzalkowski, S. C. Censky, and R. W. Prost, *Magn. Reson. Med.* **43,** 845 (2000).

[32] R. M. Weisskoff, *Magn. Reson. Med.* **29,** 553 (1993).

[33] J. Dennie, J. B. Mandeville, J. L. Boxerman, S. D. Packard, B. R. Rosen, and R. M. Weisskoff, *Magn. Reson. Med.* **40,** 793 (1998).

[34] A. P. Pathak, S. D. Rand, and K. M. Schmainda, *J. Magn. Reson. Imaging* **18,** 397 (2003).

blood rheological factors such as hematocrit. Consequently, the sensitivity to blood volume differences between tumor and normal brain tissue may be lessened when using gradient–echo susceptibility contrast agent methods.

High Molecular Weight Contrast Agents. Quantitative determination of parameters of tumor vasculature with low molecular weight contrast agents is complicated by fast extravasation of the contrast agent from leaky tumor vessels. The availability of high molecular weight contrast agents such as albumin–GdDTPA (alb–GdDTPA) complexes or synthetic compounds such as polylysine–GdDTPA and Gadomer-17 provide an opportunity for quantitative determination of tumor vascular volume and vascular permeability surface area product (PS) for molecules of comparable sizes.[35] The relatively slow leakage of these agents from the vasculature results in a long half-life time and complete equilibration of plasma concentrations within the tumor, independently of blood flow. Assuming fast exchange of water between all the compartments in the tumor (plasma, interstitium, cells), the concentration of the contrast agent within any given voxel is proportional to changes in relaxation rate $(1/T_1)$ before and after administration of the contrast. Relaxation rates can be measured either directly using fast single-shot quantitative T_1 methods[36] or from T_1-weighted steady-state experiments,[37] which provide better temporal resolution but are susceptible to experimental artifacts caused by variations in T_2 and T_2^* relaxation times. Pixel-wise maps can be generated from the acquired data and processed with the appropriate model to obtain spatial maps of tumor vascular volume and vascular permeability surface area product.

A simple linear compartment model, describing uptake of the contrast agent from plasma, postulates a negligible reflux of the contrast agent from the interstitium back to the blood compartment. Blood concentrations of the contrast agent can be approximated to be constant for the duration of the MR experiment, and under these conditions, contrast uptake is a linear function of time (Fig. 4).[38,39] On a plot of contrast agent concentration versus time, the slope of the line provides the parameter PS, and the intercept of the line with the vertical axis at time zero provides the vascular volume (Fig. 4). For absolute values of these parameters, the change in

[35] M. D. Ogan, U. Schmiedl, M. E. Mosley, W. Grodd, H. Paajanen, and R. C. Brasch, *Invest. Radiol.* **22,** 665 (1987).

[36] C. Schwarzbauer, J. Syha, and A. Haase, *Magn. Reson. Med.* **29,** 709 (1993).

[37] R. Brasch, C. Pham, D. Shames *et al., J. Magn. Reson. Imaging* **7,** 68 (1997).

[38] C. S. Patlak, R. G. Blasberg, and J. D. Fenstermacher, *J. Cereb. Blood Flow Metab.* **3,** 1 (1983).

[39] H. C. Roberts, T. P. Roberts, R. C. Brasch, and W. P. Dillon, *AJNR* **21,** 891 (2000).

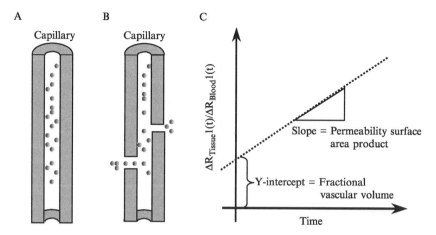

FIG. 4. Schematic illustrating (A) a blood vessel in which the macromolecular contrast agent is confined to the intravascular space. (B) In the case of a tumor vessel, due to elevated permeability, the contrast agent extravasates into the adjoining interstitial space. (C) Initially, the bulk of the T_1 relaxation effect is proportional to the intravascular space, since the contrast agent is confined to this space. The ratio of the change in relaxation rate of the tissue to that in the blood yields the fractional blood volume in that voxel. Over time, as the contrast agent extravasates into the adjoining tissue, the rate of change of the relaxation rate becomes proportional to the permeability surface area product for that vessel.

relaxation rate of the blood must be quantified. Changes in blood T_1 can be obtained separately from blood samples taken before injection of the contrast agent and at the end of the experiment, or may be obtained noninvasively.[39a]

For macromolecular agents such as alb–GdDTPA (MW $\approx$ 90 kD), blood concentrations equilibrate within 2–3 minutes and do not change for at least 40 minutes after an intravenous bolus injection. Tissue concentrations of the agent for a time period starting 5 minutes and up to 40 minutes after the bolus injection increase linearly with time. Therefore, the simple linear model is preferable for analysis of intrinsically noisy relaxation data because it is much more stable in comparison with nonlinear fitting algorithms required for the two compartment models discussed earlier. An example of vascular volume and permeability maps derived with this approach is shown in Fig. 5.[40] This linear-model approach was employed to detect vascular differences for metastatic versus nonmetastatic breast

[39a] A. P. Pathak, D. Artemov, and Z. M. Brujwalla, *Mag. Res. in Med.* **51,** 612 (2004).

[40] Z. M. Bhujwalla, D. Artemov, K. Natarajan, M. Solaiyappan, P. Kollars, and P. E. Kristjansen, *Clin. Cancer Res.* **9,** 355 (2003).

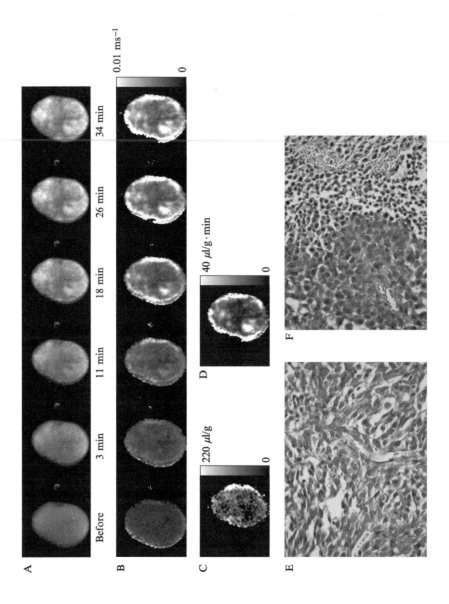

and prostrate cancer xenografts.[41] These studies not only showed that regions of high vascular volume were significantly less leaky compared with regions of low vascular volume, but that although invasion was necessary, without adequate vascularization it was not sufficient for metastasis to occur.

The accuracy of the measurement of tissue vascular volume using this approach does, however, depend on the water exchange rate between the vascular and extracellular compartments. Using a simplified model of fast exchange, where there may be intermediate to slow exchange, can lead to significant underestimation of vascular volume. Experimental approaches to minimize these errors are based on observations that the initial slope of the relaxation curve is independent of the exchange rate.[42]

Large molecular weight contrast agents may also potentially be used to measure tumor blood flow by detecting the first pass of the agent through tumor vasculature, similar to the method described in Ostergaard et al.,[43] although this approach may not be feasible when the heartbeat is very rapid, as for rodents.

Cellular and Molecular Imaging

Magnetic Resonance Detection of Cellular Targets

The development of targeted MR contrast agents has significantly increased the versatility of MR methods to detect receptor expression and specific cellular targets in cancer and other diseases. In the past, the intrinsically low signal to noise ratio (SNR) of MRI and MRS has limited the applications of MR-based methods for imaging targets at low concentrations.

[41] Z. M. Bhujwalla, D. Artemov, K. Natarajan, E. Ackerstaff, and M. Solaiyappan, *Neoplasia* **3**, 1 (2000).
[42] K. M. Donahue, R. M. Weisskoff, D. A. Chesler, K. K. Kwong, A. A. Bogdanov, J. B. Mandeville, and B. R. Rosen, *Magn. Reson. Med.* **36**, 858 (1996).
[43] L. Ostergaard, R. M. Weisskoff, D. A. Chesler, C. Gyldensted, and B. R. Rosen, *Magn. Reson. Med.* **36**, 715 (1996).

FIG. 5. (A) Raw 1 s saturation recovery images obtained from a single slice from a MatLyLu tumor, at different time points. (B) Corresponding relaxivity maps derived for this slice (using 100 ms, 500 ms, 1 s, and 7 s saturation recovery intervals) at different time points. Maps of (C) vascular volume and (D) permeability surface area product derived from the relaxivity maps for this slice. High-magnification photomicrographs from (E) viable, high-vascular volume and low-permeability regions, and (F) dying, low-vascular volume and high-permeability regions, obtained from a 5 μm-thick hematoxylin and eosin-stained section obtained from the same slice. From Bhujwalla et al.[40] (See color insert.)

Here we review strategies to improve the sensitivity of MR detection, the design and application of targeted relaxation contrast agents, and the use of signal amplification using enzymatic reactions and water exchange.

Sensitivity and Typical Target Concentration. The sensitivity of MR detection is a complex parameter that depends on various factors, including the resonant nucleus, the strength of the main magnetic field B_0, intrinsic relaxation times, and magnetic field inhomogeneities within the sample that are often determined by microarchitecture and macroarchitecture of the tissue being studied. *In vivo* proton (^{1}H) MRS performed using magnetic field strengths with B_0 in the range of 4.7–11 T can detect protons in the millimolar (mM) range. This level of sensitivity can only be obtained with optimized detection schemes, radiofrequency (RF) coils, and B_0 inhomogeneity corrections (shimming) for a spatial resolution of about 10^{-2} cm^3. Proton MRS is often complicated by the background signals of water and lipids that have to be suppressed using editing sequences.[44] In certain cases, MRS of different nuclei such as ^{19}F or ^{31}P is more appropriate for target-specific MRS.

As mentioned earlier, ^{1}H MRI detects signals from bulk water in the sample (proton concentration of about 90 M) and can provide spatial resolutions of about 10^{-5} cm^3. The specificity of MRI can be significantly increased by the use of MR contrast agents (CA). A single molecule of CA affects a large number of surrounding water molecules and induces substantial changes in the detected water signal. Shortening of the relaxation time by a contrast (or relaxation) agent can be detected with an appropriately weighted MRI method. CA are generally classified as "T_1 agents" or "T_2 agents." Reduction in T_1 results in increased intensities in T_1-weighted MR images (so-called positive contrast), whereas T_2 shortening produces a negative contrast or reduced brightness in T_2-weighted MR images. CA are characterized by T_1 or T_2 relaxivity, which is the reciprocal of the change in relaxation time (T_1 or T_2) per unit of concentration. Most T_1 CA are based on different chelate complexes of Gd, and T_2 CA usually incorporate a solid iron oxide core embedded in various polymer coatings. Therefore it is convenient to express the relaxivity of the CA per unit concentration of the metal as (m$M \cdot$ s)$^{-1}$. T_1 and T_2 relaxivities of CA are complex functions of the magnetic field B_0, temperature, and correlation times (rotational, translational, and exchange) that in turn depend on the molecular size and structure of the CA complexes. Typical T_1 relaxivity values of Gd-based CA are in the range of 5–20 (m$M \cdot$ s)$^{-1}$ [up to 80 (m$M \cdot$ s)$^{-1}$ for MS-325 bound to albumin].[45] Iron oxide–based superparamagnetic

---

[44] I. Tkac, Z. Starcuk, I. Y. Choi, and R. Gruetter, *Magn. Reson. Med.* **41**, 649 (1999).

CA have high T_2 relaxivities reaching 200 $(mM \cdot s)^{-1}$.[46] To generate detectable contrast in MR images, the average concentration of [Gd] and [Fe] should be above 10 μM (10% shortening of the relaxation times for intrinsic T_1 of 1 s and T_2 of 100 ms, respectively).

The minimum detectable concentration of the target will depend upon the efficiency of labeling of the CA, or MR probes. To estimate the upper limit of detection, it is important to consider typical *in vivo* concentrations of molecular targets that are most abundant and are of biological and medical significance. Cell surface receptors represent such a class of molecular targets, and for highly expressed receptors (10^6 per cell), the concentration of binding sites per unit volume is about 10^{15}/ml or ~ 1 μM, providing complete accessibility of the target to the MR probes. To enable MR detection, these targets have to be labeled with a significant number ($>>10$) of Gd or Fe atoms; for MRS, each target should be associated with more than 10^3 of the specific chemical probes.

Different strategies are currently available to perform target-specific or molecular MRI. One strategy is to use macromolecular carriers labeled with a large number of Gd^{3+} ions, or iron oxide nanoparticles that incorporate thousands of Fe^{3+} ions, and to direct them to receptors using highly specific high-affinity probes such as monoclonal antibodies (mAb). An alternative approach is to use enzymatic signal amplification, where the activity of an endogenous or exogenous enzyme significantly amplifies the number of MR reporter molecules. Yet another approach uses the exchange between bulk water and exchangeable groups on the surface of a reporter protein or a polymer probe. This chemical exchange can be detected as a decreased intensity of the water resonance, when the specific group is irradiated with an RF field. As in the case of relaxation agents, a significant increase in sensitivity is possible because one exchangeable group affects the signal of the abundant water molecules. The following section briefly describes some of these strategies and their application to the study of cells and preclinical tumor models.

T_1 Contrast Agents. As discussed earlier, the unpaired electrons of paramagnetic metal ions such as Gd^{3+} (III), Mn^{2+} (II), or Fe^{3+} (III) generate strong magnetic moments that induce efficient relaxation of water molecules. The chelates of these metals can be used as efficient T_1 CA. Because of the optimum electron relaxation time and high stability of the

[45] P. Caravan, N. J. Cloutier, M. T. Greenfield, S. A. McDermid, S. U. Dunham, J. W. Bulte, J. C. Amedio, Jr., R. J. Looby, R. M. Supkowski, W. D. Horrocks, Jr., T. J. McMurry, and R. B. Lauffer, *J. Am. Chem. Soc.* **124**, 3152 (2002).

[46] E. Strable, J. W. Bulte, B. M. Moskowitz, K. Vivekanandan, M. Allen, and T. Douglas, *Chem. Mater.* **13**, 2201 (2001).

complex, GdDTPA chelates are used most frequently. Early attempts to conjugate GdDTPA to a specific mAb for target-specific MRI were reported as early as 1985 by Unger *et al.*[47] and subsequently by Matsumura *et al.*[48] and Shahbazi-Gahrouei *et al.*[49] Because only a limited number of Gd chelates can be conjugated to a single mAb without significantly reducing its binding affinity,[50] in these studies the contrast generated was insufficient for MR detection.[51]

To increase the relaxivity per target site, multiple Gd chelates can be attached to a polymer carrier molecule such as a protein. In comparison with GdDTPA, these high molecular weight complexes have a longer circulation time and increased relaxivity. Several macromolecular CA platforms based on Gd chelates have been designed and used, including albumin–GdDTPA conjugates,[41,52] poly-L-lysine,[50,53] avidin,[54] and poly-amidoamine (PAMAM) dendrimers of different generations.[55,56] Large molecular size CA include cross-linked liposomes and nanoparticle emulsions labeled with a large number ($\sim$50,000) of Gd^{3+} ions.[57–59] The typical molecular size of these CA varies from about 6–8 nm for protein-based agents to $\sim$200 nm for polymerized liposomes and nanoparticle emulsions. Interestingly, the relaxivity of a complex is a linear function of the number

[47] E. C. Unger, W. G. Totty, D. M. Neufeld, F. L. Otsuka, W. A. Murphy, M. S. Welch, J. M. Connett, and G. W. Philpott, *Invest. Radiol.* **20,** 693 (1985).

[48] A. Matsumura, Y. Shibata, K. Nakagawa, and T. Nose, *Acta Neurochir.* **60**(Suppl.), 356 (1994).

[49] D. Shahbazi-Gahrouei, M. Williams, S. Rizvi, and B. J. Allen, *J. Magn. Reson. Imaging* **14,** 169 (2001).

[50] S. Gohr-Rosenthal, H. Schmitt-Willich, W. Ebert, and J. Conrad, *Invest. Radiol.* **28,** 789 (1993).

[51] W. T. Anderson-Berg, M. Strand, T. E. Lempert, A. E. Rosenbaum, and P. M. Joseph, *J. Nucl. Med.* **27,** 829 (1986).

[52] U. Schmiedl, M. Ogan, H. Paajanen, M. Marotti, L. E. Crooks, A. C. Brito, and R. C. Brasch, *Radiology* **162,** 205 (1987).

[53] A. A. Bogdanov, Jr., R. Weissleder, H. W. Frank, A. V. Bogdanova, N. Nossif, B. K. Schaffer, E. Tsai, M. I. Papisov, and T. J. Brady, *Radiology* **187,** 701 (1993).

[54] D. Artemov, N. Mori, R. Ravi, and Z. M. Bhujwalla, *Cancer Res.* **63,** 2723 (2003).

[55] L. H. Bryant, Jr., M. W. Brechbiel, C. Wu, J. W. Bulte, V. Herynek, and J. A. Frank, *J. Magn. Reson. Imaging* **9,** 348 (1999).

[56] H. Kobayashi, S. Kawamoto, T. Saga, N. Sato, A. Hiraga, T. Ishimori, Y. Akita, M. H. Mamede, J. Konishi, K. Togashi, and M. W. Brechbiel, *Magn. Reson. Med.* **46,** 795 (2001).

[57] D. A. Sipkins, D. A. Cheresh, M. R. Kazemi, L. M. Nevin, M. D. Bednarski, and K. C. Li, *Nat. Med.* **4,** 623 (1998).

[58] S. A. Anderson, R. K. Rader, W. F. Westlin, C. Null, D. Jackson, G. M. Lanza, S. A. Wickline, and J. J. Kotyk, *Magn. Reson. Med.* **44,** 433 (2000).

[59] P. M. Winter, S. D. Caruthers, X. Yu, S. K. Song, J. Chen, B. Miller, J. W. Bulte, J. D. Robertson, P. J. Gaffney, S. A. Wickline, and G. M. Lanza, *Magn. Reson. Med.* **50,** 411 (2003).

of Gd^{3+} ions attached to the polymer carrier[52] and increases with the size of the complex. It is not clear if relaxivity is saturated when standard chelate chemistry is used for sequestering Gd^{3+} ions.

Large paramagnetic Gd complexes, such as paramagnetic polymerized liposomes[57] and Gd–perfluorocarbon nanoparticles,[58,59] were successfully used to image $\alpha_v\beta_3$ integrin receptors expressed on angiogenic endothelium. These nanoparticles were targeted to the receptor by mAb either covalently bound, or attached via biotin–avidin linkers. Although these CA were large, the molecular target was intravascular and therefore easily accessible.

Smaller CA are necessary to image molecular targets, which require extravasation of the CA in solid tumors because of the reduced permeability and interstitial diffusion of large CA. MRI of mucin-like protein expressed in many types of gastrointestinal carcinomas was reported, using anti-mucin mAb covalently conjugated to poly-L-lysine–GdDTPA with a labeling ratio of 65 Gd^{3+} ions per molecule.[50] The molecular weight of the conjugate was about 200 kD. Folate-conjugated GdDTPA magnetic dendrimers (fourth generation, up to 64 complexing sites for Gd^{3+}) were used for targeted MRI of human folate receptors in ovarian tumor xenografts.[60]

Recently, a two-step labeling strategy was used for *in vivo* MRI of the HER-2/*neu* receptor. The two-step labeling allows separation of the targeted CA complex into two components with relatively low molecular weight (160 kDa for mAb, and 70 kDa for avidin), which improves the delivery of the CA. Receptors were prelabeled with biotinylated anti-HER-2/*neu* mAb, and probed with an GdDTPA–avidin conjugate 12 h later. A maximum labeling efficiency of about 50 Gd per receptor was estimated, from an average of 12.5 Gd^{3+} per avidin and 4 biotins per mAb. Positive contrast in HER-2/*neu* overexpressing tumor xenografts was detected in T_1-weighted MR images using this approach.[54]

A novel concept in MR molecular imaging is enzyme-specific amplification of the T_1 relaxivity of the targeted CA. One such approach relies on the β-galactosidase–mediated catalytic removal of the protecting sugar cap from a caged Gd contrast agent.[61] This change in chemical structure of the compound enables a large number of water molecules to be directly coordinated by the metal, dramatically reducing water T_1 relaxation time due to inner sphere effects.[62]

[60] S. D. Konda, M. Aref, M. Brechbiel, and E. C. Wiener, *Invest. Radiol.* **35,** 50 (2000).
[61] A. Y. Louie, M. M. Huber, E. T. Ahrens, U. Rothbacher, R. Moats, R. E. Jacobs, S. E. Fraser, and T. J. Meade, *Nat. Biotechnol.* **18,** 321 (2000).
[62] S. H. Koenig and R. D. Brown, 3rd, *Magn. Reson. Med.* **1,** 437 (1984).

Another approach is the enzymatic polymerization of small molecular weight Gd chelates, which results in a significant increase in T_1 relaxivity of the polymer product. Magnetic oligomers produced by polymerization of hydroxyphenol-modified GdDOTA monomers have been proposed as sensitive probes for MRI of nanomolar concentrations of oxidoreductases (peroxidase).[63] Low-relaxivity [3.75 $(\text{m}M\cdot\text{s})^{-1}$] monomeric substrates (AH) are oxidized by the peroxidase (E), thereby reducing peroxide (H_2O_2) and producing activated moeties (A^*) that self-polymerize to form high-relaxivity [11.5 $(\text{m}M\cdot\text{s})^{-1}$] magnetic oligomers (A_n):

$$2AH + [E \bullet H_2O_2] \rightarrow 2A^* + 2H_2O + E \qquad nA^* \rightarrow A_n \qquad (19)$$

This method was used for MRI of E-selectin expressed on the surface of endothelial cells using sandwich constructs of anti-E-selectin $F(ab')_2$ conjugated to peroxidase through digoxigenin (DIG)–anti-DIG Ab linkers.[63]

T_2 *Contrast Agents.* The large combined magnetic moment (so-called Curie spin) of superparamagnetic iron oxide nanoparticles makes them effective T_2 contrast agents.[64] For biocompatibility and chemical stabilization, the monocrystalline (MION) and polycrystalline (SPIO) iron oxide cores, which have diameters of $\sim$5 nm and $\sim$30 nm, respectively, are coated with a protective layer composed of biopolymers such as dextran or other polysaccharide, or a unilamellar vesicle (e.g., magnetoliposomes),[65] resulting in diameters of 17–50 nm. Magnetodendrimers consist of a superparamagnetic iron oxide core encapsulated within a dendrimer superstructure.[46,66] These compounds typically demonstrate high T_2 relaxivity, often up to and above 200 $(\text{m}M \cdot \text{s})^{-1}$ at $B_0 = 1.5$ T. In comparison, the corresponding T_1 relaxivity of these compounds is typically below 10 $(\text{m}M \cdot \text{s})^{-1}$. The superparamagnetism of the iron oxide core depends upon the alignment of individual electron spins. The T_2 relaxivity of the ultrasmall monocrystalline superparamagnetic iron oxide nanoparticles (MION-46L), with a magnetic core diameter of 4.6 nm, is close to 20 $(\text{m}M \cdot \text{s})^{-1}$ at 1.5 T magnetic field and 25°.[67] For larger SPIO particles, with a core diameter of 16 nm, the T_2 relaxivity at 1.5 T is 240 $(\text{m}M \cdot \text{s})^{-1}$. These

[63] A. Bogdanov, Jr., L. Matuszewski, C. Bremer, A. Petrovsky, and R. Weissleder, *Mol. Imaging* **1,** 16 (2002).

[64] P. Gillis, A. Roch, and R. A. Brooks, *J. Magn. Reson.* **137,** 402 (1999).

[65] J. W. Bulte, R. A. Brooks, B. M. Moskowitz, L. H. Bryant, Jr., and J. A. Frank, *Magn. Reson. Med.* **42,** 379 (1999).

[66] J. W. Bulte, T. Douglas, B. Witwer, S. C. Zhang, E. Strable, B. K. Lewis, H. A. Zywicke, B. Miller, P. van Gelderen, B. M. Moskowitz, I. D. Duncan, and J. A. Frank, *Nat. Biotechnol.* **19,** 1141 (2001).

[67] J. W. Bulte, M. de Cuyper, D. Despres, and J. A. Frank, *J. Magn. Reson. Imaging* **9,** 329 (1999).

nanoparticles produce even larger T_2^* effects by inducing local disturbances in the magnetic field. These "long range" effects cause dephasing of signal from areas much larger than the size of the nanoparticle.[68,69]

The high T_2 relaxation rate of these compounds makes them an attractive choice for MRI of cellular targets. Cells nonspecifically labeled with iron oxide nanoparticles by phagocytosis and pinocytosis can be detected by MRI at iron concentrations as low as 1.7×10^{-15} g/cell or 8.5×10^4 particles/cell.[70] MRI of a single T-cell, loaded with SPIO, was demonstrated by Dodd et al.[69] Stem cells loaded with magnetodendrimers (9–14 pg iron/cell) were successfully imaged both in vitro and in vivo after transplantation.[66] Iron oxide–based CA can be used for MRI of cellular targets with an expression level of about 10^5 cell^{-1}.

As with Gd-based CA, iron oxide CA can be targeted to molecular epitopes on the cell surface by chemical conjugation with mAb (or mAb fragments). Cultured lymphocytes were imaged with a mAb directed against the lymphocyte common antigen conjugated to SPIO particles.[68] The conjugation was achieved using a streptavidin linker connecting the biotinylated mAb and the biotinylated dextran-coated SPIO nanoparticles. E-selectin expressing human endothelial cells were imaged using the F(ab)$_2$ fragment of anti-human E-selectin mAb conjugated to CLIO (cross-linked iron oxide) nanoparticles.[71] Only cells incubated with IL-1β cytokine demonstrated a negative contrast in T_2-weighted MR images. HER-2/neu expressing malignant breast cancer cells were detected using two-step labeling with biotinylated Herceptin mAb and streptavidin–SPIO nanoparticles.[72] An example of two-step labeling of HER-2/neu cell surface receptors with streptavidin–SPIO particles and biotinylated anti-HER-2/neu mAb (Herceptin) is shown in Fig. 6. Cell lines expressing high levels of the receptor (AU-565) generate strong negative T_2 contrast as shown in Fig. 6B. Data from these studies demonstrated a linear dependence between the concentration of the target sites and T_2 relaxivity generated by SPIO CA.[72]

The in vivo applications of iron oxide–targeted CA include imaging of inflammation sites with human polyclonal immunoglobulin G (IgG)

[68] J. W. Bulte, Y. Hoekstra, R. L. Kamman, R. L. Magin, A. G. Webb, R. W. Briggs, K. G. Go, C. E. Hulstaert, S. Miltenyi, and T. H. The, Magn. Reson. Med. 25, 148 (1992).

[69] S. J. Dodd, M. Williams, J. P. Suhan, D. S. Williams, A. P. Koretsky, and C. Ho, Biophys. J. 76, 103 (1999).

[70] R. Weissleder, H. C. Cheng, A. Bogdanova, and A. Bogdanov, J. Magn. Reson. Imaging 7, 258 (1997).

[71] H. W. Kang, L. Josephson, A. Petrovsky, R. Weissleder, and A. Bogdanov, Jr., Bioconjug. Chem. 13, 122 (2002).

[72] D. Artemov, N. Mori, B. Okollie, and Z. M. Bhujwalla, Magn. Reson. Med. 49, 403 (2003).

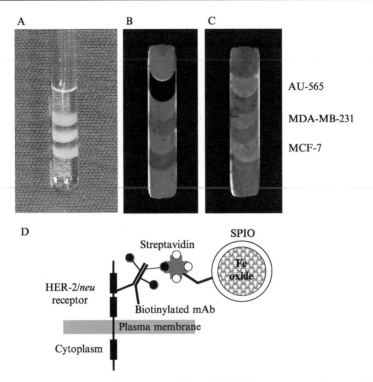

FIG. 6. MR images of AU-565, MDA-MB-231, and MCF-7 breast cancer cells embedded in agarose gel in a 5 mm NMR tube. (A) Layout of the cell sample. Cells were pretargeted with biotinylated Herceptin and a nonspecific biotinylated mAb (negative control), and probed with streptavidin SPIO microbeads as shown in (D). T_2 maps of the cell samples were reconstructed from eight T_2-weighted images acquired with a relaxation delay of 8 s and TE in the range 20 to 250 ms. A T_2 map of a cell sample probed with Herceptin is shown in (B), and the control cell sample treated with a nonspecific biotinylated mAb is shown in (C). Expression level of the HER-2/*neu* receptor was 2.7×10^6 for AU-565, 8.9×10^4 for MCF-7, and 4×10^4 for MDA-MB-231 cells, respectively. (See color insert.)

attached to MION particles,[73] and imaging of apoptosis in solid tumor models exposed to a chemotherapeutic agent.[74] In the latter study, SPIO particles were conjugated to the C2 domain of the protein synaptotagmin, which selectively binds to phosphatidylserine residues that relocate onto the outer leaflet of the plasma membrane in apoptotic cells.[74]

[73] R. Weissleder, A. S. Lee, A. J. Fischman, P. Reimer, T. Shen, R. Wilkinson, R. J. Callahan, and T. J. Brady, *Radiology* **181,** 245 (1991).

[74] M. Zhao, D. A. Beauregard, L. Loizou, B. Davletov, and K. M. Brindle, *Nat. Med.* **7,** 1241 (2001).

Weissleder et al.[75] have demonstrated a novel strategy for MR signal enhancement by the active transport of CA into cells using a transporter system. 9L glioma cells were engineered to express the modified transferrin receptor, ETR, with a knocked-down negative feedback regulation domain. These cells were loaded with MION nanoparticles conjugated to human holotransferrin (TF), which is a substrate for ETR, and imaged both in situ and in vivo using T_2^*-weighted MRI. Within an hour, up to 8×10^6 of the TF-targeted CA nanoparticles were internalized by the cells.

A significant advantage of the iron oxide nanoparticles compared with Gd-based CA is their high T_2/T_2^* relaxivity, which produces strong negative MR contrast at nanomolar concentrations of the CA. However, the large molecular size ($\sim$20–30 nm) of the superparamagnetic iron oxide–based CA can be a potential problem for in vivo applications, since delivery may be limited. The successful applications of these CA for target-specific MRI were probably a result of highly permeable vasculature at the sites of inflammation or in treated apoptotic tumors, which permitted efficient delivery of these CA into the interstitium. The long circulation time of these agents and an efficient amplification of the label by cell internalization are other possible reasons for contrast uptake within the tumor.

MR Spectroscopy and Water Exchange. Of all the stable magnetic nuclei, ^{1}H MRS provides the highest sensitivity, although strong water and lipid signals pose a problem for in vivo detection. Another magnetic isotope that can be detected by MRS with about 80% of proton receptivity is ^{19}F. The complete absence of the natural fluorine background and a large range of chemical shifts are important advantages of the use of this isotope as a reporter for in vivo MR studies.

An example of the use of ^{19}F to detect a reporter enzyme is the conversion of 5-fluorocytosine (5-FC) to 5-fluorouracil (5-FU) by the bacterial or yeast enzyme, cytosine deaminase (CD). CD is not present in mammalian cells, and therefore the formation of 5-FU from 5-FC occurs only in regions where CD is expressed. The conversion of 5-FC to 5-FU was detected, in vivo, in solid tumors derived from cancer cells transfected with yeast CD.[76] Because the chemical shift difference between 5-FC and 5-FU is approximately 1.2 ppm, and approximately 3.5 ppm between 5-FC and the fluorinated nucleotide–nucleoside products of 5-FU, all three compounds are easily resolved in vivo. In a similar approach, CD was covalently conjugated to mAb specific for the L6 antigen, which is expressed on the surface

[75] R. Weissleder, A. Moore, U. Mahmood, R. Bhorade, H. Benveniste, E. A. Chiocca, and J. P. Basilion, Nat. Med. 6, 351 (2000).

[76] L. D. Stegman, A. Rehemtulla, B. Beattie, E. Kievit, T. S. Lawrence, R. G. Blasberg, J. G. Tjuvajev, and B. D. Ross, Proc. Natl. Acad. Sci. USA 96, 9821 (1999).

of the human lung adenocarcinoma cell line, H2981.[77] *In vivo* ^{19}F MRSI of mice bearing H2981 tumors demonstrated localization of 5-FU signals within the tumor region. Although a relatively high magnetic field strength of 4.7 T was used in these studies, the available sensitivity only allowed spectroscopic images of 5-FU and 5-FC with an in-plane resolution of 6.25 × 6.25 mm, with a 20-mm-thick slice to be acquired within 35 min. The sensitivity and resolution of the method is also reduced by the rapid diffusion, clearance, and metabolic degradation of the compounds.

Another example of using MRS to detect an enzyme reporter is the detection of phosphocreatine produced from creatine *in vivo* in liver trans-fected with the murine creatine kinase enzyme (CK-B).[78] The low sensitiv-ity of ^{31}P MRS (~7% of photon) used in this study limits the application of this method for *in vivo* studies.

Chemical-exchange saturation transfer (CEST) permits MR detection of exchangeable protons with high sensitivity by detecting changes in the bulk water signal. Technically, changes in the water intensity are measured with and without RF irradiation at the chemical shift of the exchangeable group.[79,80] Hydroxyl, amine, and amide protons have been used for CEST detection.[81] Magnetization transfer from the exchangeable proton to nu-merous water protons provides a sensitivity enhancement factor of over 5000 for specially optimized probes such as polyuridilic acid (imino protons).[82] The proton transfer enhancement factor (PTE) is defined as:

$$PTE = 2 \bullet [H_2O]/[probe](1 - S_{sat}/S_0) \qquad (20)$$

where S_{sat} and S_0 are the water signal intensities with and without RF ir-radiation. The number of exchangeable protons and the exchange rate con-stant are the key parameters that define the efficiency of the CEST probe. The proton exchange rate should be low enough to enable spectral sepa-ration of the MR signal of the exchangeable group and, on the other hand, should be sufficiently fast to allow multiple water molecules to exchange protons with the exchangeable group.[80] Increasing the chemical shift differ-ence between the exchangeable group and water enables use of the higher exchange rates with correspondingly higher PTE. This can be achieved

[77] E. O. Aboagye, D. Artemov, P. Senter, and Z. M. Bhujwalla, *Cancer Res.* **58,** 4075 (1998).
[78] A. Auricchio, R. Zhou, J. M. Wilson, and J. D. Glickson, *Proc. Natl. Acad. Sci. USA* **98,** 5205 (2001).
[79] V. Guivel-Scharen, T. Sinnwell, S. D. Wolff, and R. S. Balaban, *J. Magn. Reson.* **133,** 36 (1998).
[80] N. Goffeney, J. W. Bulte, J. Duyn, L. H. Bryant, Jr., and P. C. van Zijl, *J. Am. Chem. Soc.* **123,** 8628 (2001).
[81] S. Mori, P. C. van Zijl, and D. Shortle, *Proteins* **28,** 325 (1997).
[82] K. Snoussi, J. W. Bulte, M. Gueron, and P. C. van Zijl, *Magn. Reson. Med.* **49,** 998 (2003).

with paramagnetic shift reagents that introduce a large shift to the resonance frequency of the neighboring spins.[83] CEST agents provide sensitivity similar or higher than paramagnetic CA (for the same number of exchangeable groups and Gd chelates). The sensitivity gain provided by CEST suggests that dedicated CEST contrast agents may be used as probes for targeted MRI.

Although the MRI probes and methods discussed here can be successfully used to image molecular targets in isolated cells, the translation of these techniques *in vivo* is not straightforward. The imaging properties of different targeted MR agents and potential areas of their application for *in vivo* MRI/MRS are summarized in Table I. The translation of these techniques into the clinic will require a substantial improvement in both CA chemistry and MRI technology.

Here we discuss strategies based on contrast agent chemistry to achieve gain in signal. In the following paragraphs we outline technical approaches to improve the sensitivity of MR methods.

Technical Strategies for MR Microscopy

Traditionally, cell microscopy is performed using optical techniques. The promise of noninvasively detecting molecular events with high spatial localization and sufficient sensitivity drives the effort in MR microscopy. High-resolution MR spectroscopy has been utilized in achieving spectra of volume-limited samples[84–86] and in performing localized spectroscopy on picoliter-scale samples.[87,88] High-resolution imaging has been implemented in the microscopic investigation of cell and tissue structures,[89,90] of large single cells,[91] and to detect single mammalian cells.[70] Recent work

[83] S. Zhang, P. Winter, K. Wu, and A. D. Sherry, *J. Am. Chem. Soc.* **123,** 1517 (2001).

[84] D. L. Olson, T. L. Peck, A. G. Webb, R. L. Magin, and J. V. Sweedler, *Science* **270,** 1967 (1995).

[85] T. L. Peck, R. L. Magin, and P. C. Lauterbur, *J. Magn. Reson. B* **108,** 114 (1995).

[86] R. Subramanian, M. M. Lam, and A. G. Webb, *J. Magn. Reson.* **133,** 227 (1998).

[87] R. A. Wind, K. R. Minard, G. R. Holtom, P. D. Majors, E. J. Ackerman, S. D. Colson, D. G. Cory, D. S. Daly, P. D. Ellis, N. F. Metting, C. I. Parkinson, J. M. Price, and X. W. Tang, *J. Magn. Reson.* **147,** 371 (2000).

[88] K. R. Minard and R. A. Wind, *J. Magn. Reson.* **154,** 336 (2002).

[89] Z. H. Cho, C. B. Ahn, S. C. Juh, J. M. Ja, R. M. Friedenberg, S. E. Fraser, and R. E. Jacobs, *Philos. T. Roy. Soc. A* **333,** 469 (1990).

[90] P. M. Glover, R. W. Bowtell, G. D. Brown, and P. Mansfield, *Magn. Reson. Med.* **31,** 423 (1994).

[91] J. B. Aguayo, S. J. Blackband, J. Schoeniger, M. A. Mattingly, and M. Hintermann, *Nature* **322,** 190 (1986).

TABLE I

SUMMARY OF TARGETED CA CURRENTLY AVAILABLE, THEIR APPLICATIONS AND POTENTIAL PROBLEMS

Imaging agent	Relative sensitivity of MR detection	Potential molecular targets	MRI/MRS experimental methods	Potential problems
Iron oxide-based agents (MION, SPIO, CLIO)	High, $\sim10^5$ target sites per cell	Cell surface receptors *in situ*, cell transporters	T_2- and T_2^*-weighted and T_2 quantitative MRI	Large size of nanoparticles (~30 nm) restricts *in vivo* delivery
Gd-loaded liposomes and nanoparticle emulsions	High, $\sim10^4$ targets per cell	Cellular markers on the endothelium in blood vessel lumen.	T_1-weighted and T_1 quantitative MRI	Very large molecular size, slow extravasation
Gd-labeled macromolecule complexes (proteins, dendrimers)	Moderately high, $\sim10^6$ receptors per cell	Cell targets in solid tumors with leaky vasculature	T_1-weighted and T_1 quantitative MRI	Pharmacokinetics and stability of chelate complexes *in vivo*
MRS probes (5-FC)	Low, millimolar concentrations of the probe	Reporter enzymes, expressed in the cell or delivered exogenously	^{1}H, ^{19}F, ^{31}P MRS	Low sensitivity results in low spatial resolution
Novel probes (polymerized, CEST)	Potentially can be very high	Reporter enzymes and cell surface markers	T_1-weighted MRI, proton exchange MRI	No data

in single cell tracking[92] and the study of compartmental diffusion of isolated single cells[93] provides an avenue to noninvasively study disease progression and regression, response of individual cells or small cellular structures to external perturbation, immune attack, and gene therapy. Therefore MR microscopy has promise in the noninvasive time-course study of precancerous events, early stages of tumor development, diagnosis of small tumors, and distinction of benign tumors from malignant ones.

Although MR has limited sensitivity, refinement in microfabrication and nanofabrication techniques, electronic circuitry, and pulse sequences make MR microscopy possible. The information provided from MR-derived parameters provide excellent means of obtaining high-resolution biochemical, functional, and morphologic information from cancer cells and tumors. Although the fundamental principles governing conventional MR and MR microscopy are the same, there are certain challenges distinct to microscopy. These challenges arise from molecular diffusion, low SNR resulting from small voxel size and RF coil insensitivity to the sample, and local magnetic field inhomogeneity. Several of these issues are discussed here.

Technical Aspects of Microscopy

SIGNAL TO NOISE RATIO. One of the principal impediments in microscopy is low SNR. MR is inherently insensitive because its signal amplitude relies on a small nuclear population difference between two energy states. For example, out of 2,000,009 protons at 1.5 T, only 9 protons contribute to the MR signal. Furthermore, the acquisition time varies inversely with the square of SNR. To maintain SNR, as the resolution improves from 1 mm $\times$ 1 mm $\times$ 1 mm to 100 μm $\times$ 100 μm $\times$ 100 μm, for instance, the acquisition will take 1 million times longer. Therefore microscopy efforts focus heavily on enhancing signal amplitude and detection sensitivity.

Signal enhancement may be achieved by increasing the static magnetic field strength, although high field systems are expensive and field inhomogeneity can be a potential drawback at high field. Another approach to enhance signal is polarization of the sample. As mentioned earlier, only a handful of nuclei contribute to the MR signal. The ratio of signal-generating nuclei to the total number of nuclei can be increased by polarizing the sample. Polarization in MR is most commonly achieved through optical pumping of the noble gas isotopes ^{3}He and ^{129}Xe. Hyperpolarized gas has

[92] K. A. Hinds, J. M. Hill, E. M. Shapiro, M. O. Laukkanen, A. C. Silva, C. A. Combs, T. R. Varney, R. S. Balaban, A. P. Koretsky, and C. E. Dunbar, *Blood* **102**, 867 (2003).
[93] S. C. Grant, D. L. Buckley, S. Gibbs, A. G. Webb, and S. J. Blackband, *Magn. Reson. Med.* **46**, 1107 (2001).

proven to be valuable in imaging pulmonary microstructure and function.[94] Hyperpolarized gases also induce cross-relaxation, thereby increasing signals from other nuclei within the environment, and they hold significant potential for MRI and MRS studies of the biologic microenvironment.

High-temperature superconducting coils also increase SNR by eliminating resistance in the RF sensor.[95a] Cryo-cooled probes and preamplifiers reduce thermal noise in the system, thereby increasing SNR. These noise reduction–elimination techniques are often cumbersome and expensive, and have yet to be fully optimized. The bulk of the effort in increasing sensitivity is directed toward designing and manufacturing sensitive room temperature RF coils. Improving RF sensitivity is less expensive than increasing field strength and does not require sample polarization or special cooling or superconducting circuitry. RF coils can be tailored to the sample under investigation to provide optimal SNR for that sample, providing flexibility in achieving high performance with a variety of samples.

RF SENSITIVITY. A principal challenge of MR microscopy is to increase SNR, which is proportional to RF sensitivity[95b,c]:

$$\text{SNR} \propto \frac{\omega_0^2 \bullet \left(\dfrac{B_1}{i}\right) \bullet v_s}{V_{noise}} \tag{21}$$

where ω_0 (rad/s) is the nuclear precession frequency, B_1/i (T/A) is the transverse magnetic field generated by the coil per unit current, v_s (m^3) is the sample volume, and V_{noise} (V) is the noise voltage, expressed as:

$$V_{noise} = \sqrt{4 \bullet k_B \bullet T \bullet R_{noise} \bullet \Delta f} \tag{22}$$

where k_B (1.38×10^{-23} J/K) is Boltzman's constant, T is the temperature in K, R_{noise} (Ω) is the noise owing to the sample and coil, and Δf (Hz) is the spectral bandwidth. In the microcoil regime, sample losses are negligible and the total resistance is dominated by the coil resistance.[95d] The $(B_1/i) \bullet v_s$ term is closely correlated with the filling factor of the coil. The most effective way to increase RF sensitivity when imaging volume-limited samples is to match the coil size to the sample size. When the filling factor of the coil

[94] H. E. Moller, X. J. Chen, B. Saam, K. D. Hagspiel, G. A. Johnson, T. A. Altes, E. E. de Lange, and H. U. Kauczor, *Magn. Reson. Med.* **47**, 1029 (2002).

[95a] R. D. Black, T. A. Early, P. B. Roemer, O. M. Mueller, A. Mogro-Campero, L. G. Turner, and G. A. Johnson, *Science* **259**, 793 (1993).

[95b] D. I. Holt and R. E. Richards, *J. Magn. Res.* **94**, 71 (1976).

[95c] T. L. Peck, R. L. Magin, and P. C. Lauterbur, *J. Magn. Res.* **108**, 114 (1995).

[95d] D. I. Hoult and P. C. Lauterbur, *J. Magn. Res.* **34**, 425 (1979).

increases, SNR increases. Because acquisition time varies inversely with the square of SNR, increased SNR permits the investigation of biologic systems in short, physiologically relevant times. Therefore reduction in coil size and in coil resistance, both from miniaturization and other geometric optimization, have become the foci in microcoil design.

To address the sensitivity requirements of microscopy, a new regime of RF coils called "microcoils" has been developed.[84,86,91,96–98] The challenges in developing microcoils are those of fabrication and sample positioning. Volume coils must be fabricated with thin conductors and on very small capillaries that are capable of providing mechanical stability to the coil, without compromising the coil's sensing volume. With advances in microfabrication techniques, planar coils on the order of tens to hundreds of microns can be patterned to submicron resolution[99] and integrated with microfluidic systems[100] for accurate sample placement and replacement.

Volume microcoils: Volume coils are conducive to investigating samples whose geometry is principally three-dimensional, and for applications requiring high field homogeneity (narrow spectral linewidth). Birdcage, saddle, solenoid, and scroll configurations are examples of volume coil geometries. We have focused on solenoid coils because they are most widely used, and on the novel "scroll" geometry because of its potential for miniaturization.

Classical analysis of on-axis sensitivity of a solenoid of numerous turns is:

$$\frac{B_1}{i} = \frac{\mu_0 \bullet n}{d \bullet \sqrt{1 + \left(\frac{l}{d}\right)^2}} \tag{23}$$

where μ_0 ($4\pi \times 10^{-7}$ H/m) is the permeability of free space, n is the number of turns of the solenoid, d(m) is the coil diameter, and l(m) is the length of the coil. Therefore, for a fixed length to diameter ratio, as the coil diameter decreases, sensitivity increases. The previous equation was derived for the direct current (DC) approximation, where the injected current is distributed uniformly across the cross-section of the solenoid wire. Upon

[96] A. G. Webb, *Prog. Nucl. Mag. Res. Sp.* **31,** 1 (1997).

[97] R. Subramanian and A. G. Webb, *Anal. Chem.* **70,** 2454 (1998).

[98] S. C. Lee, K. Kim, J. Kim, S. Lee, J. Han Yi, S. W. Kim, K. S. Ha, and C. Cheong, *J. Magn. Reson.* **150,** 207 (2001).

[99] C. Massin, S. Eroglu, F. Vincent, B. S. Gimi, P.-A. Besse, R. L. Magin, and R. S. Popovic, *Transducers* 967 (2003).

[100] J. D. Trumbull, I. K. Glasgow, D. J. Beebe, and R. L. Magin, *IEEE Trans. Biomed. Eng.* **47,** 3 (2000).

application of an alternating current (AC), the analysis becomes complicated owing to eddy currents generated in the coil wire governed by the Faraday induction law, which effectively push the conductive current out toward the wire perimeter. The conductive current penetration in the wire cross-section is characterized by the skin depth

$$\delta = \frac{1}{\sqrt{\mu \bullet \pi \bullet \sigma \bullet f}} \qquad (24)$$

where μ (H/m) is the permeability of the wire, σ (mho/m) is its conductivity, and f (Hz) is the operating frequency. As the operating frequency increases, the skin depth decreases and current crowding at the coil perimeter increases the coil resistance. Additionally, if the coil windings are in close proximity to each other, current crowding due to the interactions of the coil windings will further increase coil resistance owing to the phenomenon of "proximity effect," also attributable to the Faraday induction law. With the wire diameter far exceeding skin depth, proximity effects and skin depth dominate resistance, whereas for wire dimensions close to skin depth, the resistance of the coil closely approximates the DC case. Therefore fabricating microcoils with conductor thickness on the order of skin depth is advantageous to SNR, with certain provisos that are beyond the scope of this chapter but are detailed elsewhere.[85]

Also, solenoid microcoils have been routinely fabricated by winding thin wire on a small-diameter capillary. Although the spacing between the windings is difficult to control and reproduce,[101] it is a critical factor in optimizing microcoil performance.[85] Solenoid microcoils have electrical leads of lengths equivalent to the length of the coil wire, and therefore losses in the leads owing to resistance and parasitic capacitance are not negligible. To minimize lead losses, a capacitor should be placed across the leads very close to the microcoil, effectively shortening the leads. However, even a nonmagnetic capacitor cannot be placed so close to the microcoil as to cause field distortion owing to magnetic susceptibility effects. The lead losses are an important part of the circuit and should be accounted for in any electrical model characterization of microcoil performance. Solenoid microcoils suffer from scalability and difficulty in fabrication. Wire thickness is a limiting factor in coil miniaturization, and multilayered solenoids are very difficult to wind. Scroll microcoils were developed to overcome these limitations.[102] A scroll microcoil is a conductor ribbon wound

[101] D. A. Seeber, R. L. Cooper, L. Ciobanu, and C. H. Pennington, *Rev. Sci. Instrum.* **72,** 2171 (2001).
[102] B. Gimi, S. C. Grant, R. L. Magin, A. Fienerman, E. Frolova, and G. Friedman, *Experimental Nuclear Magnetic Resonance Conference* (2000).

cylindrically. To generate multiple sensing layers, the conductor is laminated with a dielectric and this ribbon is wound upon itself. Because scroll microcoils can be fabricated from sheets of very thin conductor, their dimensions are not limited by wire diameter. A thick dielectric layer makes scrolls robust and easy to wind. This microcoil geometry has the advantage of achieving very small microcoil dimensions, and with conductor thickness on the order of skin depth, which reduces resistive losses as explained earlier.

Surface microcoils: Although volume coils provide a homogenous RF field and are conducive to three-dimensional geometry, there are microscopy applications where the requirements of sample geometry, loading, and positioning make surface coils more desirable. Because volume coils are wound around a capillary for mechanical stability, the capillary wall thickness can compromise valuable sensing volume for very small coils. Additionally, precise sample positioning is often not possible when using volume coils. Surface coils, on the other hand, provide a higher filling factor for certain samples, and permit greater flexibility in sample positioning and the sample's access to nutrients, drugs, and perfusates. Additionally, planar surface microcoils are well suited to investigate principally two-dimensional geometries such as cell cultures, because they provide high localized SNR in an excitation–acquisition plane close to the plane of the coil.

Current microfabrication technology is well suited for planar geometries and less amenable to geometries with high-aspect ratio (length to diameter ratio), making surface coils easier to microfabricate than volume coils. Therefore microfabrication of surface coils offer marked flexibility in geometric parameters. A common surface configuration is that of a spiral microcoil.[103,104] In contrast to a single loop coil, a spiral provides additional field sensing/focusing turns, thereby increasing the transverse magnetic field near the region of the coil. However, as the number of turns of the spiral increase, the coil resistance increases as well. After a certain number of turns, the contribution of the coil's resistance to SNR outbalances the advantage from an increased number of sensing/focusing turns. Therefore there is an optimal number of turns of a spiral coil for maximum SNR. When the turns of the spiral are far apart, the contribution of the outer turns to the field in the central region of the coil diminishes. When the turns of the spiral are close to each other, proximity effects play a role in the

[103] C. Massin, G. Boero, F. Vincent, J. Abenhaim, P.-A. Besse, and R. S. Popovic, *Sensors and Actuators A: Physical* **97,** 280 (2002).
[104] S. Eroglu, B. Gimi, B. Roman, G. Friedman, and R. L. Magin, *Concepts in Magnetic Resonance Imaging Part B: Magnetic Resonance Engineering* **17B,** 1 (2003).

coil's performance and induce losses in the coil. An optimum interturn spacing must be selected based on electrical modeling of inductive and capacitive interturn coupling. An increased number of turns of the spiral will increase the coil inductance, and the required capacitance to achieve RF resonance would be very small, making the RF circuit difficult to tune.

Surface coils provide very high localized SNR, but their SNR advantage over volume coils decreases rapidly with increasing imaging distance from the plane of the coil.[104] Surface coils generate radiant magnetic fields that create field inhomogeneities, which results in spectral broadening, and are frequently not the preferred configuration where narrow spectral linewidth is important. However, more than one surface transceiver microcoil can be used in generating the transverse magnetic field and receiving RF signal from the sample. For instance, a Helmholtz configuration would have a higher sensing region and field homogeneity than a single planar coil. To achieve high local SNR while still imaging a large field of view, several surface microcoils can be used in configurations such as a phased array.[105] Parallel imaging techniques[106] such as SENSE, which relies on arrays of mutually coupled coils,[107] and SMASH[108] are now frequently used. Detailed discussion of these techniques is beyond the scope of this chapter but may be found elsewhere.[109,110]

Another aspect of microcoil design is that of susceptibility effects on magnetic field homogeneity. A microcoil conductor such as copper would create static field distortions owing to its diamagnetic nature. Susceptibility-compensated wire (Doty Scientific Inc., Columbia, SC) reduces this diamagnetic effect on static field distortion. For example, a composite of an aluminum core within a copper tube will reduce the wire magnetism to approximately 2% of pure copper.

Technological advances in microfabrication and electronics, development of improved contrast agents and targeted molecular probes, and the emergence of new technology such as mechanical detection of MR[111,112] are likely to carry microscopy into the realm where subcellular detection

[105] P. B. Roemer, W. A. Edelstein, C. E. Hayes, S. P. Souza, and O. M. Mueller, *Magn. Reson. Med.* **16,** 192 (1990).

[106] D. K. Sodickson and C. A. McKenzie, *Med. Phys.* **28,** 1629 (2001).

[107] S. M. Wright, R. L. Magin, and J. R. Kelton, *Magn. Reson. Med.* **17,** 252 (1991).

[108] D. K. Sodickson and W. J. Manning, *Magn. Reson. Med.* **38,** 591 (1997).

[109] K. P. Pruessmann, M. Weiger, M. B. Scheidegger, and P. Boesiger, *Magn. Reson. Med.* **42,** 952 (1999).

[110] D. K. Sodickson, C. A. McKenzie, M. A. Ohliger, E. N. Yeh, and M. D. Price, *Magma* **13,** 158 (2002).

[111] J. A. Sidles and D. Rugar, *Phys. Rev. Lett.* **70,** 3506 (1993).

[112] A. Schaff and W. S. Veeman, *J. Magn. Reson.* **126,** 200 (1997).

is routine. The push toward implantable coils,[113] combining MR with optical imaging modalities,[87] and integrating microfluidics with MR systems[100] will lead to improved cellular and subcellular detection of physiological events *in vivo,* and the detection of molecular events *in vitro.*

Metabolic and Physiologic Spectroscopy and Spectroscopic Imaging with MRS and MRSI

MRS and MRSI provide a wealth of information on tumor physiology and metabolism. Different metabolic information can be derived depending upon the nucleus (i.e., ^{1}H, ^{13}C, or ^{31}P) examined.

^{1}H MRS

The high sensitivity of ^{1}H MRS permits proton spectra to be obtained with high spatial and temporal resolution. This is important because one of the main characteristics of tumor blood flow is its heterogeneity, which results in a heterogeneous oxygen, pH, and metabolite distribution. The ability to obtain spatially localized spectra from small voxels is necessary not only to localize the measurement to tumor tissue and minimize signal contribution from normal tissue, but also to determine the spatial distribution of metabolites within the tumor to better characterize the tumor, as well as its response to therapy. Unedited proton spectra are dominated by signal from water, methyl, and methylene protons. Water suppression methods are routinely used to eliminate the signal from water during acquisition, using techniques such as VAPOR,[44] CHESS,[114] or band selective refocusing.[115,116] Although the intense methyl and methylene signals originating from mobile lipids can provide useful information,[117–119] since these signals dominate the spectrum, resonances from metabolites such as lactate and alanine, which appear near the lipid region, are obscured. The use of a long echo time (TE) of 272 ms can significantly reduce signal from

[113] X. Silver, W. X. Ni, E. V. Mercer, B. L. Beck, E. L. Bossart, B. Inglis, and T. H. Mareci, *Magn. Reson. Med.* **46,** 1216 (2001).

[114] A. Hause, J. Frahm, W. Hanicke, and D. Mattaei, *Phys. Med. Biol.* **30,** 341 (1985).

[115] J. Star-Lack, S. J. Nelson, J. Kurhanewicz, L. R. Huang, and D. B. Vigneron, *Magn. Reson. Med.* **38,** 311 (1997).

[116] D. C. Shungu and J. D. Glickson, *Magn. Reson. Med.* **32,** 277 (1994).

[117] R. Callies, R. M. Sri-Pathmanathan, D. Y. Ferguson, and K. M. Brindle, *Magn. Reson. Med.* **29,** 546 (1993).

[118] N. M. Al-Saffar, J. C. Titley, D. Robertson, P. A. Clarke, L. E. Jackson, M. O. Leach, and S. M. Ronen, *Br. J. Cancer* **86,** 963 (2002).

[119] I. Barba, M. E. Cabanas, and C. Arus, *Cancer Res.* **59,** 1861 (1999).

the mobile lipids. Localized presaturation methods are also used when the lipid signal in the tumor is localized to a peripheral region.[116,120] Recent methods for lipid suppression have employed gradient filtering of lactate multiple-quantum coherences[121,122] and the application of adiabatic pulses for spectral editing.[123] However, coherence selection may not be optimal, and two-dimensional experiments, difference spectra, or additional phase cycling steps may be required to obtain uncontaminated spectra. He et al.[124,125] have shown that a homonuclear gradient-coherence transfer method, combined with a frequency selective pulse, is very effective in suppressing both lipid and water in a single scan.

Typically, proton spectra obtained from tumors contain resonances from taurine, total choline (choline, Cho; phosphocholine, PC; and gly-cerophosphocholine, GPC), total creatine (phosphocreatine and creatine), and lactate.[126] Figure 7 displays representative ^{1}H MR spectra obtained from an invasive and metastatic human breast cancer xenograft model, MDA-MB-231. Shown in Fig. 7A is a high-resolution ^{1}H MR spectrum of a perchloric acid extract of MDA-MB-231 cells. A diffusion-weighted, water-suppressed ^{1}H MR spectrum of live perfused cells is shown in Fig. 7B, and a spectrum obtained from a representative 1 mm $\times$ 1 mm $\times$ 4 mm voxel of a ^{1}H chemical shift imaging (CSI) data set, obtained from a MDA-MB-231 tumor in a SCID mouse is shown in Fig. 7C. The CSI data set was acquired using the BASSALE sequence.[116] High-resolution ^{1}H MRS resolves the $^3J_{H-H}$ coupling in lactate at 1.33 ppm and the N-$(CH_3)_3$ resonances of GPC, PC, and Cho (Fig. 7A insert). In perfused isolated cells, as well as in tumor xenografts, an unresolved signal of total choline-containing compounds is detected at 3.2 ppm, and a combined signal of lactate and triglycerides is detected at 1.3 ppm (Fig. 7B and C).

^{1}H MR spectra of tumors typically exhibit elevated total choline and lactate levels.[127–129] The high levels of lactate are consistent with high

[120] Z. M. Bhujwalla, D. C. Shungu, and J. D. Glickson, *Magn. Reson. Med.* **36,** 204 (1996).

[121] C. H. Sotak, *Magn. Reson. Med.* **7,** 364 (1988).

[122] R. E. Hurd and D. Freeman, *NMR Biomed.* **4,** 73 (1991).

[123] R. A. de Graaf, Y. Luo, M. Terpstra, and M. Garwood, *J. Magn. Reson. B* **109,** 184 (1995).

[124] Q. He, Z. M. Bhujwalla, R. J. Maxwell, J. R. Griffiths, and J. D. Glickson, *Magn. Reson. Med.* **33,** 414 (1995).

[125] Q. He, D. C. Shungu, P. C. van Zijl, Z. M. Bhujwalla, and J. D. Glickson, *J. Magn. Reson. B* **106,** 203 (1995).

[126] F. A. Howe, R. J. Maxwell, D. E. Saunders, M. M. Brown, and J. R. Griffiths, *Magn. Reson. Q* **9,** 31 (1993).

[127] W. Negendank, *NMR Biomed.* **5,** 303 (1992).

[128] I. S. Gribbestad, S. B. Petersen, H. E. Fjosne, S. Kvinnsland, and J. Krane, *NMR Biomed.* **7,** 181 (1994).

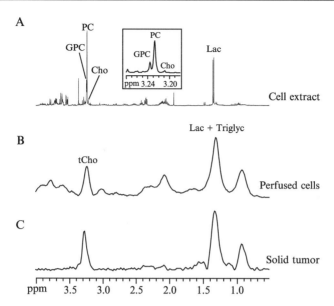

FIG. 7. ^{1}H MR spectra from (A) perchloric acid cell extracts, (B) intact cells, and (C) a 1 mm × 1 mm × 4 mm solid tumor voxel of the metastatic human breast cancer xenograft MDA-MB-231. The ^{1}H MR spectrum shown in (A) is a high-resolution spectrum of cell extracts. The insert displays a zoomed region at 3.2 ppm to demonstrate that human breast cancer cells exhibit low GPC levels, high PC levels, and high levels of total choline-containing metabolites. The diffusion-weighted, water-suppressed ^{1}H MR spectrum shown in panel (B) was acquired from intact cells perfused in our MR-compatible cell perfusion system. The ^{1}H MR spectrum shown in (C) is a representative 1 mm × 1 mm × 4 mm voxel obtained from a CSI data set acquired using the BASSALE sequence. CSI data were acquired with a TE of 272 ms and TR of 1 s within a total acquisition time of 25 minutes. Assignments made in the ^{1}H MR spectra are: Cho, free choline; GPC, glycerophosphocholine; PC, phosphocholine; Lac, lactate; tCho, total choline-containing metabolites; Lac + Triglyc, lactate + triglycerides.

glycolytic rates and poor blood flow associated with tumors.[130] The high levels of total choline detected in breast, prostate, and different types of brain tumors primarily arise from increased PC levels in tumor cells, as confirmed by high-resolution ^{1}H MRS studies of cell extracts.[131–133]

[129] F. A. Howe, S. J. Barton, S. A. Cudlip, M. Stubbs, D. E. Saunders, M. Murphy, P. Wilkins, K. S. Opstad, V. L. Doyle, M. A. McLean, B. A. Bell, and J. R. Griffiths, *Magn. Reson. Med.* **49,** 223 (2003).

[130] P. Vaupel, F. Kallinowski, and P. Okunieff, *Cancer Res.* **49,** 6449 (1989).

[131] K. K. Bhakoo, S. R. Williams, C. L. Florian, H. Land, and M. D. Noble, *Cancer Res.* **56,** 4630 (1996).

[132] E. O. Aboagye and Z. M. Bhujwalla, *Cancer Res.* **59,** 80 (1999).

[133] E. Ackerstaff, B. R. Pflug, J. B. Nelson, and Z. M. Bhujwalla, *Cancer Res.* **61,** 3599 (2001).

Molecular alterations underlying the increased PC levels observed in cancer cells include increased expression and activity of choline kinase,[134,135] a higher rate of choline transport,[136] and increased phospholipase D[137] and phospholipase A2[138] activity. Thus an increased membrane degradation or turnover combined with elevated choline kinase activity appear to cause the elevated total choline levels observed in tumor ^{1}H MR spectra, where both membrane phospholipid precursors and breakdown products contribute to the total choline signal.[132] Clinically, the total choline signal has been employed for proton MRSI of cancer. MRSI is typically performed in conjunction with high-resolution anatomic MRI, and it can significantly improve the diagnosis and the assessment of cancer location and aggressiveness. Pre- and post-therapy studies have demonstrated the potential of combined MRI and MRSI to provide a direct measure of the presence and spatial extent of cancer, as well as the time course and mechanism of therapeutic response.[139,140] The use of elevated choline levels to detect cancer with MRS or MRSI has been demonstrated for prostate,[140] brain,[141] breast,[142] and other cancers.[127]

^{13}C MRS

^{13}C MRS is uniquely suited for studying glycolysis and other metabolic pathways such as choline metabolism in cancer cells and solid tumors. The flux of metabolites through various pathways can be measured through the use of labeled substrates and metabolic modeling. Two applications of ^{13}C MRS are shown in Fig. 8. [1,2-^{13}C]-choline can be utilized to follow the production of the water-soluble ^{13}C-labeled choline phospholipid metabolites PC and GPC in human MDA-MB-231 breast cancer cells as shown in

[134] A. Ramirez de Molina, R. Gutierrez, M. A. Ramos, J. M. Silva, J. Silva, F. Bonilla, J. J. Sanchez, and J. C. Lacal, *Oncogene* **21,** 4317 (2002).

[135] A. Ramirez de Molina, A. Rodriguez-Gonzalez, R. Gutierrez, L. Martinez-Pineiro, J. Sanchez, F. Bonilla, R. Rosell, and J. Lacal, *Biochem. Biophys. Res. Commun.* **296,** 580 (2002).

[136] R. Katz-Brull and H. Degani, *Anticancer Res.* **16,** 1375 (1996).

[137] D. Y. Noh, S. J. Ahn, R. A. Lee, I. A. Park, J. H. Kim, P. G. Suh, S. H. Ryu, K. H. Lee, and J. S. Han, *Cancer Lett.* **161,** 207 (2000).

[138] C. J. Guthridge, M. R. Stampfer, M. A. Clark, and M. R. Steiner, *Cancer Lett.* **86,** 11 (1994).

[139] M. O. Leach, M. Verrill, J. Glaholm, T. A. Smith, D. J. Collins, G. S. Payne, J. C. Sharp, S. M. Ronen, V. R. McCready, T. J. Powles, and I. E. Smith, *NMR Biomed.* **11,** 314 (1998).

[140] J. Kurhanewicz, D. B. Vigneron, and S. J. Nelson, *Neoplasia* **2,** 166 (2000).

[141] X. Li, Y. Lu, A. Pirzkall, T. McKnight, and S. J. Nelson, *J. Magn. Reson. Imaging* **16,** 229 (2002).

[142] I. S. Gribbestad, B. Sitter, S. Lundgren, J. Krane, and D. Axelson, *Anticancer Res.* **19,** 1737 (1999).

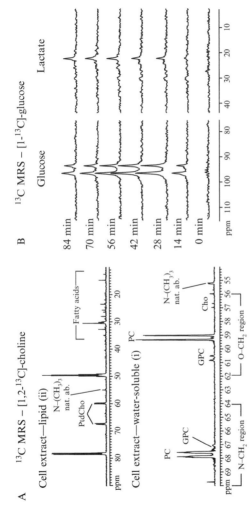

FIG. 8. (A) Representative ^{13}C MR spectra of the water-soluble (i) and the lipid fraction (ii) of MDA-MB-231 human breast cancer cells. Spectra were obtained from MDA-MB-231 cells that were labeled with 100 μM [1,2-^{13}C]-choline for 24 + 3 h and extracted using a dual phase extraction method. ^{13}C label was detected in the water-soluble metabolites glycerophosphocholine (GPC), phosphocholine (PC), and free choline (Cho), as well as in membrane phosphatidylcholine (PtdCho). (B) In vivo ^{13}C spectroscopy of a RIF-1 tumor obtained at 400 MHz using heteronuclear cross-polarization. The animal was injected with 900 mg/kg ^{13}C-labeled D-glucose. The specific glycolytic rate of the tumor can be determined from the kinetic analysis of ^{13}C-lactate buildup.

Fig. 8A (i). Using a dual-phase extraction method, the lipid fraction can be recovered and measured separately to assess incorporation of $[1,2\text{-}^{13}\text{C}]$-choline into membrane phosphatidylcholine as shown in Fig. 8A (ii). Shown in Fig. 8B are ^{13}C MR spectra obtained *in vivo* from a RIF-1 tumor at 400 MHz using heteronuclear cross-polarization, showing the appearance of signals from ^{13}C glucose and ^{13}C lactate with time. These data can be used to determine the glycolytic rate of the tumor.[143]

Although the sensitivity of ^{13}C MRS is relatively low, indirect detection methods[144] permit the detection of the ^{13}C label with a sensitivity approaching that of the proton nucleus, and greatly increase the sensitivity of detecting ^{13}C-labeled metabolites *in vivo*. Artemov *et al.*[145] have also demonstrated the use of heteronuclear cross-polarization transfer to increase the sensitivity of direct ^{13}C detection. One major advantage of direct ^{13}C detection is its relative insensitivity to motion compared with inverse detection methods.

^{13}C MRS studies (direct and indirect ^{13}C detection) of tumors have shown that in poorly differentiated transplanted tumors, $[1\text{-}^{13}\text{C}]$-labeled glucose is metabolized to lactate.[143,146–148] Ronen *et al.*[149] have shown that a well-differentiated rat hepatoma (H4IIEC3) exhibited metabolic behavior similar to that of normal hepatocytes, mainly utilizing alanine as a substrate and resorting to glucose only under conditions of nutrient deprivation. When studying perchloric acid extracts of tumors or organs from animals infused with $[1\text{-}^{13}\text{C}]$- or $[U\text{-}^{13}\text{C}]$-labeled glucose, the labeling pattern of metabolites can provide insight into metabolic compartmentalization, shuttling of metabolites between cell types or organs, and metabolic fluxes in general.[150]

^{13}C-labeled lactate can be utilized as a metabolic marker for poor blood flow and oxygenation in unperturbed tumors, although this is complicated by several factors.[151] Glycolysis is tightly regulated by pH, ADP levels,

[143] D. Artemov, Z. M. Bhujwalla, U. Pilatus, and J. D. Glickson, *NMR Biomed.* **11,** 395 (1998).

[144] P. C. van Zijl, A. S. Chesnick, D. DesPres, C. T. Moonen, J. Ruiz-Cabello, and P. van Gelderen, *Magn. Reson. Med.* **30,** 544 (1993).

[145] D. Artemov, Z. M. Bhujwalla, and J. D. Glickson, *Magn. Reson. Med.* **33,** 151 (1995).

[146] I. Constantinidis, J. C. Chatham, J. P. Wehrle, and J. D. Glickson, *Magn. Reson. Med.* **20,** 17 (1991).

[147] Z. M. Bhujwalla, I. Constantinidis, J. C. Chatham, J. P. Wehrle, and J. D. Glickson, *Int. J. Radiat. Oncol. Biol. Phys.* **22,** 95 (1992).

[148] D. G. Schupp, H. Merkle, J. M. Ellermann, Y. Ke, and M. Garwood, *Magn. Reson. Med.* **30,** 18 (1993).

[149] S. M. Ronen, A. Volk, and J. Mispelter, *NMR Biomed.* **7,** 278 (1994).

[150] A. K. Bouzier, B. Quesson, H. Valeins, P. Canioni, and M. Merle, *J. Neurochem.* **72,** 2445 (1999).

and inorganic phosphate.[152,153] As mentioned before, tumors can form lactate in the absence of oxygen. Poor delivery of ^{13}C-labeled glucose to areas of low blood flow and inhibition of phosphofructokinase at low pH may confound the interpretation of ^{13}C MR spectra of tumors to obtain indices of blood flow and oxygenation from the metabolism of ^{13}C-labeled glucose.[151] This can be accounted for by measuring blood flow (or at least the relative heterogeneity of blood flow) to the tumor. The rate of clearance of ^{13}C-labeled lactate may also provide a measure of blood flow, provided lactate transport out of cells does not vary and lactate is not metabolized. On the other hand, an acute reduction of tumor blood flow will result in a significant increase of ^{13}C-labeled lactate. For most tumors, lactate levels seem to be equilibrated by an interplay of forces involving hemodynamics, substrate supply, hypoxia, venous clearance, glucose supply, extent of necrosis, and degree of inflammatory cell infiltrate.[154] A clear correlation between decreasing tumor oxygenation and increasing glycolytic rate was observed in a murine mammary carcinoma model, studied by volume-localized ^{13}C MRS with [1]H-^{13}C cross-polarization to detect the conversion of [1-^{13}C]-glucose to [3-^{13}C]-lactate.[155] Human breast cancer cells incubated with [1-^{13}C]-glucose exhibited a diminished mitochondrial energy generation, as assessed by the reduction in the flux of pyruvate utilized for mitochondrial energy generation compared with pyruvate used to replenish tricarboxylic acid cycle intermediates, which correlated with the degree of malignancy of the breast cancer cells.[156] ^{13}C MRS methods can also be used to detect the effect of treatment aimed at the selective inhibition of glycolysis in tumors.[157] Some of these strategies have already been investigated with ^{31}P MRS.[158,159]

[1,2-^{13}C]-choline (Fig. 8A) or other ^{13}C-labeled lipid precursors, such as [3-^{13}C]-serine, [1,2-^{13}C]-ethanolamine, or [^{13}C-methyl]-methionine, can be

[151] Z. M. Bhujwalla, D. C. Shungu, J. C. Chatham, J. P. Wehrle, and J. D. Glickson, *Magn. Reson. Med.* **32,** 303 (1994).

[152] B. Trivedi and W. H. Danforth, *J. Biol. Chem.* **241,** 4110 (1966).

[153] M. Erecinska and I. A. Silver, *J. Cereb. Blood Flow Metab.* **9,** 2 (1989).

[154] M. Terpstra, W. B. High, Y. Luo, R. A. de Graaf, H. Merkle, and M. Garwood, *NMR Biomed.* **9,** 185 (1996).

[155] F. U. Nielsen, P. Daugaard, L. Bentzen, H. Stodkilde-Jorgensen, J. Overgaard, M. R. Horsman, and R. J. Maxwell, *Cancer Res.* **61,** 5318 (2001).

[156] S. Singer, K. Souza, and W. G. Thilly, *Cancer Res.* **55,** 5140 (1995).

[157] A. Floridi, M. G. Paggi, M. L. Marcante, B. Silvestrini, A. Caputo, and C. De Martino, *J. Natl. Cancer Inst.* **66,** 497 (1981).

[158] G. S. Karczmar, J. M. Arbeit, B. J. Toy, A. Speder, and M. W. Weiner, *Cancer Res.* **52,** 71 (1992).

[159] R. L. Stolfi, J. R. Colofiore, L. D. Nord, J. A. Koutcher, and D. S. Martin, *Cancer Res.* **52,** 4074 (1992).

utilized for ^{13}C MRS studies to characterize the phospholipid metabolism of cells.[160–162] These lipid precursors can be utilized for high-resolution ^{13}C studies of cell extracts, for real-time monitoring studies of perfused isolated cells, or for infusion studies of animal tumor models. [1,2-^{13}C]-choline has been applied to further elucidate the aberrant choline phospholipid metabolism in cancer cells (Fig. 8A). ^{13}C MRS studies employing [1,2-^{13}C]-choline in a cell perfusion system demonstrated that breast cancer cells exhibit enhanced choline transport and choline kinase activity.[136] This study also revealed that the rate of choline phosphorylation was much faster than the choline transport rate.[136]

^{31}P MRS

^{31}P MRS studies of solid tumors were among the first MR studies of solid tumors to be performed[163] *in vivo*. Metabolites detected in ^{31}P MR spectra of solid tumors are nucleoside triphosphates (NTP), phosphocreatine (PCr), inorganic phosphate (P_i), phosphodiesters (PDE), and phosphomonesters (PME). The chemical shift of the P_i resonance yields the pH of the tumor (discussed subsequently).

Representative ^{31}P MR spectra from perchloric acid cell extracts, intact perfused cells, and solid tumors derived from an invasive and metastatic human breast cancer xenograft, MDA-MB-231, are shown in Fig. 9. High-resolution ^{31}P MRS resolves the typical coupling pattern in the NTP and nucleoside diphosphate (NDP) signals (Fig. 9A), since divalent cations were masked with ethylenediaminetetraacetic acid (EDTA) in the extract spectra. In perfused isolated cells and tumor xenografts, unresolved NTP and NDP signals are detected (Fig. 9B and C). Detection of PC and GPC, and potentially PE and GPE, in ^{31}P MR spectra renders ^{31}P MRS useful for the study of cellular phospholipid metabolism. The P_i resonance in live cells (Fig. 9B) and tumors (Fig. 9C) reflects the overall pH of the cells and the interstitium. In isolated perfused cells, two signals representing intracellular and extracellular P_i can be resolved, given that the extracellular P_i concentration does not exceed physiologic values of 1 mM.

Early results obtained from tumors suggested that levels of NTP, PCr, and P_i would reflect the efficiency of flow and oxygenation within a

[160] S. M. Ronen and H. Degani, *Magn. Reson. Med.* **25,** 384 (1992).
[161] R. J. Gillies, J. A. Barry, and B. D. Ross, *Magn. Reson. Med.* **32,** 310 (1994).
[162] R. M. Dixon, *Anticancer Res.* **16,** 1351 (1996).
[163] J. R. Griffiths, A. N. Stevens, R. A. Iles, R. E. Gordon, and D. Shaw, *Biosci. Rep.* **1,** 319 (1981).

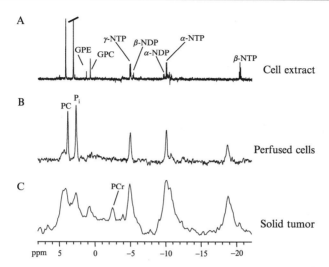

FIG. 9. ^{31}P MR spectra from (A) perchloric acid cell extracts, (B) intact cells, and (C) a solid tumor of the metastatic human breast cancer xenograft MDA-MB-231. The ^{31}P MR spectrum shown in (A) is a high-resolution spectrum of cell extracts. The ^{31}P MR spectrum shown in panel (B) was acquired from intact cells perfused in our MR-compatible cell perfusion system. The ^{31}P MR spectrum shown in (C) is from a solid MDA-MB-231 tumor inoculated in a SCID mouse. Assignments made in the ^{31}P MR spectra are: NDP, nucleoside diphosphate; NTP, nucleoside triphosphate; GPC, glycerophosphocholine; GPE, glycerophosphoethanolamine, PC, phosphocholine; PCr, phosphocreatine; P_i, inorganic phosphate.

tumor.[164] It is likely that at a given time, a ^{31}P spectrum from a tumor will depend on glucose and oxygen consumption rates, cell density, and nutritive blood flow. However, since both glucose and oxygen delivery are dependent on blood flow, energy metabolism should be tightly coupled to blood flow. Indeed, most studies show that within a single tumor line, progressive tumor growth and the associated decline of blood flow result in a decrease of high-energy metabolites such as adenosine triphosphate (ATP) and PCr and increase of P_i.[165,166] Vaupel *et al.*[130,167] measured changes in ^{31}P MRS parameters with tumor growth and, in parallel experiments, measured tumor oxygen tensions with oxygen electrodes. A strong

[164] W. T. Evanochko, T. C. Ng, and J. D. Glickson, *Magn. Reson. Med.* **1,** 508 (1984).

[165] P. G. Okunieff, J. A. Koutcher, L. Gerweck, E. McFarland, B. Hitzig, M. Urano, T. Brady, L. Neuringer, and H. D. Suit, *Int. J. Radiat. Oncol. Biol. Phys.* **12,** 793 (1986).

[166] S. J. Li, J. P. Wehrle, S. S. Rajan, R. G. Steen, J. D. Glickson, and J. Hilton, *Cancer Res.* **48,** 4736 (1988).

[167] P. Vaupel, P. Okunieff, F. Kallinowski, and L. J. Neuringer, *Radiat. Res.* **120,** 477 (1989).

correlation was observed between ^{31}P MRS parameters PCr/P$_i$ and NTP/P$_i$ and tumor oxygenation. Rofstad et al.[168] investigated the dependence of ^{31}P MRS parameters on the radiobiologic hypoxic fraction, a parameter of importance for radiation therapy, for two tumor lines with widely different hypoxic fractions. Within a given cell line the (NTP + PCr)/P$_i$ ratio decreased with tumor volume and radiobiologic hypoxic fraction. However, when ^{31}P MRS parameters were related to hypoxic fractions across cell lines, the tumor line with the higher hypoxic fraction did not have a lower (NTP + PCr)/P$_i$ ratio. A dissociation of the energy metabolism and the radiobiologic hypoxic fraction can exist if cells are highly glycolytic and do not require oxygen. In fact, it is possible that highly glycolytic tumors such as the 9L glioma may have a low hypoxic fraction.[169] ^{31}P MRS parameters may, however, be useful in detecting blood flow–mediated changes in tumor reoxygenation[170] during a course of radiation therapy.[171]

Acute changes in tumor blood flow produce more dramatic changes in high-energy phosphates. As mentioned before, the action of vasoactive agents that either improve or decrease tumor blood flow has been studied over the past decade with the aim of creating hypoxia and thereby increasing response to bioreductive drugs targeted towards hypoxic cells, or improving tumor blood flow and oxygenation and sensitizing tumors to radiation.[172] ^{31}P MR spectra of experimental tumors show a prompt decrease in NTP, PCr, and pH and an increase in P$_i$ following delivery of agents known to reduce tumor blood flow.[173,174] However, the direct action of some of these agents on cellular metabolism may also contribute to the metabolic changes observed.[175] The dependence of NTP levels on tumor blood flow may have a threshold that may be characteristic of each tumor type, depending upon its energy requirements; blood flow rates above this threshold may not further elevate NTP levels.[174]

^{1}H MRS studies of choline phospholipid metabolism in cancer cells and solid tumors have been complemented by a vast array of ^{31}P MRS investigations performed in vivo and in vitro (reviewed in[127,176,177]).

[168] E. K. Rofstad, P. DeMuth, B. M. Fenton, and R. M. Sutherland, Cancer Res. 48, 5440 (1988).

[169] J. E. Moulder and S. Rockwell, Int. J. Radiat. Oncol. Biol. Phys. 10, 695 (1984).

[170] R. F. Kallman, Radiology 105, 135 (1972).

[171] G. M. Tozer and J. R. Griffiths, NMR Biomed. 5, 279 (1992).

[172] J. Denekamp, J. Natl. Cancer Inst. 85, 935 (1993).

[173] P. Okunieff, F. Kallinowski, P. Vaupel, and L. J. Neuringer, J. Natl. Cancer Inst. 80, 745 (1988).

[174] Z. M. Bhujwalla, G. M. Tozer, S. B. Field, R. J. Maxwell, and J. R. Griffiths, Radiother. Oncol. 19, 281 (1990).

[175] G. M. Tozer, R. J. Maxwell, J. R. Griffiths, and P. Pham, Br. J. Cancer 62, 553 (1990).

Phospholipid metabolites have been monitored in cancer, using ^{31}P MRS during mutant *ras*-oncogene transformation,[178] drug treatment with anti-microtubule drugs,[179] or the nonsteroidal anti-inflammatory agent indo-methacin.[180,181] These data indicate that diverse molecular alterations and treatments arrive at common endpoints in choline phospholipid metabolism of cancer cells.

Tumor pH

^{31}*P MRS.* MR measurements of pH are obtained from the chemical shift difference between P_i and an endogenous reference such as PCr or α-NTP.[182] The P_i signal in the spectrum originates from the intracellular and extracellular compartment. Tumor interstitial fluid has P_i concentrations of 1–2 mM, which is very close to the plasma P_i concentration, suggesting that pH as measured by MRS is weighted toward intracellular pH by at least 70% for tumors with extracellular volumes less than 50%.[183] Although large necrotic areas do not accumulate P_i,[171] it is likely that cells in areas with poor blood flow may have a substantially higher P_i concentration than cells in well-perfused regions of the tumor. Therefore, in such cases, the chemical shift of the P_i signal may be biased toward the pH of the cells with poor blood flow. Gillies *et al.*[184a] have reported the use of an exogenous compound, 3-aminopropylphosphonate (3-APP), to obtain extracellular pH (pH$_e$) *in vivo* from a ^{31}P spectrum. Results obtained from a transplanted tumor confirm predictions, based on electrode measurements, of an acidic extracellular environment and a neutral to alkaline intracellular pH.

The more acidic (mainly extracellular) pH measured by electrodes and the neutral to alkaline (mainly intracellular) pH measured by MRS suggest that tumor cells may have increased H$^+$-transporting activity relative to normal tissue. Gillies *et al.*[184b] have shown that Ehrlich ascites cells can maintain intracellular pH values of 7.1 in the presence of an extracellular

[176] J. D. de Certaines, V. A. Larsen, F. Podo, G. Carpinelli, O. Briot, and O. Henriksen, *NMR Biomed.* **6,** 345 (1993).

[177] F. Podo, *NMR Biomed.* **12,** 413 (1999).

[178] S. M. Ronen, L. E. Jackson, M. Beloueche, and M. O. Leach, *Br. J. Cancer* **84,** 691 (2001).

[179] M. Sterin, J. S. Cohen, Y. Mardor, E. Berman, and I. Ringel, *Cancer Res.* **61,** 7536 (2001).

[180] K. Natarajan, N. Mori, D. Artemov, and Z. M. Bhujwalla, *Neoplasia* **4,** 409 (2002).

[181] K. Glunde, E. Ackerstaff, K. Natarajan, D. Artemov, and Z. M. Bhujwalla, *Magn. Reson. Med.* **48,** 819 (2002).

[182] R. B. Moon and J. H. Richards, *J. Biol. Chem.* **248,** 7276 (1973).

[183] M. Stubbs, Z. M. Bhujwalla, G. M. Tozer, L. M. Rodrigues, R. J. Maxwell, R. Morgan, F. A. Howe, and J. R. Griffiths, *NMR Biomed.* **5,** 351 (1992).

[184a] R. J. Gillies, Z. Liu, and Z. Bhujwalla, *Am. J. Physiol.* **267,** C195 (1994).

[184b] R. J. Gillies, T. Ogino, R. G. Shulman, and D. C. Ward, *J. Cell Biol.* **95,** 24 (1982).

pH environment of 6.8. The gradient collapsed in the absence of oxygen and glucose, suggesting that the proton extrusion mechanisms were energy driven. Several *in vivo* MR studies have also demonstrated the close dependence of intracellular pH on blood flow and substrate supply. An acute reduction of blood flow usually results in a significant decrease of intracellular pH.[175,184] Similarly, the P_i signal-derived pH usually decreases with progressive growth of the tumor.[130,166–168] However, a human ovarian xenograft did not exhibit this decrease with growth,[185] suggesting that different types of tumors may have different proton extrusion capabilities. Studies on the effects of hyperglycemia on tumor metabolism also suggest that significant decreases in the P_i signal-derived pH occur mainly when hyperglycemia produces a decrease of tumor blood flow or energy metabolism.[186] In fact, an increase in glucose delivery to the tumor can increase pH and energy levels.[187]

¹H MRS. ¹H MRS is more sensitive than ³¹P or ¹³C MRS, but currently there are no ¹H-detectable endogenous markers to quantitatively measure tumor pH *in vivo*. ¹H MRS approaches to measure tumor pH have relied on compounds, such as imidazoles and aromatics, that resonate far downfield of endogenous metabolites. Ballesteros and her colleagues[188] have developed an imidazole compound, (+/−)2-imidazole-1-yl-3-ethoxycarbonylpropionic acid (IEPA), which has been used to measure extracellular pH using ¹H MRS of breast tumor xenografts with a spatial resolution of $1 \times 1 \times 2$ mm³.[189] These pH maps showed pH values ranging from 6.4 to 6.8, which were consistent with those measured with ³¹P MRS of 3-APP. By combining MRSI of IEPA and vascular MRI using albumin–GdDTPA, Bhujwalla *et al.*[190] have demonstrated the feasibility of obtaining coregistered maps of vascular volume, permeability, and extracellular pH.

¹H MRSI techniques employing proton chemical-exchange-dependent saturation transfer (CEST) provide another strategy to measure pH *in vivo*. Magnetization transfer (MT) has been employed to evaluate proton and other nuclide chemical exchange to study chemical reactions.[191]

[185] E. K. Rofstad and R. M. Sutherland, *Int. J. Radiat. Oncol. Biol. Phys.* **15,** 921 (1988).

[186] J. L. Evelhoch, S. A. Sapareto, D. E. Jick, and J. J. Ackerman, *Proc. Natl. Acad. Sci. USA* **81,** 6496 (1984).

[187] P. Okunieff, P. Vaupel, R. Sedlacek, and L. J. Neuringer, *Int. J. Radiat. Oncol. Biol. Phys.* **16,** 1493 (1989).

[188] S. Gil, P. Zaderenzo, F. Cruz, S. Cerdan, and P. Ballesteros, *Bioorg. Med. Chem.* **2,** 305 (1994).

[189] R. van Sluis, Z. M. Bhujwalla, N. Raghunand, P. Ballesteros, J. Alvarez, S. Cerdan, J. P. Galons, and R. J. Gillies, *Magn. Reson. Med.* **41,** 743 (1999).

[190] Z. M. Bhujwalla, D. Artemov, P. Ballesteros, S. Cerdan, R. J. Gillies, and M. Solaiyappan, *NMR Biomed.* **15,** 114 (2002).

[191] J. R. Alger and R. G. Shulman, *Q. Rev. Biophys.* **17,** 83 (1984).

Potential exogenous CEST–contrast agents have been tested, demonstrating the feasibility of a CEST-based MRI contrast agent.[192] Recently, the use of amide protons of endogenous mobile cellular proteins and peptides (chemical shift of 8.3 ± 0.5 ppm), which exchange with water for CEST, has been investigated by ^{1}H MRS and MRI.[193] To achieve amide proton CEST contrast, amide protons were selectively irradiated with radiofrequency at 8.3 ppm, and water was imaged after several seconds of transfer, thereby calculating an MT ratio asymmetry parameter.[193] However, pH maps based on the amide proton transfer (APT) contrast are difficult to quantify, since variation of several parameters, such as the concentration of amide protons and water, can affect the actual measurement. Although APT contrast images seem to provide relative pH maps, it would be desirable to achieve absolute quantification. For pH imaging of solid tumors, it would also be necessary to distinguish between intracellular and extracellular pH, which is not (yet) possible using this new ^{1}H MRS-based pH imaging technique.

Examples of Integrated Imaging and Spectroscopy Approaches to Studying Cancer

MR can be used for an integrated imaging and spectroscopic imaging approach, which is extremely useful in studying a disease as complex as cancer. In this section we have included two examples of such an approach in studying cancer cell invasion, and the vascularization and metabolism of solid tumors.

MR Metabolic Boyden Chamber for Studying Cancer Cell Invasion

The ability of cancer cells to invade and metastasize is one of the most lethal aspects of cancer. Cancer cells invade by secreting enzymes that degrade basement membrane. This invasive potential is commonly assayed by determining the penetration of cells into reconstituted basement membrane gel (Matrigel or extracellular matrix [ECM] gel). Invasion is then quantified by counting the number of cells that invade ECM gel-coated filters over a period of 5–72 h. However, these methods do not permit evaluation of the metabolic state of tumor cells, nor do they allow invasion to be measured dynamically in the same sample, under controlled environmental conditions, or following therapeutic interventions.

[192] K. M. Ward, A. H. Aletras, and R. S. Balaban, *J. Magn. Reson.* **143,** 79 (2000).
[193] J. Zhou, J. F. Payen, D. A. Wilson, R. J. Traystman, and P. C. van Zijl, *Nat. Med.* **9,** 1085 (2003).

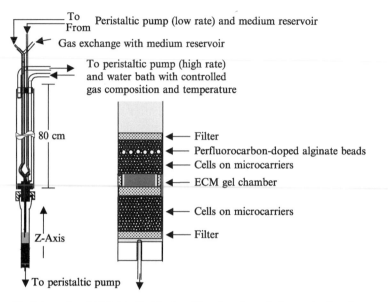

To Peristaltic pump (low rate) and medium reservoir
From

Gas exchange with medium reservoir

To peristaltic pump (high rate)
and water bath with controlled
gas composition and temperature

80 cm

← Filter
← Perfluorocarbon-doped alginate beads
← Cells on microcarriers
← ECM gel chamber

← Cells on microcarriers

← Filter

Z-Axis

To peristaltic pump

FIG. 10. Schematic of "Metabolic Boyden Chamber Assay" demonstrating the reproducible layer of ECM gel and perfluorocarbon-doped alginate beads placed within the cell layers. (Adapted from Pilatus et al.[194])

We therefore developed an invasion assay system, termed the Metabolic Boyden Chamber (MBC) assay,[194] to dynamically track the invasion of cancer cells into ECM gel and simultaneously characterize oxygen tensions and physiologic and metabolic parameters. In this assay, ^{1}H, ^{31}P, and ^{19}F MR experiments of cancer cells continually perfused with medium are performed on a GE Omega 400 MHz MR spectrometer with 130 G/cm shielded gradients. A schematic describing the system is shown in Fig. 10. The current design achieves a reproducible thickness of ECM gel while allowing free perfusion through the tube. By replacing Norprene tubing with less permeable Viton tubing, we obtain oxygen tensions under 1.5%, which is necessary to evaluate the impact of the hypoxic tumor environment on cancer cell invasion. We incorporate perfluorotripropylamine (FTPA)-doped alginate beads in the NMR tube and embed FTPA into the ECM gel to directly measure oxygen tensions in the sample using ^{19}F MR relaxometry.[195] An example of the oxygen tensions achieved in this system is shown in Fig. 11.

[194] U. Pilatus, E. Ackerstaff, D. Artemov, N. Mori, R. J. Gillies, and Z. M. Bhujwalla, *Neoplasia* **2,** 273 (2000).

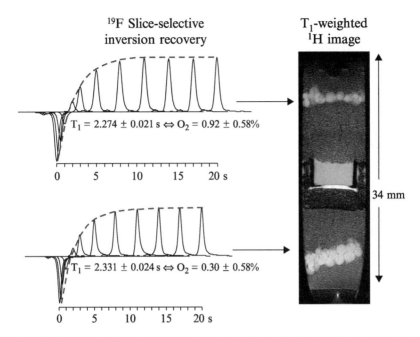

FIG. 11. Characterization of oxygen tensions in the Metabolic Boyden Chamber perfusion system.

T_1-weighted ^{1}H MR images are used to visualize the geometry of the sample, including the ECM gel, as well as changes in the integrity of the ECM gel due to invasion and degradation. The 1D distribution of invading cells in the sample is obtained from the profile of intracellular water measured with 1D imaging along the length of the sample with a spatial resolution of 0.031 mm (zero-filled to 0.016 mm). A quantitative index of invasion is obtained from these cellular profiles. We routinely obtain localized proton spectra of intracellular water and metabolites with a slice thickness of 0.31 mm. With the current hardware, experiments with a spatial resolution of 0.312 mm (128 phase-encoding steps, 40 mm field of view [FOV]) can be performed within 9 hours. The energy status of the cells, pH, and levels of choline phospholipid metabolites are obtained from ^{31}P MR spectra.

Typical invasion and metabolic data obtained with the MBC assay for the noninvasive prostate cancer cell line, DU-145, and the invasive prostate

[195] K. A. McGovern, J. S. Schoeniger, J. P. Wehrle, C. E. Ng, and J. D. Glickson, *Magn. Reson. Med.* **29,** 196 (1993).

cancer cell line, MatLyLu, at comparable time points are displayed in Fig. 12. Representative T_1-weighted ^{1}H MR images demonstrating the different rates of degradation of ECM gel by three prostate cancer cell lines, MatLyLu, PC-3, and DU-145, are presented in Fig. 13. Quantitative differences in the invasion index for three different prostate cancer cell lines are summarized in Fig. 14, and demonstrate the differences in the invasion index for the three cell lines.

Multinuclear MRI and MRSI of Preclinical Models of Cancer

Over the past decade we have established several MRI and MRSI techniques to study preclinical models of cancer. From the earlier sections it is apparent that a unique aspect of MR is its potential ability to investigate the relationships between tumor vascularization, physiology, and metabolism. Although it is useful to study these as separate characteristics, the ability to relate metabolism and vascularization within the same region of interest adds a new dimension to our understanding of how one impacts on the other for tumor models with different vascular and metastatic characteristics, or for those overexpressing a selected gene or receptor, or for tumors with different drug resistance. We have also developed combined vascular, metabolic, and extracellular pH imaging, and can perform combined vascular, metabolic, and optical imaging or positron-emission tomography (PET).

For combined metabolic and vascular imaging of tumor models, both proton spectroscopic images (SI) and vascular maps are obtained using a ^{1}H RF coil on our GE Omega 4.7-T spectrometer during a single experiment, without disturbing the position of the animal or the tumor (Fig. 15). During the experiment, the body temperature of the animal is maintained at 37° by means of a heating blanket. Typically we first obtain a spectroscopic imaging (SI) data set using the BASSALE pulse sequence.[116,190,196] Images of total choline and lactate–lipid are generated from the SI acquired from a 4-mm-thick slice, with an in-plane resolution of 2 mm (zero-filled to 1 mm), FOV = 32 mm, matrix size = 16 × 16 × 512, 4 scans per phase encode step, TE = 136 ms, TR = 1 s. After SI, which is completed within 30 minutes, measurements of vascular volume and permeability surface area product are obtained. Vascular imaging is done as follows. Multislice relaxation rates (T_1^{-1}) of the tumor are obtained by a saturation recovery method combined with fast T_1 SNAPSHOT–FLASH

---

[196] Z. M. Bhujwalla, D. Artemov, and M. Solaiyappan, "Combined Vascular and Metabolic Characterization of Orthotopically Implanted Prostate Cancer Xenografts." International Society for Magnetic Resonance in Medicine, Honolulu, 2002.

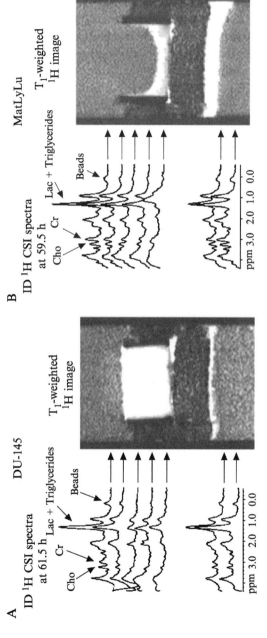

FIG. 12. Expanded ^1H MR image of sample showing the ECM gel region for (A) DU-145 and (B) MatLyLu cells at comparable time points. Images were obtained with TR 1 s, TE 30 ms, FOV 40 mm, slice thickness 2 mm, and an in-plane resolution of 0.078 mm. Corresponding ^1H MR spectra in the figure are from every third 0.31-mm-thick slice localized within the sample. Metabolites assigned are Cho (total choline), Cr (total creatine), Lac+Triglycerides (lactate and triglycerides). Signal from the beads to which the cancer cells are attached is also observed in the spectra. (Adapted from Pilatus et al.[194])

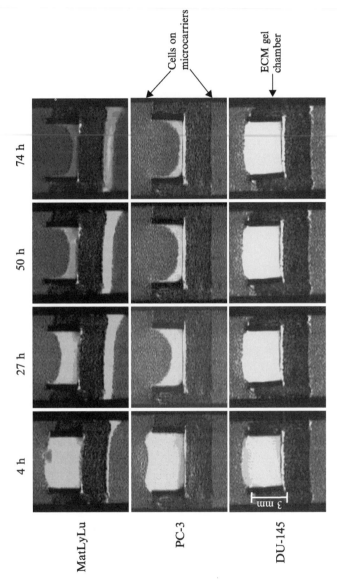

FIG. 13. "Metabolic Boyden Chamber Assay" demonstrating differences in invasive characteristics of three prostate cancer cell lines. The T_1-weighted ^{1}H MR images show the bright ECM gel layer, which is significantly degraded by MatLyLu and PC-3 cells, but not by DU-145 cells.

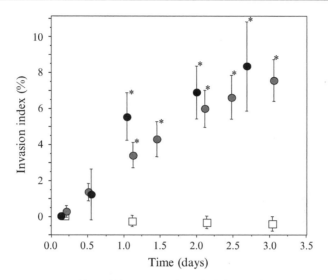

FIG. 14. Invasion Index (I) for (●) MatLyLu (n = 6), (◉) PC-3 (n = 4), and (□) DU-145 (n = 4) prostate cancer cells. Values are mean ± S.E. * represents a statistically significant difference between MatLyLu and DU-145, and between PC-3 and DU-145 cells ($p < 0.01$).

imaging (flip angle = 10°, TE = 2 ms). Images of 4 slices matching the SI data (slice thickness of 1 mm) are acquired with an in-plane spatial resolution of 0.125 mm (matrix = 128 × 128, FOV = 16 mm, NS = 16) for 3 relaxation delays (100 ms, 500 ms, and 1 s) for each of the slices. An M_0 map with a recovery delay of 7 s is acquired once at the beginning of the experiment. Images are obtained before intravenous (IV) administration (via the tail vein) of 0.2 ml of 60 mg/ml albumin–GdDTPA in saline (dose of 500 mg/kg) and repeated every 5 minutes, starting 3 minutes after the injection, for up to 30 minutes.

Relaxation maps are computed from the data sets for three different relaxation times and the M_0 data set on a voxel-wise basis.[41] Vascular volume and permeability product surface area (PS) maps are generated from the ratio of $\Delta(1/T_1)$ values in the images to that of blood. As described earlier (see Fig. 4), the slope of $\Delta(1/T_1)$ ratios versus time in each pixel is used to compute PS (Fig. 15B), whereas the y-intercept is used to compute vascular volume (Fig. 15A). Thus vascular volumes are corrected for permeability of the vessels. Blood T_1 is obtained from a sample withdrawn from the inferior vena cava in terminal experiments, or the tail vein when not sacrificing the animal. A special microcoil can be used to determine T_1 of 20-μl blood samples (two drops of blood) at 4.7 T, the magnetic field used for all

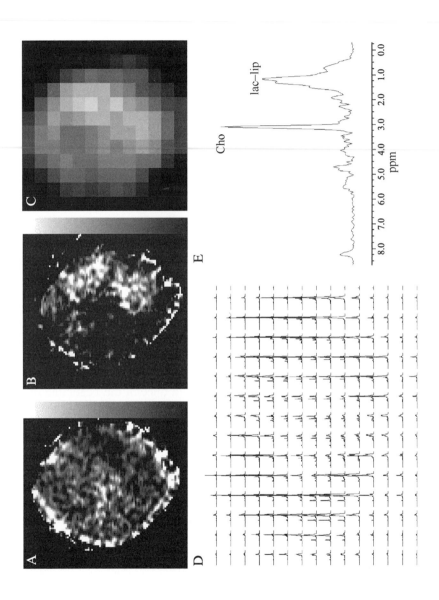

the *in vivo* studies presented here. Because blood T_1 can be determined from a couple of drops of blood from the tail vein without sacrificing the animal, it allows us to perform longitudinal measurements of vascular volume and permeability in the same animal. We can also use an *in vivo* tail coil that allows us to obtain blood T_1 s by using voxels localized on a tail vein blood vessel.[7] T_1 maps of the tail are acquired with short echo time spin–echo imaging with saturation recovery preparation and variable recovery delays. Acquisition parameters are typically TE = 5 ms, TR = 250 ms, RD = 50–200 ms, FOV = 4 mm, ST = 1 mm, matrix size = 64 × 40 for an in-plane resolution of 62.5 μm. This tail coil system has been validated with a flow phantom and provides reproducible T_1 measurements for flow rates of up to 0.5 cm/s, and its high in-plane resolution minimizes volume-averaging effects.

This approach of combined vascular and metabolic imaging can be extended to determining the relationship between any (three) colocalized parameters such as vascular volume, permeability, extracellular pH, total choline, or lactate–lipid. By displaying images for each parameter through a unique color channel (e.g., vascular volume as red, vascular permeability as green, and extracellular pH or choline or lactate as blue), it is feasible to visually inspect the interrelationship between these by fusing the color maps and determining the fractional volumes of the patterns for each tumor.[190] Parametric maps of vascular volume, permeability, and pH$_e$ can also be used to generate three-dimensional, volumetric histogram matrices.[190] This display extends the conventional histogram plot to a volumetric histogram. Each voxel in the 3D display corresponds to an entry in a three-dimensional matrix of vascular volume, permeability, and pH$_e$. The intensity of color of the voxel represents the frequency of occurrence (histogram count) of the entry from the sampled data set. Although not presented here, these approaches of acquiring and analyzing multiparametric data can be extended to perform combined MR molecular and functional imaging to visualize molecular markers, and understand their role in tumor vascularization and metabolism.

In conclusion, with its enormous versatility, MR is certainly a method capable of effectively investigating a disease as complex as cancer. As with

---

Fig. 15. Coregistered maps of (A) vascular volume, (B) permeability surface area product, and (C) total choline obtained from the central slice of a transgenic PC-3 tumor (270 mm^3). Vascular volume ranged from 0 to 126 μl/g and PSP from 0 to 3.4 μl/g.min. Proton CSI spectra obtained from 1 mm × 1 mm × 4 mm voxels used to generate the total choline map are shown in (D), and the spectrum obtained from a single 1 mm × 1 mm × 4 mm voxel with large signals from total choline and lactate–lipid is shown in (E).

most methods, however, it is imperative to identify the questions most appropriate for the method. It is hoped that this chapter has succeeded in describing some of the current applications and advances of MR methods in functional and molecular imaging of cancer.

Acknowledgments

Work from the authors' program was supported through funding by NIH grants 2R01 CA73850, 1R01 CA90471, 1R01 CA82337, P20 CA86346, and P50 CA 103175.

[2] A Modified Transorbital Baboon Model of Reperfused Stroke

By Anthony L. D'Ambrosio, Michael E. Sughrue, J. Mocco, William J. Mack, Ryan G. King, Shivani Agarwal, and E. Sander Connolly, Jr.

Introduction

Stroke is the third leading cause of death in the United States, with 750,000 deaths per year, a number exceeded only by the millions of people who are left permanently debilitated or institutionalized by the disease.[1] Extensive research efforts have been undertaken in the realm of therapeutics to reduce the immense morbidity and mortality associated with this devastating disease.

Acute stroke can be subdivided into two categories: ischemic and hemorrhagic. Acute cerebral ischemia is far more common and can occur via atherosclerotic plaque rupture or vessel occlusion secondary to an embolus.[2] This acute event results in an ischemic core, where tissue damage is permanent and irreversible, and a surrounding region called the penumbra.[3,4] Tissues in the penumbra are believed to remain viable and could be salvaged if the appropriate measures are implemented in a timely fashion. Recent research has implicated several deterimental event cascades that are set in motion after an acute ischemic event, including leukocyte

[1] D. M. Bravata, S. Y. Ho, L. M. Brass, J. Concato, J. Scinto, and T. P. Meehan, *Stroke* **34,** 699 (2003).

[2] E. Wraige, C. Hajat, W. Jan, K. R. Pohl, C. D. Wolfe, and V. Ganesan, *Dev. Med. Child Neurol.* **45,** 229 (2003).

[3] I. F. Pestalozza, S. Di Legge, M. Calabresi, and G. L. Lenzi, *Clin. Exp. Hypertens.* **24,** 517 (2002).

[4] N. Bonaffini, M. Altieri, A. Rocco, and V. Di Piero, *Clin. Exp. Hypertens.* **24,** 647 (2002).

Copyright 2004, Elsevier Inc.
All rights reserved.
0076-6879/04 $35.00

infiltration, upregulation of inflammatory mediators, platelet activation, complement upregulation, free radicals, apoptosis, and microvascular failure.[5] The implication of these multiple event cascades in cerebral ischemia has given researchers several promising targets for the development of novel neuroprotective strategies.

Unfortunately, although more than 49 neuroprotective agents have been tested recently in approximately 114 stroke trials, none have shown efficacy.[6] Although many of these novel agents demonstrated significant neuroprotection in preclinical rodent trials, few were tested in nonhuman primate models of cerebral ischemia. Many investigators believe that this repeated failure of promising neuroprotective agents to demonstrate efficacy in clinical trails is rooted, at least in part, in species-specific differences in anatomy, physiology, and specific responses to ischemic injury between the rodent and the human.[7]

Animal Models of Stroke

Rodent models are the most commonly used animal models of cerebral ischemia used in preclinical testing today. Across laboratories, rodent models have considerable variation in terms of technique, species, methods of anesthesia, and duration of ischemia. Earlier models of cerebral ischemia in the rat utilized permanent middle cerebral artery occlusion (MCAo) via a craniotomy.[8] However, as we gained a better understanding of the critical role played by reperfusion injury and progressive microvascular failure in the pathophysiology of cerebral ischemia, many investigators moved away from models of permanent ischemia and developed reperfusion models that are believed to be more physiologic.

The most commonly used model of ischemia–reperfusion (I–R) injury involves insertion of a silicon-coated suture into the internal carotid artery and advancing it intracranially to occlude MCA blood flow.[9] The suture is subsequently removed to allow for cerebral reperfusion. More recently, models of embolic cerebral ischemia are rapidly gaining favor. Embolic models involve the injection of an autologous clot into the MCA via an intracarotid catheter.[10,11] Other less utilized models include photothrombotic models[12] and vessel occlusion using macrospheres.[13]

[5] W. D. Heiss, *J. Neural Transm.* **63**(Suppl.), 37 (2002).

[6] D. J. Gladstone, S. E. Black, and A. M. Hakim, Heart and Stroke Foundation of Ontario Centre of Excellence in Stroke. *Stroke* **33**, 2123 (2002).

[7] T. J. DeGraba and L. C. Pettigrew, *Neurol. Clin.* **18**, 475 (2000).

[8] S. T. Chen, C. Y. Hsu, E. L. Hogan, H. Maricq, and J. D. Balentine, *Stroke* **17**, 738 (1986).

[9] E. Z. Longa, P. R. Weinstein, S. Carlson, and R. Cummins, *Stroke* **20**, 84 (1989).

[10] Z. Zhang, M. Chopp, R. L. Zhang, and A. Goussev, *J. Cereb. Blood Flow Metab.* **17**, 1081 (1997).

Primate Models

A variety of large animals, including dogs[14] and rabbits,[15] have been utilized in several models of focal cerebral ischemia. However, nonhuman primate models provide the closest representation of human pathophysiology by reducing neuroanatomic variability and physiologic responses to ischemic insult.[16,17] Furthermore, the primary endpoint of most clinical stroke trials is improvement of functional outcome. Primate behavior is more complex, and as a consequence, functional deficits are more similar to those seen in humans. For this reason, outcome measures are more comparable and improvements in outcome are more clinically applicable.

In an effort to address these issues, we developed a model of reperfused nonhuman primate stroke that utilized, via a transorbital approach, unilateral internal carotid artery (ICA) and bilateral anterior cerebral artery (ACA) occlusion.[18] We describe specific details of our model, including several refinements made over a 5-year experience with primate reperfused stroke.

Preoperative Care

Obtaining Animals and Supplies

Test subjects are young adult male baboons, *(Papio anubis)*, weighing 14–23 kg (Buckshire Farms, Perkasie, PA). Animals are individually housed in stainless steel cages and uniquely identified with a body tattoo and a cage card. Upon arrival, each subject is thoroughly assessed as per our standard Institutional Review Board (IRB) protocol by the animal care team. All animals are acclimated to laboratory conditions for a minimum of 45 days prior to study initiation.

Preparing for a transorbital procedure can be an involved process. This requires advance planning to coordinate staff to provide anesthesia support, procurement of appropriate supplies (Fig. 1), use of MRI facilities, and 24-hour care overnight following the procedure.

[11] R. L. Zhang, M. Chopp, Z. G. Zhang, Q. Jiang, and J. R. Ewing, *Brain Res.* **766,** 83 (1997).

[12] P. Wester, B. D. Watson, R. Prado, and W. D. Dietrich, *Stroke* **26,** 444 (1995).

[13] T. Gerriets, F. Li, M. D. Silva, X. Meng, M. Brevard, C. H. Sotak, and M. Fisher, *J. Neurosci. Meth.* **122,** 201 (2003).

[14] G. Brenowitz and H. Yonas, *Surg. Neurol.* **33,** 247 (1990).

[15] M. A. Yenari, A. de Crespigny, J. T. Palmer, S. Roberts, S. L. Schrier, G. W. Albers, M. E. Moseley, and G. K. Steinberg, *J. Cereb. Blood Flow Metab.* **17,** 401 (1997).

[16] A. R. Lake, I. J. Van Niekerk, C. G. Le Roux, T. R. Trevor-Jones, and P. D. De Wet, *Am. J. Anat.* **187,** 277 (1990).

[17] J. S. Meyer, M. Yamamoto, L. A. Hayman, F. Sakai, S. Nakajima, and D. Armstrong, *Neurol. Res.* **2,** 101 (1980).

[18] J. Huang, J. Mocco, T. F. Choudhri, A. Poisik, S. J. Popilskis, R. Emerson, R. L. DelaPaz, A. G. Khandji, D. J. Pinsky, and E. S. Connolly, Jr., *Stroke* **31,** 3054 (2000).

Materials and Suppliers

Medications	Supplier
ketamine HCl	Fort Dodge Animal Health
propofol	Zeneca Pharmaceutical
0.5% isoflurane	Baxter
Balanced nitrous oxide	Tech Air
fentanyl	Elkins-Sinn
cefazolin	Bristol/Meyers Squibb
vecuronium	Organon
midazolam	Roche
phenylephrine hydrochloride	Gensia Laboratories
lidocaine with epinephrine 1:100,000	Abbott Laboratories
pentobarbital	Veterinary Laboratories
mannitol	AmVet

Anesthesia Equipment	Supplier
Ohmeda 7000 ventilator	Ohmeda
Arterial blood pressure monitor	Datascope
Femoral vein catheter	Arrow International
Transurethral foley catheter	Baxter
Esophageal probe	Datascope
Parenchymal sensor	Neuromonitor, Codman
Laser doppler probe (Model PF2B)	Perimed, Inc.

Surgical Supplies	Supplier
Adjustable stereotactic frame	Stoelting
Hand-held twist drill	Neurocare
Self-retaining lid retractor	Codman
Bipolar electrocautery	Valley Forge Scientific
Operating microscope	Model F170, Zeiss
High-speed pneumatic drill/diamond bit	Midas Rex
3 micro-yasargil aneurysm clip	Aesculap
18-gauge needles	Becton-Dickinson
10, 11, and 15 scalpel blades	Henry-Schein
Gel foam	Pharmacia and Upjohn
Radiolucent methylmethacrylate	Codman Cranioplastic, J&J

Pathology Supplies	Supplier
2% 2,3,5-triphenyltetrazolium in 0.9% PBS	Sigma

FIG. 1. Comprehensive list of materials and manufacturers.

Anesthesia and Preparation

On the morning of surgery, the study animal is brought into a preoperative holding area and anesthetized with ketamine (Fort Dodge Animal Health, Ft. Dodge, IA) at an intramuscular (IM) dose of 5 mg/kg. The head, neck, forearm, and femoral regions are shaved with an electric clipper. Two 18- or 20-gauge peripheral intravenous (IV) catheters are placed, and 0.9% normal saline (NS) is started. Propofol (Zeneca Pharmaceutical, Macclesfield, UK) is then given as a bolus infusion prior to oropharyngeal intubation using a size 6 or 7 endotracheal tube. The animal is then transported to the operating room where assisted ventilation is initiated (Ohmeda 7000 ventilator) with an inhalation mixture composed of 0.5% isoflurane (Baxter, Deerfield, IL) and balanced nitrous oxide (Tech Air, West Plains, NY) and oxygen. In anticipation of the placement of additional monitoring devices, an IV bolus infusion of fentanyl (Elkins-Sinn, Cherry Hill, NJ) at 50 μg/kg is administered, followed by a continuous fentanyl infusion of 50 to 70 μg/kg/hr. The concentration of isoflurane in the inhalation anesthetic agent is maintained between 0% and 0.6%. IV cefazolin (Bristol/Meyers Squibb, New York, NY) is administered as prophylaxis and continued for 48 hours postoperatively.

Before final positioning in the head frame (Stoelting, Wooddale, IL), a continuous IV infusion of vecuronium (Organon, Roseland, NJ) is started at 0.04 mg/kg/hr. Additionally, a 0.1 mg/kg IV bolus of midazolam (Roche, Indianapolis, IN) is given every 30 minutes. Once the transorbital approach has begun, the fentanyl infusion rate is increased to 70 to 100 μg/kg/hr, while the isoflurane rate is decreased to <0.5%.

Physiologic Monitoring

An intra-arterial catheter is introduced into the femoral artery for continuous systemic blood pressure monitoring to facilitate multiple specimen collections. Blood pressure is monitored (Datascope, Paramus, NJ) to maintain a mean arterial pressure (MAP) of 60- to 80-mm Hg. Hypotensive episodes can be treated with injections of phenylephrine hydrochloride (Gensia Laboratories, Irvine, CA). Central venous pressure (CVP) is monitored using a femoral vein catheter (Arrow International, Reading, PA) and sustained at 5 ± 2 mm Hg. An indwelling, transurethral Foley catheter (Baxter, Deerfield, IL) allows accurate urine output monitoring to guide management of fluid balance and CVP.

Arterial blood gas (ABG) analysis is performed at regular intervals (Stat profile 3, Nova Biomedical, Waltham, MA) while the respiratory rate and tidal volume are continuously adjusted to maintain a carbon dioxide pressure (P_{CO_2}) between 35 and 40 mm Hg. Continuous core body temperature is monitored using an esophageal probe (Datascope, Montvale, NJ). Continuous intracranial pressure (ICP) monitoring is accomplished with a parenchymal sensor (Neuromonitor, Codman, Piscataway, NJ). Sustained ICP of >20 mm Hg for >5 minutes is treated with a bolus administration of IV mannitol at 0.5 g/kg.

Before intubation and the administration of anesthesia, baseline complete peripheral blood cell counts are performed. Systemic blood pressure, CVP, cerebral perfusion pressure (CPP), and core body temperature are maintained at constant levels throughout the procedure and for the first 24 hours postoperatively. During ischemia, ventilation is adjusted to maintain a P_{CO_2} similar to those at baseline.[18]

Operative Technique

Positioning

After the insertion of all indwelling catheters and before the placement of the ICP monitor and the laser Doppler probe, the animal is positioned prone, in an adjustable stereotactic frame, with two-point head fixation via the external auditory canals. The cranial base is positioned parallel to the floor, with the anterior skull base elevated approximately 15° and turned slightly to the right. Dependent pressure points are adequately padded to prevent tissue necrosis.

Placement of ICP Monitor and Laser Doppler Probe

A left frontal approach via a paramedian linear skin incision and two burr holes is used for the insertion of an intraparenchymal ICP monitor and a straight laser Doppler probe (Model PF2B, Perimed, Inc., Piscataway, NJ). A hand-held twist drill (Neurocare, San Diego, CA) is used to create the burr holes. The dura is cauterized and sharply incised to allow passage of the fiberoptic pressure sensor and the laser Doppler probe. The laser Doppler probe is lowered 10 mm below the inner table of the calvarium into the cortex. The probe is immobilized with a plastic burr hole cover and secured to the cranium with contact cement. The skin incision is closed with 3-0 interrupted nylon sutures. Intraoperative measurements of ICP and local cerebral blood flow (CBF) are continually recorded throughout the procedure.

Transorbital Approach

Infiltration of the medial and lateral canthi of the left orbit with 0.5% lidocaine with epinephrine 1:100,000 (Abbott Laboratories, Abbott Park,

IL) is performed before making an incision in the plane along the palpebral fissure. A self-retaining lid retractor is then placed. An 18-gauge needle is carefully inserted into the anterior and posterior chambers of the globe for aspiration of the vitreous and aqueous humors. Internally decompressing the globe in this manner allows it and the periorbital soft tissue to be circumferentially dissected from the walls of the orbit. The globe can be removed after transection of the optic nerve and ophthalmic artery. Care should be taken at this point to completely cauterize the ophthalmic artery to prevent unnecessary blood loss. The residual soft tissue is removed with bipolar cauterization and curettage.

The operating microscope (Model F170, Zeiss, Thornwood, NJ) is used for the remainder of the procedure. A high-speed pneumatic drill with a coarse diamond bit (Midas Rex, Fort Worth, TX) is used to remove the bone of the posteromedial orbit. The dura of the anterior cranial fossa is then incised with a No. 11 blade and the edges cauterized to reveal the anterior circle of Willis.

Vessel Occlusion

Microsurgical technique is used to identify the cerebral arteries and clear them from the surrounding arachnoid membrane. After reconfirmation of the stability of the physiologic variables, vessel occlusion is accomplished through the sequential placement of three micro-Yasargil (Aesculap, Center Valley, PA) aneurysm clips: (1) on the proximal segment of the left anterior cerebral artery, proximal to the anterior communicating artery (left A1), (2) on the proximal right anterior cerebral artery (right A1), and (3) across the left ICA at the level of the anterior choroidal artery so that the clip incorporates and occludes the anterior choroidal artery (left ICA). Aneurysm clip placement and operative exposure are illustrated in Fig. 2.

Previous model protocols utilized a test occlusion period to confirm ischemic cerebral blood flow using both motor evoked potential (MEP) monitoring and laser Doppler flowmetry. This test occlusion period lasted 5 minutes, followed by 15 minutes of reperfusion and then 60 minutes of permanent occlusion.[19] After the 60-minute occlusion period, the aneurysm clips were removed and the brain was allowed to reperfuse. This process is represented in Fig. 3A. In a currently submitted publication, we demonstrated excellent correlation between MEP dropout and laser Doppler flowmetry tracings during permanent occlusion.[19] Because laser Doppler flowmetry (Fig. 3B) is a direct measure of CBF, we have since abandoned

---

[19] C. J. Wintree, W. J. Mack, D. Hoh, R. King, A. F. Ducruet, A. L. D'Ambrosio, M. E. Sughrue, J. McKinnell, and E. S. Connolly, Jr., *Acta Neurochir (Wien)* **145,** 1105 (2003).

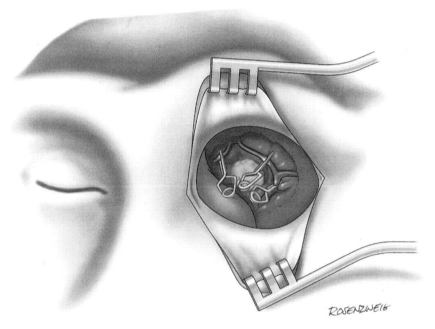

FIG. 2. Transorbital approach demonstrating placement of occlusive aneurysm clips on the internal carotid artery (ICA) and both anterior cerebral arteries (A1 segments). Reproduced with permission from Huang *et al.*[18]

MEP monitoring and no longer utilize the 5-minute test occlusion period. In recent trials, we have used permanent occlusion times of 60, 75, and 90 minutes and found the 75-minute occlusion time to give the most reliable results (unpublished data).

After 75 minutes of occlusion, the clips are removed in reverse order to permit reperfusion. A layer of gel foam (Pharmacia and Upjohn, Kalamazoo, MI) is placed over the dural defect, and the retractor is removed. Radiolucent methylmethacrylate (Codman Cranioplastic, Johnson & Johnson, Piscataway, NJ) is used to fill the orbital defect, and the eyelid is sutured closed with a running 3-0 nylon suture.

Postoperative Care

Postoperative Monitoring

At the conclusion of the surgical procedure, the animal is removed from the surgical head frame and placed supine on a padded mattress with 30° of head elevation. The fentanyl infusion is subsequently lowered to 20 μg/kg/hr, and the isoflurane is increased to 0.1–0.6%. The nitrous oxide is

A

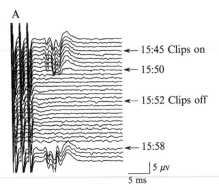

←— 15:45 Clips on

←— 15:50

←— 15:52 Clips off

←— 15:58

| 5 μv

5 ms

B

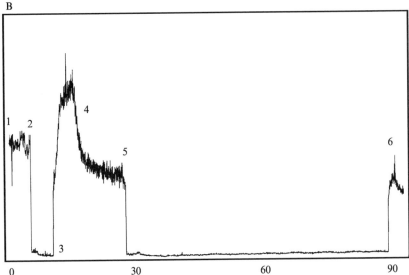

FIG. 3. (A) Representative motor evoked potential (MEP) tracing demonstrating electrophysiologic evidence of cerebral blood flow decrement and restoration with clip placement and removal, respectively. Reproduced with permission from Huang *et al.*[18] (B) Representative tracing of laser Doppler dropout following middle cerebral artery (MCA) territory occlusion. Time point 1 represents baseline cerebral blood flow (CBF). Time point 2 marks the beginning of a 5-minute test occlusion period. Time point 3 represents a 15-minute reperfusion period after the three temporary clips have been removed. Time point 4 demonstrates a brief period of hyperperfusion after the test occlusion period. Time point 5 represents the beginning of a 60-minute permanent occlusion period. Note the immediate drop-off of CBF signal. Time point 6 represents the final removal of all three aneurysm clips and the reconstitution of CBF. Modified with permission from Winfree *et al.*[19]

replaced with a balanced air and oxygen mixture. Vecuronium and midazolam are continued. The animal remains intubated and sedated with continuous physiologic monitoring and close regulation for the first 18 hours of reperfusion. Sustained ICP elevations of >20 mm Hg for >5 minutes are treated with IV infusions of mannitol (0.5 mg/kg/dose) as needed. Aggressive pulmonary toilet is achieved through frequent suctioning and chest physical therapy. Eighteen hours after reperfusion, the inhalation anesthetic and narcotic agents are tapered off, and the animal is permitted to regain consciousness. If the ABGs demonstrate satisfactory gas exchange, all of the indwelling catheters and monitors are removed to allow extubation and return to housing cages for observation in the animal intensive care unit. Blood samples are drawn through an arterial catheter several times throughout the postoperative period. If extubated, the animal is anesthetized prior to drawing blood with IM ketamine. Animals are continuously monitored for their ability to self-care, eat, and drink. All wounds are examined regularly for signs of breakdown and/or infection.

Criteria for Sacrifice

Clinical examinations are performed several times daily. If the veterinarian determines that continued survival would be unethical secondary to devastating functional impairment prior to postoperative day 3, the animal is reintubated until day 3, at which time an MRI scan is obtained. If at any point after postoperative day 3 the animal fails neurologic examination, it is euthanized with an IV injection of pentobarbital (Veterinary Laboratories, Winnipeg, MB). Animals are survived to a maximum of 10 days, at which point they are sacrificed.

Data Collection and Analysis

Neurologic Evaluation

Daily neurologic assessments are performed by two investigators who are blinded to all imaging data. Originally, we used the 100-point neurologic scale developed by Spetzler and associates,[20] with higher scores reflecting better neurologic function. With the NIH Stroke Scale[21] as a template, we designed a task-oriented neurologic scoring scale that demonstrated less interobserver variability and better correlation with radiographic infarct volume.[22] This neurologic scoring system is shown in Fig. 4.

[20] R. F. Spetzler, W. R. Selman, P. Weinstein, J. Townsend, M. Mehdorn, D. Telles, R. C. Crumrine, and R. Macko, *Neurosurgery* **7**, 257 (1980).
[21] T. Brott, H. P. Adams, Jr., C. P. Olinger, J. R. Marler, W. G. Barsan, J. Biller, J. Spilker, R. Holleran, R. Eberle, and V. Hertzberg, *Stroke* **20**, 864 (1989).

Task-Oriented Baboon Neurologic Scale		
Behavior (50 points)	**Motor (50 points)**	
Level of Awareness (30 points)	*Tone and Posture (20 points)*	
Alive 5	Sits upright with support	4
Movement/reaction to tactile stimuli 5	Sits upright without assistance	4
Aware of examiner's presence/turns head to noise 5	Stands in cage	4
Interacts with examiner/maintains attention 5	Walks around cage	4
Attempts slow movement as examiner approaches 5	Moves/jumps around cage at full speed	4
Shows teeth, growls, aggressive/defensive movements 5		
Ability to Self-care (20 points)	*Distal Strength/Coordination (30 points)*	
Swallows if water is placed in mouth 5	Lifts unaffected arm against gravity	3
Chews if food is placed in mouth 5	Attempts to grasp cage/bar with unaffected arm	3
Feeds from examiner's hand 5	Grasps cage/supports with unaffected arm	3
Feeds from cage 5	Shakes cage/pulls self to stand with unaffected arm	3
	Lifts affected arm against gravity	4
	Attempts to grasp cage/bar with affected arm	4
	Grasps cage/supports with affected arm	4
	Shakes cage/pulls self to stand with affected arm	4
	Preserves coordination in both arms	2

FIG. 4. Task-oriented neurologic scale. Neurologic function is assessed on a task-oriented 100-point scoring system with 50 points each given to motor and behavioral tasks. Reproduced with permission from Mack et al.[22]

Animals are generally quite self-sufficient postoperatively. Most animals are able to move around their cage with a mild spastic limp. Some climb, but most do not. Additionally, most animals are capable of sitting upright without assistance. Those that cannot are assisted with a cloth sling wrapped around the upper chest and under each armpit, which is then secured to the cage. This provides adequate arm mobility in the good extremity while effectively supporting the injured arm. Animals are usually capable of using the injured arm proximally, allowing comfortable positioning during routine daily activities and sleep. Most can feed themselves adequately with the left upper extremity but generally have little useful strength on the right. Animals incapable of self-feeding are deemed unable to self-care.

Radiographic Imaging

After 72 hours of reperfusion, the animal is anesthetized with ketamine and sedated with an IV pentobarbital bolus and propofol infusion titrated to allow independent respiratory function for up to 6 hours while the

[22] W. J. Mack, R. G. King, D. J. Hoh, A. L. Coon, A. F. Ducruet, J. Huang, J. Mocco, C. J. Winfree, A. L. D'Ambrosio, M. N. Nair, R. R. Sciacca, and E. S. Connolly, Jr., *Neurol. Res.* **25,** 280 (2003).

airway is maintained with an endotracheal tube. Brain MRI is then performed at this "early" time point, with the acquisition of axial T_2-weighted, gradient echo, diffusion–perfusion, fluid activation inversion recovery (FLAIR), and MR angiography sequences (Fig. 5). The T_2-weighted images are acquired with a slice thickness of 3 mm and without intervening space between images. The decision for animal survival past 72 hours is a clinical judgment by the veterinarian and is based on, among other things, level of alertness, respiration, and ability to self-care. If the animal is deemed able to self-care, it is survived out to day 10, at which time a "late" MRI is obtained before the animal is sacrificed.

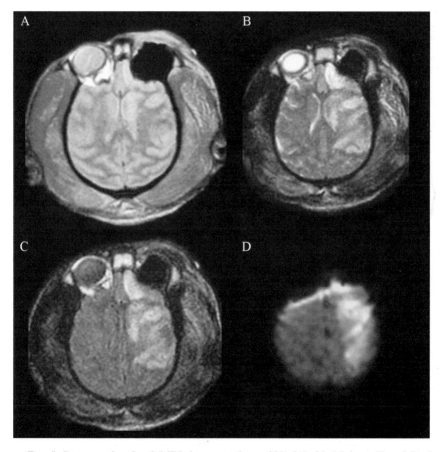

Fig. 5. Postoperative day 3 MRI demonstrating a 30% left-sided infarct. T_1-weighted (A) and T_2-weighted (B) axial images demonstrating anterior cerebral artery (ACA) and middle cerebral artery (MCA) signal changes consistent with an acute infarct. FLAIR (C) and diffusion-weighted images (D) further delineate the lesion.

Infarct Volume

For histologic confirmation of infarct location and correlation with MRI, the brain is removed after sacrifice with the surrounding dura intact. Three coronal sections of 5-mm thickness are collected from the ischemic ipsilateral and stereoanatomic equivalent, normal contralateral hemispheres. The first section is obtained from the medial portion of the most posterior aspect of the precentral gyrus and immersed in a solution of 2% 2,3,5-triphenyltetrazolium (TTC) (Sigma, St. Louis, MO) in 0.9% phosphate buffered saline. Additional sections are obtained immediately anterior and posterior to the first section and embedded in Tissue-Tek compound for further histologic processing. Infarcted tissue is visualized as nonstained portions of brain.[23,24]

Infarct volume is determined by two blinded observers. Areas of ischemic damage demonstrate high signal intensity in T_2-weighted MRI. Using commercially available graphics software, infarct volume can be quantified by planimetric analysis and expressed as the percentage of the total volume of the ipsilateral hemisphere.

Model Application: HuEP5C7

We have used our transorbital model to test the efficacy of a variety of therapeutic interventions on infarct volume and neurologic outcome. One such study investigated a humanized monoclonal antibody, HuEP5C7, which competitively binds E-selectin and P-selectin, critical adhesion molecules implicated in both microvascular failure and inflammation following cerebral ischemia.[25] Eighteen adult baboons were randomized to receive either HuEP5C7 (20 mg/kg IV, $n = 9$) or placebo ($n = 9$) immediately following a 1-hour ischemic period. Serum levels of HuEP5C7, E-selectin, and P-selectin were determined at 30 min, 120 min, 24 h, 48 h, 72 h, and 9 days postoperatively. MRI scans and neurologic assessments were performed as previously outlined. Histologic examination of the infarct was analyzed to determine tissue levels of P- and E-selectin.

This study demonstrating trends toward reduced polymorphonuclear leukocyte (PMN) infiltration into the ischemic cortex, reduced infarct volumes, improved neurologic score, and improved ability to self-care. This study suggests that HuEP5C7 is a good candidate for evaluation in clinical trials and may be a safe and effective neuroprotective agent in the future.

[23] J. B. Bederson, L. H. Pitts, S. M. Germano, M. C. Nishimura, R. L. Davis, and H. M. Bartkowski, *Stroke* **17,** 1304 (1986).

[24] C. Dettmers, A. Hartmann, T. Rommel, S. Kramer, S. Pappata, A. Young, S. Hartmann, S. Zierz, E. T. MacKenzie, and J. C. Baron, *Neurol. Res.* **16,** 205 (1994).

[25] J. Huang, T. F. Choudhri, C. J. Winfree, R. A. McTaggart, S. Kiss, J. Mocco, L. J. Kim, T. S. Protopsaltis, Y. Zhang, D. J. Pinsky, and E. S. Connolly, Jr., *Stroke* **31,** 3047 (2000).

Conclusion

Our current model of temporary aneurysm clip occlusion of both A1 segments and the ICA at the level of the choroidal artery for 75 minutes, together with laser Doppler flowmetry confirmation of ischemia, allows for large and consistent hemispheric infarcts in baboons. Importantly, our animals are survivable and often recover their ability to self-care,[18,25] making postoperative neurologic assessments possible. Nonhuman primate models of cerebral I–R injury are essential in the preclinical testing of potential neuroprotective strategies for stroke. Such models provide a testing ground for novel neuroprotective agents without putting patients at risk for potential negative side effects.

[3] MRI Investigations of Graft Rejection Following Organ Transplantation Using Rodent Models

By YI-JEN L. WU, KAZUYA SATO, QING YE, and CHIEN HO

Introduction

Organ transplantation is now the preferred modality of treatment for patients with end-stage organ failure. With improved immunosuppressive therapy, the survival rate and the quality of life for patients with organ transplantation have improved greatly. However, acute allograft rejection remains the leading cause of mortality in transplant patients.[1] The occurrence of acute rejection episodes often is the most predictive factor for the later development of chronic rejection[2] and graft failure in heart[3] and in kidney[4,5] transplantation. Thus early diagnosis of acute rejection is critical before the condition progresses to higher grades. For a general discussion on mechanisms, pathology, and diagnosis of solid organ transplant rejection, see Solez et al.[6]

[1] C. H. Knosalla, M. Hummel, J. Muller, O. Grauhan, R. Ewert, and R. Hetzer, Curr. Opin. Transplant. 5, 118 (2000).

[2] S. G. Tullius and N. L. Tilney, Transplantation 59, 313 (1995).

[3] S. H. Kubo, D. C. Naftel, R. M. Mills, Jr., J. O'Donnell, R. J. Rodeheffer, G. B. Cintron, J. L. Kenzora, R. C. Bourge, and J. K. Kirklin, J. Heart Lung Transplant. 14, 409 (1995).

[4] J. M. Cecka, Clin. Transplant. 8, 324 (1994).

[5] A. Tejani, D. Stablein, S. Alexander, R. Fine, and W. Harmon, Transplantation 59, 500 (1995).

[6] K. Solez, L. C. Racusen, and M. E. Billingham, "Solid Organ Transplant Rejection: Mechanisms, Pathology, and Diagnosis." Marcel Dekker, Inc., New York, 1996.

Copyright 2004, Elsevier Inc.
All rights reserved.
0076-6879/04 $35.00

When acute rejection occurs, immune cells, such as T-cells and macrophages, infiltrate the rejection sites, with accompanying loss of organ functions. For biopsy, the current clinical "gold standard" for diagnosing rejection, finite tissue samples from the grafts are harvested for histologic examination of immune cell infiltration and other pathologic changes. Although widely utilized, biopsy is not free of limitations and pitfalls. It is not only invasive, associated with finite risks of morbidity and mortality, but also is prone to sampling errors due to limited size and/or locations of tissue available, particularly in the pediatric age group. In addition, the interpretation of a biopsy is often objective, leading to errors in diagnosis,[7] and, as a result, inability to predict the onset of acute rejection episode.[1] More importantly, there can be a discrepancy between biopsy-based diagnosis and rejection resulting in organ dysfunction.[8,9] In the case of humoral rejection, patients manifest severe rejection and organ dysfunction, yet lack cellular infiltration and may exhibit a negative biopsy. These patients are at high risk of dying, but are silent in the biopsy evaluation. The disassociation of biopsy grading from the evaluation of acute rejection and patient outcomes makes the role of biopsy in diagnosing and monitoring acute rejection controversial and questionable.[10–12] Since the threshold for antirejection therapy has recently been uncoupled from histologic biopsy grading, it has become more closely linked to graft dysfunction.[13] Hence it is highly desirable to have a reliable and noninvasive alternative to biopsy for detecting acute rejection and also to be able to evaluate organ function after transplantation.

Magnetic resonance imaging (MRI) can potentially be a powerful imaging modality for noninvasive evaluation of organ rejection, since it can detect organ function and morphology with high resolution, is not limited by depth of penetration of the tissue or organ, and has excellent soft tissue contrast. It could potentially have a great impact on the evaluation of rejection if noninvasive MRI-based methodologies can be established that not only can detect immune cell infiltration at the rejection sites, but also can evaluate organ dysfunction resulting from acute rejection.

[7] G. L. Winters, C. C. Marboe, and M. E. Billingham, *J. Heart Lung Transplant.* **17**, 754 (1998).

[8] R. M. Mills, D. C. Naftel, J. K. Kirklin, A. B. Van Bakel, B. E. Jaski, E. K. Massin, H. J. Eisen, F. A. Lee, D. P. Fishbein, and R. C. Bourge, *J. Heart Lung Transplant.* **16**, 813 (1997).

[9] E. Pahl, D. C. Naftel, C. E. Canter, E. A. Frazier, J. K. Kirklin, and W. R. Morrow, *J. Heart Lung Transplant.* **20**, 279 (2001).

[10] M. M. Boucek, *Pediatr. Transplant.* **4**, 173 (2000).

[11] M. Hummel, J. Muller, M. Dandel, and R. Hetzer, *Transplant. Proc.* **34**, 1860 (2002).

[12] M. R. Mehra, P. A. Uber, W. E. Uber, M. H. Park, and R. L. Scott, *Curr. Opin. Cardiol.* **17**, 131 (2002).

[13] J. A. Kobashigawa, *Transplant. Proc.* **31**, 2038 (1999).

The Scope of This Chapter

The focus of this chapter is to describe emerging noninvasive MRI methodologies for evaluating acute allograft rejection as potential alternatives for surveillance biopsy, particularly the novel two-pronged approach our laboratory has been using in our rodent transplantation models. First, we use MRI to detect immune cell infiltration at the rejection sites by monitoring the accumulation of dextran-coated ultrasmall superparamagnetic iron oxide (USPIO)-labeled immune cells *in vivo*. Second, we establish MRI parameters that are associated with organ dysfunction resulting from acute rejection.

Animal Models

Animal models are important for developing noninvasive MRI techniques for detecting graft rejection after organ transplantation. Rodent models have the advantage of having a large number of inbred strains available for genetic background control and increased experimental options, compared with other larger animal models. Rats have suitable body size for standard animal MRI instruments, without the need to implement special hardware for microimaging. In this section, we describe several kidney, heart, and lung transplantation models in rats for MRI studies on allograft rejection.

In order to study the entire rejection process with minimal physiologic deterioration of transplanted animals, we have chosen unilateral orthotopic renal transplantation and heterotopic heart and lung transplantation models. In our renal transplantation model, the recipient rat receives only one transplanted kidney, while the other native kidney remains intact. In our heterotopic heart and lung transplantation model, in addition to its own native heart and lung, the recipient rat has an additional heart and lung transplanted in the abdominal or groin area. With this design, the recipient animals do not depend on the grafts to support life, so that the grafts can be monitored until being fully rejected with massive necrosis.

It is important to choose proper donor–recipient rat strain pair combinations, because the extent of histocompatibility mismatch between donor and recipient, the alloreactive potential of the recipient, time course of surgical recovery, sensitivity to ischemic insult, and mortality are all strain dependent. Incomplete recovery from ischemic injury after transplantation surgery makes functional evaluation of the transplanted organ difficult. We have chosen inbred male Dark-Agouti (DA; $RT1^{avl}$) to Brown Norway (BN; RTI^n) transplantation, because organs are rejected later in the DA-to-BN transplantation model than in a Fisher 344-to-ACI combination in heart transplantation and it has more immune cell infiltration and less edema in kidney model. This allows better surgical recovery before extensive allograft rejection occurs. In addition, BN rats are less fatty and have larger vessels and ureters than other rats, making surgery easier. Other commonly used transplantation pairs are DA-to-Lewis, PVG-to-Lewis, and Fisher-to-ACI. In our experiments, all rats used were 8–10 weeks of age and weighed between 250 and 300 g, since larger rats have more fatty tissue that can cause difficulties in surgical dissection. Animals have received care in compliance with the *Principles of Laboratory Animal Care* and the *Guide for Care and Use of Laboratory Animals*, published by the National Institutes of Health (NIH publication No. 96–03, revised 1996). Rat transplantation was performed under general anesthesia by inhalation of isoflurane (Abbott Laboratories, North Chicago, IL) or methoxyflurane (Mallinckrodt Veterinary, Mundelein, IL) via a nose cone.

Kidney Model

The renal transplantation model was first described by Fisher and Lee.[14] In our model, an orthotopic kidney transplantation model has been established for studying renal allograft rejection. The left kidney of the recipient rat is removed (nephrectomy) and replaced with another kidney from the same strain (syngeneic transplantation) or a different strain (allogeneic transplantation), whereas the contralateral right kidney is left intact to serve as an internal control. This design was chosen to enable direct comparison between the animal's own native kidney and the transplanted kidney. More importantly, since the recipient does not need to depend on the kidney graft to support life (i.e., the contralateral native kidney remains in place to support metabolism in the recipient rat), this design makes functional study of the entire rejection process possible with minimal physiologic alterations. Left nephrectomy of the recipient is performed before the left kidney transplantation. The graft (the donor's left kidney) is flushed with lactated Ringer's solution (Abbott Laboratories) containing heparin (100 U/ml) and then stored in the same solution at 4°. It is important that harvest time for the graft kidney is approximately 15 min and that the graft not be touched during the whole procedure. The donor renal artery and vein is anastomosed end-to-side to the recipient abdominal aorta and inferior vena cava (IVC) with 10–0 sutures (Sherwood Medical, St. Louis, MO). The ureter of the recipient is anastomosed end-to-end with the donor's ureter by 10–0 sutures using four stitches.

The major causes of mortality after renal transplantation are vascular complications, ureteric complications, and hypovolemic shock. In our

[14] B. Fisher and S. Lee, *Surgery* **59,** 904 (1965).

model, vascular complications can be minimized by using an aortic segment that saves extensive dissection at the junction of the renal artery and aorta, so that the renal graft vasospasm may be avoided. When performing vessel anastomosis, a large lumen with a sufficient length is helpful to avoid thrombosis and to save time. Keeping the animal body warm and short surgery time are important factors for reducing hypovolemic shock. Ureteric reconstruction is the most problematic and least standardized procedure as discussed by Engelbrecht et al.[15] We have evaluated several techniques, including ureter-to-bladder anastomosis, bladder-to-bladder anastomosis, and end-to-end ureter anastomosis. We prefer using end-to-end ureter anasotomosis because this technique results in the fewest ureteric complications, including obstructive uropathy, calculus formation, and ureteric infection. The success rate with this renal transplantation model is >95%, which permits reliable MRI studies.

Immediately after MRI evaluation, transplanted kidneys were scored histologically for extent of rejection based on the Banff Schema (Score 0–6) with minor modification[16,17]: Score 0, normal kidney architecture, no sign of rejection; Score 1, minor interstitial lymphoplasmacytic infiltration with occasional penetration of tubular epithelium; Score 2 to 3, widespread focal interstitial lymphoplasmacytic infiltration with mild tubulitis and normal glomeruli and arteriolar vessels; Score 4 to 5, extensive interstitial lymphoplasmacytic infiltration with definite tubulitis and intimal cell prominence with subintimal vacuolation in arterial vessels; and Score 6, extensive interstitial lymphoplasmacytic infiltration with tubulitis and lymphoplasmacytic infiltration of arterial walls, sometimes accompanied by fibrinoid change.

In our renal transplantation model, most of the allogeneic kidney grafts show mild acute rejection on postoperative day (POD) 3, with light interstitial edema and some foci of lymphoplasmacytic infiltration in the perivascular space and in the surrounding interstitium, but the glomeruli are within the normal range. On POD 4, the allografts undergo moderate acute rejection with lymphoplasmacytic infiltration found over the entire interstitium. On POD 5, extensive interstitial lymphoplasmacytic infiltration is found with hemorrhage. By POD 7, all allogeneic kidney grafts have undergone irreversible severe rejection with extensive necrosis, hemorrhages, and generalized edema.

[15] G. Engelbrecht, F. Duminy, and R. Hickman, *Microsurgery* **13**, 340 (1992).

[16] K. Solez, R. A. Axelsen, H. Benediktsson, J. F. Burdick, A. H. Cohen, R. B. Colvin, B. P. Croker, D. Droz, M. S. Dunnill, P. F. Halloran et al., *Kidney Int.* **44**, 411 (1993).

[17] K. Solez, D. Battaglia, H. Fahmy, and K. Trpkov, *Curr. Opin. Nephrol. Hypertens.* **2**, 904 (1993).

Heart and Lung Models

Heterotopic heart and lung transplantation models were chosen for MRI studies. In the heterotopic heart and lung transplantation model, in addition to keeping in place the native heart and lung, the recipient rat receives another heart and lung located outside the chest. The reasons for the heterotopic transplantation models are twofold. First, this enables studies of the entire rejection process without many physiologic alterations of a transplanted animal, because the heart and lung grafts do not have a life-supporting function. Second, a cardiopulmonary bypass system is not available for rodents, so that orthotopic heart transplantation is not feasible at the present time.

Conventional intra-abdominal heterotopic heart and lung transplantation in the rat was described originally by Lee *et al.*[18] and modified by Fox and Montorsi.[19] This conventional model of transplantation is not suitable for MRI studies of lungs because of abdominal and respiratory interference. For immunologic studies with MRI, we have established a new heterotopic heart and lung transplantation model (Model 1) where the transplanted heart is placed in the inguinal region to avoid difficulties in abdominal imaging, such as susceptibility artifacts from the intestines and abdominal gas, as well as motion artifacts from the bowel movement. In addition, the lung graft preserves all the normal circulation, yet without air (i.e., collapsed), so it can be easily imaged with MRI.

In Model 1, illustrated in Fig. 1A, an *en bloc* heart and lung graft is transplanted to the groin area of the recipient. After the chest wall is opened in the donor, 500 U/kg body weight of sodium heparin is injected into the IVC, and the left lung is ligated and excised. The azygos vein with the left and right superior vena cava (SVC) is ligated and divided. The descending thoracic aorta is transected, and 10 ml of cold lactated Ringer's solution (Abbott Laboratories) is infused into the IVC until the fluid draining from the aorta is clear. The IVC is then ligated and divided. The ascending aorta is dissected and transected at the portion between the left common carotid artery and the left subclavian artery, followed by ligation and division of the right brachiocephalic artery and the left common carotid artery. The grafts are then placed in cold lactated Ringer's solution until transplantation. The left inguinal region is opened in the recipient and dissected to make enough space for the transplanted organs. The left lower part of the abdominal wall is opened in a transverse fashion from the left femoral vessels to the midline. The abdominal organs are retracted to the right, and both the aorta and the IVC just beyond the bifurcation are

[18] S. Lee, W. F. Willoughby, C. J. Smallwood, A. Dawson, and M. J. Orloff, *Am. J. Pathol.* **59,** 279 (1970).
[19] U. Fox and M. Montorsi, *J. Microsurg.* **1,** 377 (1980).

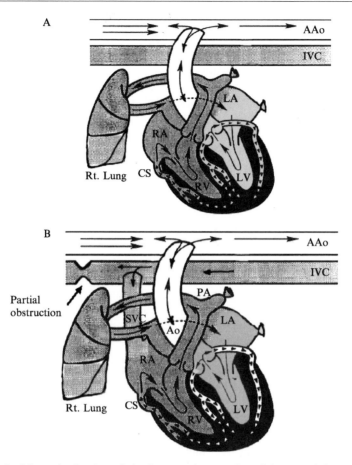

FIG. 1. Schematic drawing of the heterotopic transplanted heart and lung models. (A) Model 1, nonworking heart model in the groin area. (B) Model 2, biventricular working heart model in the abdominal area. AAo, Recipient abdominal aorta; IVC, recipient inferior vena cava; Ao, donor ascending aorta; PA, pulmonary artery; LA, left atrium; LV, left ventricle; RA, right atrium; RV, right ventricle; SVC, donor superior vena cava; Rt. Lung, donor right lung; CS, coronary sinus. Arrows indicate directions of the blood flow.

dissected. The aorta is clamped, an appropriate opening of the aorta is made to receive the aorta of the graft in an end-to-side fashion, and the aortas are sutured together with 8–0 polypropylene sutures (Ethicon Inc., Somerville, NJ). Rhythmic heartbeats commence spontaneously as the heart and lung regain circulation after removal of the clamp. The abdominal wall is then sutured with 6–0 silk (Ethicon Inc.), with care taken to avoid kinking or obstructing the aorta of the graft. Figure 1A shows a schematic drawing of circulation in this model. The total ischemic time for the transplant surgery is about 30 min.

However, this heterotopic heart and lung transplantation model (Model 1) with the transplanted organs in the inguinal region, like other conventional heterotopic heart transplantation models, is a nonworking heart model. The heart does not receive proper pressure preload, and the myocardium experiences atrophy over time. It is a very useful model for immunologic studies and MR imaging, yet it is inadequate for evaluation of cardiac function. Therefore, for functional studies, we have established a new heterotopic heart and lung transplantation model with the organs placed in the abdomen (Model 2). This is a biventricle (both left and right ventricles) working heart model. In this model, the stroke volume (SV), ejection fraction (EF), and the pressures at all chambers are close to those of native hearts. Surgically, this model is similar to Model 1, but with the following modifications (Fig. 1B): (1) The donor SVC is anastomosed to the recipient IVC. (2) The IVC of the recipient is partially obstructed to increase the preload of the transplanted heart. (3) The graft is in the abdomen instead of in the inguinal region to avoid physical constraint in the groin region. Figure 1B shows the configuration of this model. The total ischemic time for the transplant surgery is about 30 min.

Transplanted heart and lung were graded histologically according to the International Society for Heart and Lung Transplantation (ISHLT) criteria[7,20]: Grade 0, no rejection; Grade 1, mild rejection, characterized by small foci of mononuclear cell infiltration limited to the subendocardial, the perivascular, or both zones and without evidence of myocyte necrosis; Grade 2, moderate rejection, more extensive interstitial but still focal areas of mononuclear cell infiltration, with or without focal regions or myocyte necrosis; Grade 3, moderate rejection, more diffused inflammatory cell infiltration with myocyte necrosis; and Grade 4, severe rejection, widespread inflammatory cell infiltration, including polymorphonuclear leukocytes, hemorrhage, and myocyte necrosis.

In our heterotopic heart and lung transplantation models, all lung grafts developed severe Grade 4 rejection on POD 5, whereas the transplanted hearts did not develop severe rejection until POD 7.

Noninvasive Detection of Rejection with MRI by Monitoring USPIO-Labeled Immune Cells

Spatial sampling errors, as in biopsy, in probing immune cell infiltration at the transplanted organ can be avoided by monitoring immune cell accumulation in the whole graft with MRI. This is made possible by labeling

[20] M. E. Billingham, N. R. Cary, M. E. Hammond, J. Kemnitz, C. Marboe, H. A. McCallister, D. C. Snovar, G. L. Winters, and A. Zerbe, *J. Heart Transplant.* **9,** 587 (1990).

immune cells with an MRI-detectable intracellular marker, such as dextran-coated USPIO particles. We have successfully detected organ rejection by monitoring the accumulation of USPIO-labeled immune cells *in vivo*. In this section, we give a summary of our progress ranging from detecting single immune cells *in vitro* to monitoring the accumulation of immune cells *in vivo* in our rat transplantation models for kidney, heart, and lung.

Dextran-Coated USPIO Particles for Cell Labeling and Tracking

Dextran-coated superparamagnetic iron oxide (SPIO) nanoparticles are contrast agents that exhibit excellent MR relaxivity, mainly apparent spin–spin relaxation time (T_2*) effect, yet are biologically indolent or non-toxic, so that they have gained much interest for *in vivo* cell tracking and imaging. They have an iron core composed of a ferrous and ferric iron oxide complex surrounded by a dextran coat to shield the body from the toxicity of the iron crystals. The superparamagnetic characteristics of these nanoparticles[21] make them excellent cellular contrast agents because they can create a large microscopic field gradient that can affect a much larger (50 times) region than their actual size.[22–24]

Two general procedures are being used to label mammalian cells: (1) Specific cell types can be isolated from a host and labeled *ex vivo* in a culture medium. The labeled cells can then be infused into an animal or a human and their migration or accumulation monitored by MRI. (2) Cells can also be labeled *in vivo* by introducing an MRI contrast agent directly into an animal or a human through intravenous (IV) injection. Unlike the *ex vivo* labeling, the *in vivo* procedure results in the labeling of nonspecific cell types, unless the contrast agent is conjugated to a cell-specific transporter or marker.

The two most utilized dextran-coated SPIO and USPIO particles are: (1) SPIO (Feridex), which is FDA approved for clinical use, and (2) USPIO (or MION-46L), which is in Phase III clinical trials. Although both possess pronounced relaxivity, they have different behavior *in vivo*. SPIO particles (sizes ranging from >30 to 1000 nm in diameter) are cleared from the blood by the mononuclear phagocytic system (MPS) quickly, in

[21] C. W. Jung, *Magn. Reson. Imaging* **13**, 675 (1995).
[22] P. C. Lauterbur, M. L. Bernardo, Jr., M. H. Menonca Dias, and L. K. Hedges, "Microscopic NMR Imaging of the Magnetic Field around Magnetic Particles," p. 229. Proceedings of the Society for Magnetic Resonance in Medicine, 5th Annual Meeting. Montreal, Quebec, Canada, 1986.
[23] S. J. Dodd, M. Williams, J. P. Suhan, D. S. Williams, A. P. Koretsky, and C. Ho, *Biophys. J.* **76**, 103 (1999).
[24] J. W. Bulte, R. A. Brooks, B. M. Moskowitz, L. H. Bryant, Jr., and J. A. Frank, *Magn. Reson. Med.* **42**, 379 (1999).

the order of minutes. On the other hand, the much smaller USPIO particles (10–30 nm in diameter) migrate very slowly across the capillary endothelium and have a much longer half-life in blood, about 100 minutes in rats.[25–28] Thus USPIO particles are more suitable for *in vivo* cell labeling.

MRI Detection of Single Immune Cells in an In-Vitro *System.* Detection of a single immune cell labeled with dextran-coated USPIO particles has been demonstrated *in vitro* with labeled cells suspended in a gel matrix, such as agarose or gelatin. Endocytosis of USPIO particles by T-cells[23,29,30] and macrophages[28,31] leads to the accumulation of these particles in phagocytic vacuoles, resulting in an increased efficacy of the contrast agent due to the larger susceptibility effect of the combined particles. Dodd *et al.*[23] have shown that a concentration of 10×10^6 USPIO-labeled T-cells/ml with a 20% labeling efficiency estimated from transmission electron microscopy (TEM) results in a 50% reduction of the T_2 value in a 4.7-Tesla (T) magnetic field. In this sample, one would expect to find on average two USPIO-labeled T-cells for every nanoliter or $100 \times 100 \times 100\ \mu m^3$ volume, or a single cell in half that volume or $50 \times 100 \times 100\ \mu m^3$. This unit volume can be critical for MRI detection of a labeled cell.

Figure 2 shows T_2*-weighted MR images with $50 \times 50 \times 100\ \mu m$ resolution (4×10^6 voxels/ml) of USPIO-labeled and unlabeled T-cell samples with concentrations of 2.0×10^6, 1.0×10^6, and 0.25×10^6 T-cells/ml at 4.7 T using a 2D FT (two-dimensional Fourier transform) gradient–echo sequence with an echo time (TE) of 30–40 ms.[23] Comparing the images of the labeled cells in Fig. 2A, C, and E with the images of unlabeled cells in Fig. 2B, D, and F clearly shows the samples containing USPIO-labeled cells giving rise to punctuate loss of signal (dark spots), the quantity of which decreases with a decrease in the concentration of labeled cells. Each dark spot is an USPIO-labeled T-cell. The few dark spots seen in samples with unlabeled cells most likely are due to small air bubbles. Similarly, we have found that single macrophages labeled with dextran-coated USPIO particles and suspended in 5% gelatin can also be detected by MRI.[28]

[25] R. Weissleder, D. D. Stark, B. L. Engelstad, B. R. Bacon, C. C. Compton, D. L. White, P. Jacobs, and J. Lewis, *Am. J. Roentgenol.* **152,** 167 (1989).
[26] R. Weissleder, G. Elizondo, J. Wittenberg, C. A. Rabito, H. H. Bengele, and L. Josephson, *Radiology* **175,** 489 (1990).
[27] O. Hauger, C. Delalande, H. Trillaud, C. Deminiere, B. Quesson, H. Kahn, J. Cambar, C. Combe, and N. Grenier, *Magn. Reson. Med.* **41,** 156 (1999).
[28] S. Kanno, P. C. Lee, S. J. Dodd, M. Williams, B. P. Griffith, and C. Ho, *J. Thorac. Cardiovasc. Surg.* **120,** 923 (2000).
[29] T. C. Yeh, W. Zhang, S. T. Ildstad, and C. Ho, *Magn. Reson. Med.* **30,** 617 (1993).
[30] T. C. Yeh, W. Zhang, S. T. Ildstad, and C. Ho, *Magn. Reson. Med.* **33,** 200 (1995).
[31] S. Kanno, Y. J. Wu, P. C. Lee, S. J. Dodd, M. Williams, B. P. Griffith, and C. Ho, *Circulation* **104,** 934 (2001).

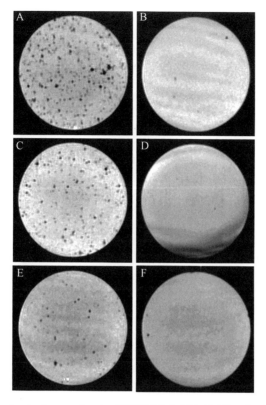

FIG. 2. Gradient–echo images showing USPIO-labeled (A, C, E) and unlabeled (B, D, F) T-cell samples. Cell concentrations were: (A) and (B), 2.0×10^6 T-cells/ml; (C) and (D), 1×10^6 T-cells/ml; (E) and (F), 0.25×10^6 T-cells/ml. MRI parameters were: TR/TE = 1000/30 ms; FA = $45°$; matrix size = 128×128; FOV = 6.4 mm; slice thickness = 100 μm; in-plane resolution = 50×50 μm; acquisition bandwidth = 20 kHz; number of averages (NA) = 8; experiment time = 17 min. In each case, the direction of the read gradient was from left to right and the phase-encoding direction was vertical. Bulk samples for high-resolution MRI studies were prepared in 1-ml plastic syringes by suspending T-cell samples in 3% gelatin. The MRI experiments were carried out in a 4.7-T Bruker AVANCE DRX MR instrument equipped with a 40-cm horizontal bore superconducting solenoid. Taken from Fig. 2 of Dodd et al.,[23] with permission.

Monitoring the Accumulation of Immune Cells Labeled with USPIO Particles at Transplanted Organs In Vivo. Labeling immune cells *in vivo* by intravenous injection of USPIO, then monitoring the accumulation or infiltration of USPIO-labeled immune cells at the transplanted organ, is a convenient way to detect rejection. In their original form without modification, USPIO particles are taken up by cells through endocytosis.[29,30] USPIO particles are incorporated into macrophages more efficiently than into other cell types. In our DA-to-BN rat transplantation models, we have found that there is excellent correlation between the signal loss of MRI due to the USPIO-labeled macrophages and the iron content of the graft samples, as well as the conventional assessments of graft rejection by histology and immunology. Thus we have found that *in vivo* labeling of macrophages with dextran-coated USPIO particles is a convenient and sensitive way to detect rejection of kidney, heart, and lung transplants by MRI.

DETECTING KIDNEY REJECTION BY MRI. To detect the acute rejection *in vivo*, USPIO particles are administered intravenously on POD 4 in our rat kidney transplantation model (DA-to-BN rat pairs), when moderate acute rejection has occurred. The animals are subjected to T_2*-weighted imaging on POD 5, 24 hr after USPIO injection, to allow ample time for USPIO incorporation and immune cell migration and accumulation at the rejection sites. Figure 3 shows T_2*-weighted gradient–echo images of allografts with different dosages of USPIO particles and isografts.[32] Allograft kidneys show darkening in the images 24 hr after USPIO injection, whereas isografts have no significant signal changes. The extent of hypointensity in allografts is proportional to the amount of USPIO administered. Based on the results, 6 mg Fe/kg body weight appears to be the optimal dosage for visualization of rejection-associated changes in the renal cortex in our rat model. Some darkening is seen in allograft medulla even without USPIO injection. It is due to hemorrhage associated with rejection. Histologic and immunohistochemical analyses of the grafts show that the iron-containing cells revealed by Perl's iron staining correlate with ED1$^+$ cells, which are macrophage lineage cells. These results indicate that detecting rat renal acute rejection with MRI following an IV injection of dextran-coated USPIO particles appears to be feasible and promising.

DETECTING LUNG REJECTION BY MRI. In our heterotopic DA-to-BN lung transplantation model, allogeneic lungs show Grade 4 rejection on POD 5. USPIO particles are administered intravenously on POD 4, and the signal changes are followed by T_2*-weighted MRI within 24 hr after USPIO injection.[28] As seen in Fig. 4, an allogeneic lung graft shows extensive darkening after USPIO injection, whereas no significant signal decrease is detected in a syngeneic lung graft. On the other hand, allografts treated with immunosuppressant cyclosporine A (CsA, 3 mg/kg/day) exhibit intermediate signal decrease between isografts and nonimmunosuppressed allografts. Figure 5 gives a summary of the MR signal changes upon USPIO administration. Histologic analysis also shows that the extent of iron staining

[32] Q. Ye, D. Yang, M. Williams, D. S. Williams, C. Pluempitiwiriyawej, J. M. Moura, and C. Ho, *Kidney Int.* **61,** 1124 (2002).

POD 4, 10 min   POD 4, 5 min   POD 5, 24 hr
before infusion   after infusion   after infusion

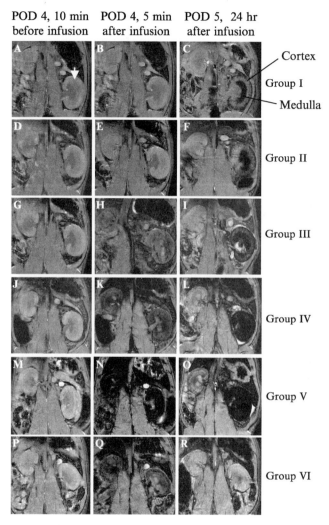

Cortex

Group I

Medulla

Group II

Group III

Group IV

Group V

Group VI

FIG. 3. Representative gradient–echo MR images showing the effect of different doses of dextran-coated USPIO particles. The transplanted kidneys appear on the right (arrow), and the native kidneys appear on the left in these panels. The MR images from immediately before, 5 min after, and 24 hr after infusion of dextran-coated USPIO particles are shown in the first, second, and third columns, respectively. Allograft images (A), (B), and (C) are from Group I without USPIO infusion (infused with PBS only); (D), (E), and (F) are from Group II infused with 1 mg Fe/kg; (G), (H), and (I) are from Group III infused with 3 mg Fe/kg; (J), (K), and (L) are from Group IV infused with 6 mg Fe/kg; and (M), (N), and (O) are from Group V infused with 12 mg Fe/kg. (P), (Q), and (R) show isograft MR images from Group VI infused with 6 mg Fe/kg. MR images of the coronal view of the grafts are shown in each case. The MRI experiments were carried out in a 4.7-T Bruker AVANCE DRX MR instrument equipped with a 40-cm horizontal bore superconducting solenoid. Taken from Fig. 3 of Ye et al.,[32] with permission.

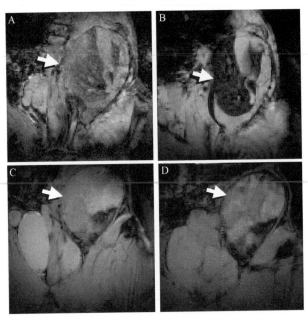

Fig. 4. Representative MR images showing the effect of dextran-coated USPIO particle infusion in allogeneic (A and B) and syngeneic (C and D) rat lung grafts. The MR images of the graft on POD 4 before USPIO infusion are at (A) and (C), and the MR images taken on POD 5, 24 hr after infusion are at (B) and (D). The lung is indicated by the arrow in each panel. The MR imaging sequence consisted of a gradient–echo sequence, triggered to the electrocardiogram (ECG) with the following parameters: TR/TE = 1000/10 ms; FA = Ernst angle; slice thickness = 1 mm, FOV = 6.0 cm; matrix size = 256 × 130 (zero-filled to 256 × 256); and scan time = 5 min. The MRI experiments were carried out in a 4.7-T Bruker AVANCE DRX MR instrument equipped with a 40-cm horizontal bore superconducting solenoid. Taken from Figs. 3 and 4 of Kanno et al.,[28] with permission.

correlates with rejection grades of the grafts. Immunohistochemical analysis shows that the Perl's iron staining mainly coincides with $ED1^+$ cells (i.e., macrophages) and less with $CD3^+$ T-cells. Taken together, these results indicate that the iron particles, which are phagocytosed by macrophages, modulate MR signal intensity at the rejection site of the allograft in accordance with the degree of acute rejection. This indicates that *in vivo* labeling of macrophages with USPIO is a sensitive way of detecting lung rejection.

DETECTING HEART REJECTION BY MRI. Cardiac grafts in our heterotopic heart and lung transplantation model exhibit similar phenomena to those seen in lung grafts upon USPIO administration, yet with a different time course and optimal USPIO dosage.[31] The allogeneic heart does not develop severe rejection until POD 7. The USPIO particles are administered

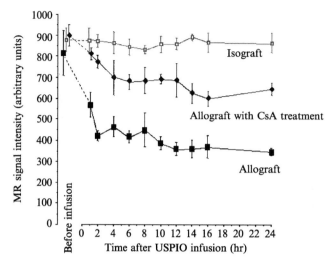

FIG. 5. Summary of the measurements of MR signal intensity (MRSI) in transplanted lungs. MRSI was measured at 5 different points of the horizontal image of each transplanted lung. Data are indicated as mean ± standard deviation (SD) of the mean, shown as arbitrary units. Allografts (solid square) show significantly lower MRSI compared with isografts (open square) at every time point after USPIO infusion. Allografts with CsA treatment (diamond) also show significantly lower MRSI compared with isografts. There is also a significant difference between allografts with or without CsA treatment. Taken from Fig. 6 of Kanno et al.,[28] with permission.

on POD 6, then the rats are imaged 1, 2, and 24 hr after USPIO injection. The optimal dosage for visualizing USPIO-labeled macrophage accumulation at the rejected organ is 1.5 mg Fe/kg for the lung, but 3 mg Fe/kg for the heart. Allografts show significant signal decrease 1 day after USPIO injection, and isografts show little decrease, whereas CsA-treated allografts show intermediate signal decrease. The distribution of iron at the cellular level correlates with ED1$^+$ macrophages.

More interestingly, after the first infusion of USPIO particles on POD 6, a group of rats with allotransplants is then given CsA for either 4 days (POD 7 to 10) or 7 days (POD 7 to 13), and USPIO particles are reinjected on POD 14 and reimaged on POD 15. The animals treated with CsA for 7 days show no significant signal reduction 24 hours after USPIO injection, whereas animals treated with CsA for 4 days show a significant decrease in MR signal intensity in the heart, indicative of graft rejection. These results suggest that repetitive evaluation with in vivo USPIO labeling of macrophages is feasible.

Taken together, the results obtained in allotransplanted kidneys, lungs, and hearts indicate that dextran-coated USPIO particles phagocytosed by macrophages modulate the MR signal at rejection sites, and this can be a sensitive and promising way of detecting organ rejection.

Using MRI to Measure Organ Function and to Detect Organ Dysfunction Resulting from Acute Allograft Rejection

Detecting Organ Dysfunction Resulting from Acute Allograft Rejection by MRI

MRI can detect subtle anatomic alterations associated with graft rejection, such as interstitial edema and mass or volume changes. Many studies have detected noticeable increases in T_1 and T_2 values, and organ volume enlargement, in rejecting allografts in heart, kidney, and liver transplantation, both in humans and in animal models. The prolonged T_1 or T_2 is presumably due to interstitial edema. However, such changes in relaxation times or volume can be nonspecific and not very sensitive for correct diagnosis of acute allograft rejection. The focus of this section is using MRI to monitor function in various animal transplantation models. These may become useful alternatives for MRI-based surveillance for managing organ rejection.

Monitoring Renal Function by MRI. In the assessment of disorders of renal function, current clinically implemented methods, including laboratory urine tests, classic dye clearance tests, radionuclide nephrography, and Doppler flowmetry, are less than ideal because they are either very invasive, lack sensitivity and specificity, lack spatial resolution, or provide poor quantitative information. For reviews, see Cho et al.[33] and Racusen et al.[34] MRI has the capacity of combining both anatomic and functional information and thus is a suitable modality for noninvasive assessment of renal function in the acute rejection process.

The kidney is the most important organ for maintaining ion and electrolyte homeostasis. Renal function is reflected by renal blood flow and glomerular filtration rate. This section focuses on using MRI for noninvasive and sensitive measurements of renal blood flow and glomerular filtration rate.

---

[33] Y. W. Cho, P. I. Terasaki, and B. Graver, *in* "Fifteen-Year Kidney Graft Survival, in Clinical Transplant" (I. Terasaki, ed.), p. 325. UCLA Tissue Typing Laboratory, 1989.

[34] L. C. Racusen, K. Solez, and J. F. Burdick, "Kidney Transplant Rejection: Diagnosis and Treatment." Marcel Dekker, Inc., New York, 1998.

Renal blood flow can be measured by perfusion MRI, which involves tracking the rate of an MRI-detectable tracer traveling through the renal vasculature. Perfusion MRI usually uses one of two methods: the steady-state arterial spin labeling (ASL) technique without application of external contrast agents, or the bolus-tracking method with administration of an exogenous contrast agent. The steady-state ASL method uses water spins in blood as endogenous tracers without application of external contrast agents. It is completely noninvasive with no toxicity, and quantization of the absolute perfusion rate is readily available. In the bolus-tracking method, a bolus of an external paramagnetic contrast agent, such as Gd-DTPA or iron oxide particles, is administered to the animal or human, then the dynamics of the contrast agent passing through the renal vascular structure is monitored by fast imaging. Only relative perfusion rates are available with this method, since it is not feasible to determine the absolute blood flow due to fundamental limitations in measuring mean transit time or mean residence time of the tracer in tissue. Nevertheless, the bolus-tracking method is still the most robust approach for assessing renal perfusion. Although absolute perfusion rate is not available with the bolus-tracking method, we have developed a sensitive autoregression (AR) model to analyze the dynamic first-pass renal perfusion signals, which make distinguishing renal allograft rejection possible.[35]

PERFUSION QUANTIZATION WITH ARTERIAL SPIN LABELING. The ASL technique uses the water in blood as an endogenous tracer, with no need of administering an external contrast agent. The ASL method makes blood water "visible" by magnetically labeling the spins of the incoming arterial blood water.[36,37] The water proton spins in an incoming major artery are inverted at the proximal labeling plane by flow-induced adiabatic fast passage. For renal perfusion, ASL is performed at the suprarenal aorta before it enters the kidney,[38–40] which is approximately 12 mm away from the perfusion imaging plane in a rat. The spin-inverted water protons carry the "negative" magnetization as they travel through the target organ. Thus the degree of signal decrease in the target organ is proportional to the

[35] D. Yang, Q. Ye, M. Williams, Y. Sun, T. C. Hu, D. S. Williams, J. M. Moura, and C. Ho, *Magn. Reson. Med.* **46,** 1152 (2001).

[36] J. A. Detre, J. S. Leigh, D. S. Williams, and A. P. Koretsky, *Magn. Reson. Med.* **23,** 37 (1992).

[37] D. S. Williams, J. A. Detre, J. S. Leigh, and A. P. Koretsky, *Proc. Natl. Acad. Sci. USA* **89,** 212 (1992).

[38] J. A. Detre, W. Zhang, D. A. Roberts, A. C. Silva, D. S. Williams, D. J. Grandis, A. P. Koretsky, and J. S. Leigh, *NMR Biomed.* **7,** 75 (1994).

[39] D. A. Roberts, J. A. Detre, L. Bolinger, E. K. Insko, R. E. Lenkinski, M. J. Pentecost, and J. S. Leigh, Jr., *Radiology* **196,** 281 (1995).

[40] D. S. Williams, W. Zhang, A. P. Koretsky, and S. Adler, *Radiology* **190,** 813 (1994).

blood flow. The absolute perfusion rate f (ml/g/min) can then be generated by the following equation[37]

$$f = [\lambda/T_{lobs} \times 2\alpha] \times [(M_c - M_l)/M_c] \tag{1}$$

where α is spin-labeling efficiency, T_{lobs} is the *in vivo* tissue water apparent spin–lattice relaxation time, and λ (ml/g) is the blood–tissue partition coefficient of water. For renal cortical perfusion, λ is a spatially constant value of 0.9 ml/g, whereas T_{lobs} and α can be measured directly.

In the orthotopic Fisher 344-to-ACI rat renal transplantation model, the left kidney of an ACI recipient rat is replaced with an allograft kidney from a Fisher 344 donor rat or with an isograft kidney from another ACI rat. The contralateral right native kidney is left untouched to serve as an internal control. The ASL perfusion evaluation is performed on POD 3, when only mild rejection has occurred (Banff Score 2 to 4), or on POD 7, when irreversible severe acute rejection has already taken place (Banff Score 6). The renal cortical perfusion rate in normal native kidney is 7.5 ± 0.8 ml/g/min. On POD 3, before marked acute rejection has occurred, the renal cortical perfusion rate is similar in syngeneic (3.3 ± 1.7 ml/g/min) and allogeneic (3.0 ± 2.4 ml/g/min) kidneys. In contrast, on POD 7, allografts with severe rejection have very low renal cortical perfusion (0.3 ml/g/min), whereas isografts preserve very good perfusion (5.2 ± 2.0 ml/g/min). There is an excellent inverse correlation between the renal cortical perfusion measured by ASL and the histologic scores for acute allograft rejection (Fig. 6). In addition to renal cortical perfusion, quantitative MRI provides other information that may be useful in assessing the health of a kidney. The renal cross-sectional area and renal cortical T_{lobs} are noticeably larger in allografts on POD 7, presumably due to extensive interstitial lymphoplasmacytic infiltration with generalized edema. As with renal cortical perfusion, these two parameters also show good correlation with histologic rejection grades. For details, see Wang *et al.*[41]

Most of the exogenous contrast agents are excreted through the kidney. Because the ASL technique uses endogenous water in the blood as an MRI-detectable tracer, there is no potential toxicity from injected tracers, so it is very desirable for renal transplantation patients who may have compromised renal function. The completely noninvasive and quantitative monitoring of renal cortical perfusion with the ASL method offers a promising new avenue for both short- and long-term management of patients with renal transplantation.

[41] J. J. Wang, K. S. Hendrich, E. K. Jackson, S. T. Ildstad, D. S. Williams, and C. Ho, *Kidney Int.* **53**, 1783 (1998).

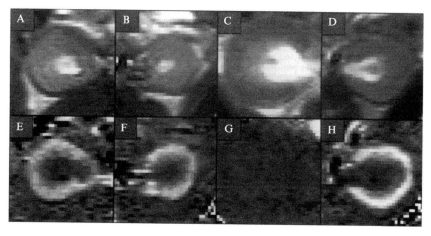

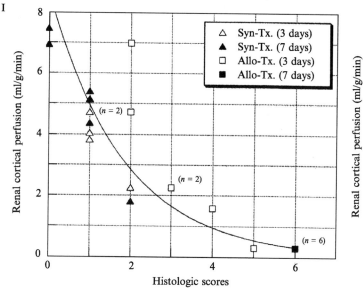

FIG. 6. (A–H) Expansions (2 cm × 2 cm) from representative MR images of rat kidney. Top rows (A–D) are anatomic images. Bottom rows (E–H) are the corresponding relative renal perfusion maps, where high perfusion rates are depicted as high intensity. The first two columns (A, E, B, and F) are from a normal rat, and the last two columns (C, G, D, and H) are from a rat with allogeneic transplantation at POD 7. (I) Shows correlation between the renal cortical perfusion rate and histologic scores in transplanted kidneys. The correlation coefficient (r) was -0.82, $p < 0.05$ ($n = 23$). One value was an outlier. Taken from Figs. 1 and 3 of Wang et al.,[41] with permission.

DYNAMIC MRI FOR FIRST-PASS RENAL PERFUSION. Renal perfusion can also be measured by the bolus-tracking method. The chemical and physical properties of the contrast agents administered, as well as the temporal resolution of imaging, are of importance for the physiologic phenomena measured by MRI. The intravascular blood pool contrast agents, which have minimal migration across the capillary endothelium, would stay in the vascular structure without interstitial diffusion, glomerular filtration, or tubular excretion; they are the most suitable agents for studying the vascular sector of the kidney. The dextran-coated iron oxide particles are such a type of blood pool agent, including SPIO and USPIO particles. However, SPIO particles have the drawback of having a very short blood half-life, whereas the prolonged blood pool effect of the USPIO particles makes them more suitable for measuring renal perfusion. On the other hand, most commonly used gadolinium (Gd)-based contrast agents, which can diffuse into interstitial space and are filtered freely by the glomerulus in kidneys without tubular excretion, seem most suitable for measuring glomerular filtration, rather than renal perfusion.

The first-pass renal perfusion is completed within a few seconds after the injection of the contrast agent in rats. Therefore it is important to have a proper temporal resolution of MRI for catching the fast dynamic behavior. To better capture the fast first-pass dynamic behavior, we have used the snapshot fast low-angle shot (FLASH) pulse sequence, which is characterized by low flip angle, very short pulse duration, short signal readout, and very short TE and repetition time (TR). The temporal resolution for the sequence is 134 msec per image, which is suitable for first-pass renal perfusion evaluation. For details, see Yang *et al.*[35]

First-pass dynamic MRI has been used to evaluate renal allograft rejection in an orthotopic renal transplantation model in rats using DA-to-BN rats.[35] With this strain combination, the allografts undergo mild/moderate acute rejection on day 4 after the transplantation. On POD 4, a bolus of dextran-coated USPIO particles is given intravenously to rats with either allografts or isografts, or with no surgery. After the USPIO bolus, 128 consecutive snapshot FLASH coronal images are taken within 43 s (i.e., 134 ms per frame). The curve shape of the first-pass dynamic MRI depends on the type and the dosage of the contrast agent administered, field strength, and the MR pulse sequences. With USPIO dosage lower than 12.1 mg Fe/kg body weight, the typical well-shaped first-pass transient is readily seen (Fig. 7A). After the USPIO bolus injection, a signal decrease, or the "wash-in" of the bolus, occurs within the cortex first, followed by a centripetal progression of the signal decrease toward the medulla. This is followed by a signal intensity increase in the cortex and the medulla because of the "wash-out" of the bolus. For USPIO dosages

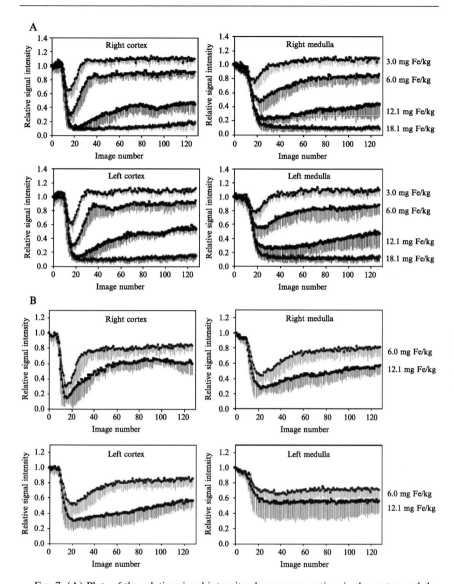

FIG. 7. (A) Plots of the relative signal intensity changes versus time in the cortex and the outer medulla in normal BN rats after a bolus injection of four different doses. Images were repeated at 334 ms. Shaded bars represent the standard deviations of the normalized signal intensities. (B) Plots of the relative signal intensity changes versus time in the cortex and the outer medulla in the allograft rats after a bolus injection of two different doses. Images were repeated at 334 ms. Shaded bars represent the standard deviations of the normalized signal intensities. Taken from Figs. 3 and 4 of Yang et al.,[35] with permission.

higher than 18.1 mg Fe/kg, the dynamic curve does not recover after the initial signal intensity reduction.

The wash-in slope, maximum signal decrease (MSD), and time of occurrence of MSD (t_{MSD}) are extracted from the dynamic curves. For BN or DA rats with no transplantation surgery, there are no significant differences in all three parameters between left and right native kidneys or between rat strains at all doses studied. In transplanted animals, isograft kidneys show a perfusion pattern similar to those of normal kidneys in the cortex and the outer medulla. Allograft kidneys, on the other hand, exhibit lower MSD, longer t_{MSD}, and lower wash-in slope compared with those in normal kidneys (Fig. 7B). This indicates impaired perfusion in the renal allografts. All these parameters examined, MSD, t_{MSD}, and wash-in slope, show differences in allografts compared with isografts or native normal kidneys, thus they all can be used as functional indexes for detecting acute renal rejection. When plotting these three parameters against one another, values from native kidneys and isografts all cluster together, which can be considered the "normal" range for renal function, whereas all allografts fall far outside these "normal" boundaries.

AUTOREGRESSIVE MODELING FOR DYNAMIC MRI FIRST-PASS RENAL PERFUSION STUDIES. Although the first-pass dynamic of the USPIO bolus tracking appears to be an excellent methodology for detecting renal allograft rejection, no absolute quantitative perfusion rate (in ml/g/min) can be obtained. All three functional indexes, MSD, t_{MSD}, and wash-in slope, are relative changes and are greatly dependent on the contrast agent chosen, the dosage used, the pulse sequences, experimental parameters, the field strength, and other factors. As a result, it is difficult to establish a universal "normal" range for evaluating renal rejection for potential clinical application. Hence we have developed a new signal processing methodology for the automatic detection of rejection.[35] Instead of looking at just one aspect of the first-pass dynamic curve, such as the wash-in slope, this signal processing methodology, which is based on AR modeling, looks at the whole time series and generates one novel parameter, the subspace distance, that best reflects the overall characteristics of the dynamic curves and can thus detect rejection readily.

Our modified AR statistical modeling method looks at the overall time series of the first-pass dynamic, conditions the signals internally with the sample mean, and then models the signal sequences as a liner combination of past samples plus a noise sample with AR model order p and a set of AR coefficients. It is known from signal processing theory that the dynamic behavior of an AR model is best characterized by its poles, the eigenvalues of the companion matrix of the AR model coefficients. Thus the poles are the set of parameters that can best reflect the overall characteristics of the time series and are the potential universal indexes for detecting renal rejection.

The combined information criterion is used to select the order of the AR model, and the AR coefficients are estimated by Burg's method, which is based on the minimization of the sum of the forward and backward squared prediction errors. It is found that the fourth-order AR model represents the best fit for the first-pass dynamic data. From each dynamic time series, four poles, or two pairs of poles, are generated from our model, and the pole configuration reflects the overall dynamic behavior of the original time series (Fig. 8A). One of the two pairs of poles shows the largest dissimilarity between the normal and allograft kidneys, so we have chosen to work with the subspaces spanned by the two eigenvectors associated with this pair of poles. The subspace distance between these two eigenvectors quantifies the dissimilarity between the left and right kidneys. Thus this single parameter, the subspace distance (Fig. 8B), captures the overall dynamic characteristics of the whole first-pass dynamic curves (including MSD, t_{MSD} and wash-in slope) in a single quantitative measure. These results demonstrate that the subspace distance can detect pattern dissimilarity in dynamic intrarenal signals and can be a candidate for clinical assessment of kidney function. For details, see Yang et al.[35]

RENOGRAPHY. In addition to renal blood flow, a common method for assessing renal allograft function is the estimation of glomerular filtration rate. However, this measure is not reliable or sensitive, and it is inconvenient due to the inadequacies of serum creatinine and clearance methods for detection of renal dysfunction in kidney transplantation patients. The most commonly used extracellular MRI contrast agents, the Gd chelates, are predominately cleared in the kidney, and filtered freely by the glomerulus without tubular excretion, reabsorption, or known nephrotoxicity. They can be used to measure the glomerular filtration rate and give a measure of the function for kidneys.[42–44] In a DA-to-Lewis orthotopic renal transplantation model,[45] a T_1-weighted FLASH pulse sequence was used to capture the dynamic behavior of a bolus of Gd-DOTA passing through the renal system in a time-resolved manner. The rejected allografts showed very little signal enhancement and very slow clearance, which indicates severely compromised function.

[42] R. K. Sharma, R. K. Gupta, H. Poptani, C. M. Pandey, R. B. Gujral, and M. Bhandari, *Transplantation* **59,** 1405 (1995).

[43] J. Taylor, P. E. Summers, S. F. Keevil, A. M. Saks, J. Diskin, P. J. Hilton, and A. B. Ayers, *Magn. Reson. Imaging* **15,** 637 (1997).

[44] D. Baumann and M. Rudin, *Magn. Reson. Imaging* **18,** 587 (2000).

[45] N. Beckmann, J. Joergensen, K. Bruttel, M. Rudin, and H. J. Schuurman, *Transpl. Int.* **9,** 175 (1996).

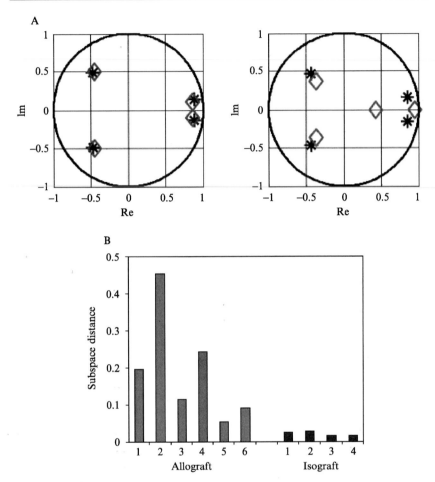

FIG. 8. (A) The poles of the AR model of the signal in the cortex for two rats: isografts on the left and allografts on the right. The stars represent the poles of the right kidney, and the diamonds represent the poles of the left kidney. Im, imaginary part; Re, real part; ∗, native kidney; ◇, transplanted kidney. (B) The subspace distances between the left and right kidneys for six allograft and four isograft rats. The left six columns represent allograft rats, and the right four columns represent isograft rats. Taken from Figs. 2 and 7 of Yang et al.,[35] with permission.

Monitoring Cardiac Function by MRI. Clinically, cardiac acute rejection is characterized by symptoms, as well as by the standard chest radiography, which may show increased heart size, pulmonary edema, pleural effusion, and decreased cardiac output. However, these parameters are nonspecific and reflect advanced stages of acute cardiac rejection, which is difficult to reverse by antirejection therapy.[1] It is important to have a non-invasive method for accessing acute heart rejection at an early stage.

MRI and MR spectroscopy (MRS) have been used for noninvasive evaluation of cardiac rejection after transplantation (Table I), and the parameters found to be correlated with the rejection mainly are prolonged T_1 and T_2 relaxation times, increase in Gd-DTPA uptake, increased heart size or mass, and reduced high-energy phosphates. Although showing good correlation with rejection noninvasively, these parameters are detecting increase in edema or necrosis, which are manifestations of advanced rejection. It will be more useful if the acute cardiac rejection can be found earlier before the irreversible rejection occurs. In our heterotopic working heart model (Model 2) in rats, we have found parameters associated with acute cardiac rejection with cine MRI and myocardial tagging.

CARDIAC FUNCTION BY CINE MRI. Electrocardiogram (ECG)- and respiration-gated cine imaging[46] was used to obtain high-resolution density-weighted short-axis images covering the entire heart volume and the whole cardiac cycle (Fig. 9). The dimensions of the left ventricle (LV) volume (LVV) and LV wall can be precisely quantified[47] by segmenting out the contours of the LV cavity and wall,[48] and the global systolic functional parameters can be obtained, including LVV, SV, EF, cardiac output, and heart mass. No significant changes in global systolic functions were detected until the cardiac allografts had suffered severe acute rejection.[49] Thus, although useful for late-stage rejection diagnosis, these global systolic functional parameters are inadequate in detecting early rejection.

In addition to the global systolic function, regional wall motion can be obtained through centerline method analysis.[50] The time-dependent hemodynamics can be measured for LVV and LV wall motion regionally throughout the whole cardiac cycle. With sufficient temporal resolution, each phase in a cardiac cycle can be identified from the volume–time relationship, including isovolumic contraction, ejection, isovolumic relaxation, rapid filling, diastasis, and atrial contraction. The relative time span and the

[46] R. I. Pettigrew, J. N. Oshinski, G. Chatzimavroudis, and W. T. Dixon, *J. Magn. Reson. Imaging* **10**, 590 (1999).

[47] R. J. van der Geest and J. H. Reiber, *J. Magn. Reson. Imaging* **10**, 602 (1999).

[48] S. Slawson, B. B. Roman, D. S. Williams, and A. P. Koretsky, *Magn. Reson. Med.* **39**, 980 (1998).

[49] Y. L. Wu, S. Kanno, and C. Ho, *J. Cardiovas. Magn. Reson.* **4**, 4 (2002).

[50] F. P. van Rugge, E. E. van der Wall, S. J. Spanjersberg, A. de Roos, N. A. Matheijssen, A. H. Zwinderman, P. R. van Dijkman, J. H. Reiber, and A. V. Bruschke, *Circulation* **90**, 127 (1994).

TABLE I
MRI AND MRS METHODS FOR DIAGNOSIS OF ACUTE HEART REJECTION

Study	Object model system	Method	Parameters investigated and outcome
Marie et al., 2001[a]; Marie et al., 1998[b]; Wiesenberg et al., 1987[c]; Lund et al., 1988[d]; Lund et al., 1987[e]; Nishimura et al., 1987[f]	Human	Black-blood MRI Anatomic MRI	Lengthened ^1H relaxation times (T_1 or T_2) correlated with advanced rejection
Smart et al., 1993[g]	Human	T_2-weighted MRI	Increased SI ratio of myocardial/skeletal muscle
Mousseaux et al., 1991[h]; Mousseaux et al., 1993[i]; Almenar et al., 2003[j]	Human	Gd-enhanced MRI	Increased signal intensity due to increased Gd uptake at the areas with necrotic myocardium
Konstam et al., 1988[k]	Heterotopic rat model (LBN to Lewis)		
Globits et al., 1997[l]	Human	Cine MRI	LV hypertrophy, reduced wall stress
Almenar et al., 2003[j]	Human	Cine MRI	No changes in LVV, EF, or hypertrophy
Johansson et al., 2002[m]	Heterotopic rat model (PVG to Wistar/Kyono)	Perfusion with USPIO	Changes in perfusion
Walpoth et al., 1998[n]	Orthotopic pig model	^{31}P MRS and ^1H MRI	Reduction in high-energy phosphates and increase in inorganic phosphates or phosphomonoester
Walpoth et al., 1995[o]	Heterotopic rat model (DA to Lewis)		
Onno van Dobbenburgh et al., 1996[p]	Heterotopic rat model (BN to Lewis)		
Donofrio et al., 1999[q]	Human infant	Tagging	Even with no rejection, heterogeneous baseline patterns for strain, twist, and radial shortening

(continued)

TABLE I   *(continued)*

Study	Object model system	Method	Parameters investigated and outcome
Nishimura *et al.*, 1989[r]	Heterotopic dog model	^{23}Na MRI	Increased intensity in ^{23}Na MRI due to increased intracellular Na in necrotic myocardium

[a] P. Y. Marie, M. Angioi, J. P. Carteaux, J. M. Escanye, S. Mattei, K. Tzvetanov, O. Claudon, N. Hassan, N. Danchin, G. Karcher, A. Bertrand, P. M. Walker, and J. P. Villemot, *J. Am. Coll. Cardiol.* **37**, 825 (2001).

[b] P. Y. Marie, J. P. Carteaux, M. Angioi, N. S. Marwan, K. Tzvetanov, J. M. Escanye, N. David, S. Mattei, N. Danchin, G. Karcher, A. Bertrand, and J. P. Villemot, *Transplant. Proc.* **30**, 1933 (1998).

[c] G. Wiesenberg, P. W. Pflugfelder, W. J. Kostuk, C. A. McKenzie, and F. S. Prato, *Am. J. Cardiol.* **60**, 130 (1987).

[d] G. Lund, R. L. Morin, M. T. Olivari, and W. S. Ring, *J. Heart Transplant.* **7**, 274 (1988).

[e] G. Lund, J. G. Letourneau, D. L. Day, and J. R. Crass, *Radiol. Clin. North Am.* **25**, 281 (1987).

[f] T. Nishimura, M. Sada, H. Sasaki, C. Yutani, T. Kozuka, H. Amemiya, T. Fujita, T. Akutsu, and H. Manabe, *Heart Vessels* **3**, 135 (1987).

[g] F. W. Smart, J. B. Young, D. Weilbaecher, N. S. Kleiman, R. E. Wendt, 3rd, and D. L. Johnston, *J. Heart Lung Transplant.* **12**, 403 (1993).

[h] E. Mousseaux, D. Farge, R. Guillemain, C. Amrein, C. Vulser, and P. Bruneval, *J. Am. Coll. Cardiol.* **17**, 326A (1991).

[i] E. Mousseaux, D. Farge, R. Guillemain, P. Bruneval, C. Vulser, J. P. Couetil, A. Carpentier, and J. C. Gaux, *J. Comput. Assist. Tomogr.* **17**, 237 (1993).

[j] L. Almenar, B. Igual, L. Martinez-Dolz, M. A. Arnau, A. Osa, J. Rueda, and M. Palencia, *Transplant. Proc.* **35**, 1962 (2003).

[k] M. A. Konstam, M. J. Aronovitz, V. M. Runge, D. M. Kaufman, B. A. Brockway, J. M. Isner, N. A. Katzen, A. R. Dresdale, J. T. Diehl, E. Kaplan *et al.*, *Circulation* **78**, III 87 (1988).

[l] S. Globits, T. De Marco, J. Schwitter, H. Sakuma, M. O'Sullivan, C. Rifkin, F. Keith, K. Chatterjee, W. W. Parmley, and C. B. Higgins, *J. Heart Lung Transplant.* **16**, 504 (1997).

[m] L. Johansson, C. Johnsson, E. Penno, A. Bjornerud, and H. Ahlstrom, *Radiology* **225**, 97 (2002).

[n] B. H. Walpoth, M. F. Muller, B. Celik, B. Nicolaus, N. Walpoth, T. Schaffner, U. Althaus, and T. Carrel, *Eur. J. Cardiothorac. Surg.* **14**, 426 (1998).

[o] B. H. Walpoth, F. Lazeyras, A. Tschopp, T. Schaffner, U. Althaus, M. Billingham, and R. Morris, *Transplant. Proc.* **27**, 2088 (1995).

[p] J. Onno van Dobbenburgh, C. Kasbergen, P. J. Slootweg, T. J. Ruigrok, and C. J. van Echteld, *Mol. Cell Biochem.* **163–164**, 247 (1996).

[q] M. T. Donofrio, B. J. Clark, C. Ramaciotti, M. L. Jacobs, K. E. Fellows, P. M. Weinberg, and M. A. Foel, *Am. J. Physiol.* **277**, R1481 (1999).

[r] T. Nishimura, M. Sada, H. Sasaki, N. Yamada, Y. Yamada, C. Yutani, H. Amemiya, T. Fujita, T. Akutsu, and H. Manabe, *Cardiovasc. Res.* **23**, 561 (1989).

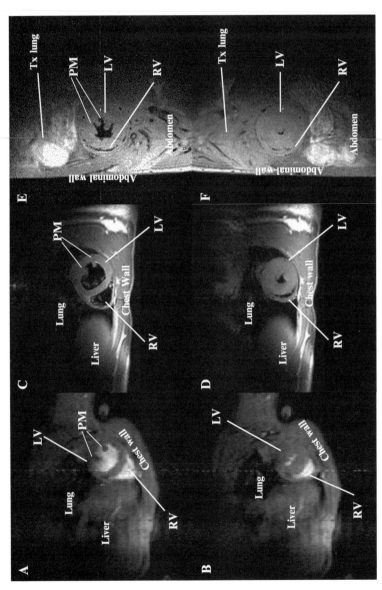

Fig. 9. Cine imaging for native and heterotopic transplanted hearts: (A, B) Short-axis images for a native mouse heart with a bright blood pulse sequence (FLASH). (C, D) Short-axis images for a native rat heart with a black-blood spin–echo pulse sequence. (E, F) Short-axis images for a transplanted rat heart in the abdomen with a black-blood spin–echo pulse sequence. The top panels (A, C, E) are images acquired at the end-diastole (ED), whereas the lower panels (B, D, F) sequence.

rate constant of each phase are extracted from the time-dependent volume or wall thickness curves.

Because our goal is to search for sensitive functional parameters that can detect cardiac rejection at earlier stages, transplanted hearts have been subjected to detailed MRI with higher temporal resolution before severe rejection occurs. In the earlier phases, isografts and allografts undergoing mild to moderate rejection largely exhibit a similar volume–time relationship. Most of the parameters examined have shown no significant changes, except the diastolic rate constant and the relative time span for the isovolumic relaxation (T_{IVR}). As rejection grades increase over time, T_{IVR} increases significantly for the allografts, which is indicative of gradual loss of diastolic function. Conversely, T_{IVR} decreases over time for the isografts, due to recovering from the ischemic injury of the surgery operation.

This result indicates that the early dysfunction of acute cardiac rejection is mainly diastolic and the relative time span for the isovolumic relaxation and the diastolic rate constant can be potential sensitive indexes for early diastolic dysfunction resulting from acute rejection.[49]

CARDIAC FUNCTION BY MYOCARDIAL TAGGING. Myocardial tagging[51] by SPAMM[52–56] or DANTE[57] schemes deposits planes of signal saturation intersecting the myocardium prior to imaging. This creates grid-like "tags" on the myocardium and provides material points for tracking regional wall motion or contractility (Fig. 10A and B). This is a powerful tool for assessing mechanical function of the heart. The tags are applied as close as to R-wave as possible to facilitate tracking the motion throughout the whole systole and diastole phases. Strain (Fig. 10E and F) and motion (Fig. 10C and D) can be extracted from the tagging data regionally at each material

[51] N. Reichek, *J. Magn. Reson. Imaging* **10**, 609 (1999).

[52] E. A. Zerhouni, D. M. Parish, W. J. Rogers, A. Yang, and E. P. Shapiro, *Radiology* **169**, 59 (1988).

[53] L. Axel and L. Dougherty, *Radiology* **172**, 349 (1989).

[54] L. Axel and L. Dougherty, *Radiology* **171**, 841 (1989).

[55] R. McVeigh and E. Atalar, *Magn. Reson. Med.* **28**, 318 (1992).

[56] A. A. Young, L. Axel, L. Dougherty, D. K. Bogen, and C. S. Parenteau, *Radiology* **188**, 101 (1993).

[57] N. V. Tsekos, M. Garwood, H. Merkle, Y. Xu, and N. Wilke, *Magn. Reson. Med.* **34**, 395 (1994).

are images acquired at the end-systole (ES). LV, left ventricle; RV, right ventricle; PM, papillary muscle; Tx, transplanted. The air-filled native lung is almost invisible in the normal MRI, whereas the heterotopic transplanted lung is readily seen because there is no air in the lung.

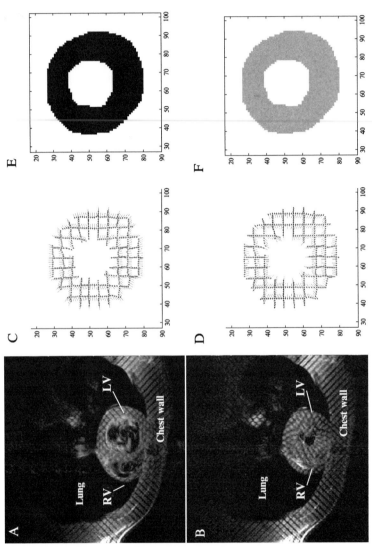

Fig. 10. Myocardial tagging for biomechanical evaluation of ventricular function: (A, B) tagging images of a native rat heart at the ED (A) and ES (B); and (C, D) displacement field of the heart. The direction of the arrowheads represents the direction of the motion, whereas the length of the arrowheads represents the extent of the movement. (E, F) Strain map of the heart. The brightness of the strain represents extent of the contraction.

point throughout the whole volume of the heart. Strain is the unit less measurement of deformation that relates the changes in myocardial wall shape. A positive strain represents elongation, whereas negative strain represents compression. Several types of motion can be portrayed by tagging (e.g., shortening, twist, or rotation). The radial shortening describes the distance changes of each material point relative to the centroid of the ventricular cavity, whereas the twist is defined as the extent of rotation or wringing motion of the heart. The displacement field (Fig. 10C and D) can be defined for each material point at each cardiac phase. The direction of the arrows represents the direction of the movement, whereas the length of the arrows represents the extent of the movement. Thus the regional contractility of the heart can be defined in detail.

In pediatric transplantation patients, it has been found that transplanted hearts exhibit inhomogeneous changes in patterns of strain, twist, and radial shortening[58] even though the hearts undergo no rejection (Grade 0) or very mild rejection (Grade 1A). The biomechanics of the transplanted ventricles is significantly different from that of native hearts, even without rejection.

In our heterotopic working heart model in rats, we have observed heterogeneous changes in strain and motion by tagging. At early rejection phases, allografts exhibit spatial heterogeneity in contractility. Figure 11 shows the strain (Fig. 11A and B) and the displacement field (Fig. 11C and D) mapped by tagging of a moderately rejected allograft. The brightness of the strain map represents the extent of stretching. The angle and the extent of the twist and radial shortening are represented by the direction and the length of the arrowheads, respectively. For this particular heart, the inferolateral (upper left) and anterior (lower right) LV wall have preserved most of the motility, whereas the anterolateral wall (lower left) has lost its contractility. The loss in mechanical function at early rejection phase is inhomogeneous, and the spatial heterogeneity of functional loss in early acute rejection can be captured by myocardial tagging. For details, see Wu et al.[58] Hence cardiac MRI is a promising tool for noninvasive and sensitive detection for early acute cardiac allograft rejection.

[58] Y. L. Wu, K. Sato, Y. Sun, J. M. F. Moura, and C. Ho, *J. Cardiovasc. Magn. Reson.* **6,** 87 (2004).

(C and E) depict the dilation motion of the heart, whereas (D and F) depict the contraction motion of the heart between two consecutive cardiac phases. Figs. 10C, D, E, and F are computer-generated strain maps and displacement fields based on modeling of a native rat heart (unpublished results of Y. Sun, J. M. Moura, J.-Y. L. Wu, K. Sato, and C. Ho).

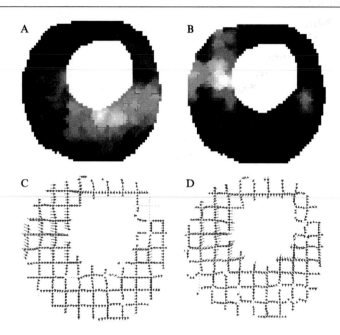

Fig. 11. Myocardial tagging for a moderately rejection-transplanted rat heart: (A, B) strain map and (C, D) displacement field of a moderately rejection-transplanted allograft. (A) and (C) depict motion at a systolic phase, whereas (B) and (D) depict motion at a diastolic phase. Figs. 11A, B, C, and D are computer-generated strain maps and displacement fields based on tagged MRI data obtained from a moderately rejected rat heart (unpublished results of Y. Sun, J. M. F. Moura, J.-Y. L. Wu, K. Sato, and C. Ho).

Summary and Future Perspectives

Our experimental results, as discussed previously, clearly indicate that our two-pronged approach to detect the accumulation of macrophages labeled with dextran-coated USPIO particles at a transplanted organ and to monitor the function of a transplanted organ in our rat models is a very promising, noninvasive way to follow the status of the transplanted organ. With further advances in designing new MR contrast agents with better sensitivity and specificity, as well as in improving the performance of MRI instruments, MRI offers an excellent modality to monitor and to manage transplant patients. A major challenge in the treatment of transplant patients is to administer the right kinds, as well as the appropriate amounts, of immunosuppressive drugs. Thus our two-pronged approach to detect the initiation of graft rejection and to monitor the function of a transplanted organ during various stages of the graft rejection process using the noninvasive MRI methodology is very attractive and merits further development.

Acknowledgments

We thank Dr. E. Ann Pratt for helpful discussions in preparing our manuscript. Our research is supported by research grants from the National Institutes of Health (P41EB-001977 and R01EB/AI-00318). The Pittsburgh NMR Center is supported by the National Institute of Biomedical Imaging and Bioengineering as a National Biomedical Research Resource Center. Y.-J. L. W is supported by a Ruth L. Kirschstein National Research Service Award (F32HL-068423).

[4] Structural and Functional Optical Imaging of Angiogenesis in Animal Models

By Richard L. Roberts and Pengnian Charles Lin

Introduction

Discovery in cancer biology has moved at an accelerated pace in recent years with a considerable focus on the transition from *in vitro* to *in vivo* models. As a result, there has been a significant increase in the need to develop noninvasive, high-resolution *in vivo* imaging approaches to characterize molecular and cellular events that lead to cancer development and progression.[1] Noninvasive molecular imaging is particularly critical for angiogenesis studies[2,3] because blood vessel formation is a multistep process, and it is strongly influenced by the microenvironment. Current *in vitro* assays cannot replicate many aspects of this dynamic process. Therefore *in vivo* imaging in animal models is critical to elucidate mechanisms that regulate angiogenesis and to identify and validate new therapeutic targets. Traditional approaches for studying tumor development have relied on tissue biopsies and histologic analysis, which only gives a snapshot of a highly dynamic, complex process. Noninvasive molecular imaging provides an opportunity to overcome these limitations and may be critical in evaluating therapeutic responses to new antiangiogenic therapies that will be an important adjunct in translating the knowledge into clinical practice. Notably, antiangiogenic therapy is intended to target the angiogenic vessels only and will potentially cause fewer side effects because normal vessels are typically quiescent. Traditional clinical trial designs looking for the maximum toxicity dose would not apply to this new type of therapy.[4]

---

[1] R. Weissleder and U. Mahmood, *Radiology* **219**, 316 (2001).

[2] J. Folkman and K. Beckner, *Acad. Radiol.* **7**, 783 (2000).

[3] D. M. McDonald and P. L. Choyke, *Nat. Med.* **9**, 713 (2003).

[4] M. Cristofanilli, C. Charnsangavej, and G. N. Hortobagyi, *Nat. Rev. Drug Discov.* **1**, 415 (2002).

Copyright 2004, Elsevier Inc.
All rights reserved.
0076-6879/04 $35.00

Instead, identifying surrogate markers for antiangiogenic therapy is imminent. Noninvasive imaging may offer the ability to optimize therapy for each treatment and for each individual.

The ability to visualize the dynamic biologic process by *in vivo* imaging is revolutionizing many areas in biology. Recent advances in the application of fluorescent proteins have permitted microscopy to move from static images to dynamic recordings of cellular behavior in living animals.[5] Tissue visualization with light is probably the most common imaging in medical research, and new optical imaging approaches offer great promise for real-time monitoring of tumor development and progression in living animal models. Microscopic optical imaging methods ranging from regular microscopy, fluorescence, confocal, and multiphoton microscopy are particularly useful for elucidating structural and functional abnormalities of tumor blood vessels. They are relatively inexpensive and provide the highest resolution. Furthermore, optical imaging complements other imaging technologies such as computed tomography (CT) and magnetic resonance imaging (MRI). In this chapter, we present two animal models for microscopic optical imaging of angiogenesis.

Intravital Microscopy and Animal Window Models

Microscopic optical imaging methods are powerful tools for dissecting the cellular and molecular features of the microvasculature. Optical imaging offers high resolution and real-time functional readouts in animal models. However, a major limitation of optical imaging is tissue penetration and light absorption. The most powerful current approach, multiphoton microscopy, only has the depth of imaging to hundreds of micrometers. Animal window chamber models are designed to overcome this limitation of optical imaging. A variety of window models have been developed that include the dorsal skin window model, mammary tumor window model, and cranial tumor window model.[6–8] The window chamber is a chronical model, which is installed on living animals (mice or rats). The tissues implanted in the transparent chambers in skin and other sites can be imaged over time by microscopy. The animal window models, coupled with intravital microscopy, have provided stunning insight into tumor host

---

[5] R. D. Phair and T. Misteli, *Nat. Rev. Mol. Cell Biol.* **2,** 898 (2001).

[6] P. Lin, P. Polverini, M. Dewhirst, S. Shan, P. S. Rao, and K. Peters, *J. Clin. Invest.* **100,** 2072 (1997).

[7] S. Shan, B. Sorg, and M. W. Dewhirst, *Microvasc. Res.* **65,** 109 (2003).

[8] R. M. Foltz, R. E. McLendon, H. S. Friedman, R. K. Dodge, D. D. Bigner, and M. W. Dewhirst, *Neurosurgery* **36,** 976; discussion 984 (1995).

interaction, tumor angiogenesis initiation, and therapeutic responses of tumor blood vessels.[9] Rate of blood flow, blood vessel diameter, vascular density, endothelium permeability, leukocyte–endothelial cell interactions, and gene expression are among the variables that can be measured by this model.

The Dorsal Skin Window Chamber Model

The most commonly used window chamber is the dorsal skin window chamber that is installed on the back skinfold of mice or rats.[6] Sterile technique is followed for this surgical procedure. The mice or rats are anesthetized, and the back of the mice or rats is shaved and has warm hair removal cream (Nair) applied for 2 minutes. After the cream and remaining hair have been wiped away, the back is alternately wiped with surgical sponges soaked in Hibiclens and then alcohol. The area is cleaned at least three times in this way. Then using sterile instruments, the skin is gathered from the back and sutured to a C-shaped holder, which functions to hold the skin tautly. A 10-mm diameter hole for rats or 8-mm diameter hole for mice is traced on the skin and is dissected away from the dorsal skin flap. This is done on both sides of the back for rats or one side of skin for mice, leaving a fascial plane with associated vasculature. The hole is then held vertically away from the body with a titanium superstructure (the "window chamber") that is sutured to both sides of the flap. Glass cover slips are attached to the center of the titanium saddle to cover the surgical site. A small piece of tumor or tumor cells can be placed onto the fascial plane and the chambers are sealed with glass cover slips. The C-shaped holder is then removed, and the skin surfaces adjacent to the implant are covered with Neosporin ointment. Finally, the mouse or rat is placed on a heating blanket to recover. The tissue within the chamber is approximately 200 μm thick and is semitransparent (Fig. 1).

Intravital Microscopy Imaging. To visualize and record tumor angiogenesis and tumor development *in vivo,* the mice or rats are anesthetized. The window chamber is placed in a special plexiglass holder to stabilize the window. The mouse or rat is then placed on a heated microscope stage for observation. One can use a regular upright microscope, fluorescent microscope, confocal microscope, or multiphoton microscope, depending on the need. Images of the vasculature of the tumor are captured by cameras and/ or recorded on videotapes for further analysis. A variety of angiogenic functions and abnormalities of blood vessels in pathologic conditions can be measured in live animals.

[9] E. B. Brown, R. B. Campbell, Y. Tsuzuki, L. Xu, P. Carmeliet, D. Fukumura, and R. K. Jain, *Nat. Med.* **7,** 864 (2001).

A   B

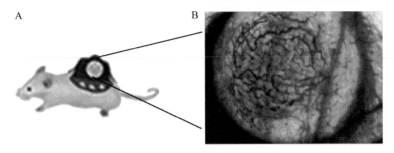

FIG. 1. Dorsal skinfold tumor window model. A small window chamber is installed on the back skinfold of mice. An 8-mm-diameter hole is dissected in the epithelium, and a fascial plane with associated vasculature remains. Tumor cells or tumor tissues can be implanted onto the fascial plane, and the chamber is sealed with glass cover slips. The chamber with tissue is approximately 1 mm thick and is semitransparent (A). Tumor growth and angiogenesis can be directly visualized and recorded in live mice by microscopy. (B) A live image of a well-vascularized tumor grown in the window chamber. (See color insert.)

Imaging Tumor Angiogenesis and Tumor Development in the Window Chamber Model

Imaging studies have identified multiple cellular and molecular abnormalities that distinguish tumor vessels from their normal vessels and help explain their unusual appearance, irregular blood flow, and vascular leakiness. Most tumor vessels display a disorganized pattern and do not fit into the conventional hierarchy of arterioles, capillaries, and venules. We have used the tumor window model to study tumor angiogenesis and tumor development. The tumors typically grow to an observable size at 5–7 days after surgery. Tumor angiogenesis can be seen at the same time. Microscopy and recording devices are used to record the development of tumor blood vessels and tumors (Fig. 1). The following vascular and tumor indexes can be measured.

Tumor Vascular Length Density. The microphotographs of live tumors in the window chamber are used to obtain measurements of tumor vascular length density as an indicator of tumor vasculature. Briefly, a square grid is superimposed over the tumor images and the number of interactions between the grid and flowing vessels are counted. The vascular length density in mm/mm^3 is calculated using the formula:

$$\text{Vascular length density} = N/(4gdt)$$

where N is the average number of intersections between vessels and grid per sheet; g is the number of blocks in the grid ($= 54$); d is the length of one grid square calibrated by a micrometer image at the same magnification;

and t is the measured depth of field through which microvessels could be discerned ($=0.2$ mm).

Vessel Dilation (VD). Vessel diameter can be measured from the images taken at each time point using the National Institutes of Health (NIH) Image 1.12 analysis program. At least 10 vessels per window and five measurements per vessel are taken. The mean of the vessel size (VD) will be used for comparison. Vessel dilation index is calculated using VD at each time point divided by VD at time 0.

Vessel Tortuosity (VT). Tumor vessels and angiogenic vessels in pathologic conditions commonly display tortuous morphology compared with straight and uniform normal blood vessels. The vascular length (L) between two points (4 mm apart) of the same vessel is measured from the images. At least 10 vessels per window are measured, and the mean of the vessel length is used for analysis. Tortuosity index is calculated as $L/4$ mm. VT is equal to 1 in normal blood vessels (straight vessels) and is larger than 1 in tortured vessels.

Vascular Permeability (VP). Leaky blood vessels are common in tumors due to defective endothelial barrier function. The vascular leakiness is important because it contributes to the high interstitial pressure in tumors and may facilitate angiogenesis by releasing plasma into tissues and forming a scaffold for endothelial cell function. To measure the vascular permeability, we have used fluorescence-labeled tracers. Extravasated dye increases the fluorescent intensity in the vicinity of any chosen blood vessel over time. The leakiness of individual tumor vessels can be measured with direct microscopic visualization. Local vascular permeability is calculated as $Fn/F0$ at the vicinity area of a chosen blood vessel at each time point. Fn is the fluorescent intensity at n time point after the injection of dye. $F0$ is the measurement prior to the injection of dye. One should keep in mind that vascular permeability correlates with molecular mass, which reflects the pore size of tumor blood vessels. Depending on tumor type and location, the cutoff size varies.

Blood Flow (BF). Blood flow velocity in each vessel can be measured by the following methods: (1) One can perform a single line scanning using confocal microscopy or multiphoton microscopy and then count the number of red blood cells (RBCs) that go by. (2) One can record blood flow velocity using a VCR recorder. Because vessel size and blood flow velocity can be easily measured, a relative index of blood flow can be calculated as $BF = V \times S$, where V is the velocity and S is the size of the vessel.

Hypoxia. Hypoxia is a common feature of tumors, an important stimulus for angiogenesis, and a potential target for imaging. Hypoxia regulates angiogenesis through induction and stabilization of a transcription factor, hypoxia inducible factor (HIF). HIF binds to an 8-base pair hypoxia response element (HRE) in the promoter region of angiogenic genes and induces angiogenic factor production. A hypoxia reporter construct has been developed (5HREhCMVmpd2EGFP), of which enhanced green fluorescent protein (EGFP) is under the control of HRE.[10] Hypoxia can be directly visualized by the expression of EGFP in tumor cells harboring this construct.

Tumor Size. To obtain the tumor size, hematoxylin and eosin (H & E) stained sections representing the largest cross-sectional area of each tumor are photographed and the thickness (t) and the diameter (d) of tumors are measured from the microphotographs. Tumor volumes, which are assumed to approximate a flat cylinder in shape, can be calculated using the formula:

$$\text{Tumor volume (mm}^3) = 3.14t(d/2)^2$$

Alternatively, one can implant fluorescent protein-labeled tumor cells in the window chamber model (Fig. 2). The fluorescence intensity of the tumor can be used as a relative index of the tumor size.

Imaging of Tumor–Host Interaction and Angiogenesis Initiation Using Fluorescent Protein-Labeled Tumor Cells

Tumor cells labeled with green fluorescent protein (GFP), or another fluorescent or bioluminescent reporter can be noninvasively localized and measured in preclinical models by optical imaging. Optical imaging has the ability to study cellular and molecular events in living animals. We have used the tumor cells labeled with GFP in a tumor vascular window model to examine the initiation of tumor angiogenesis and tumor–host interaction. A murine mammary tumor line transfected with GFP (4T1–GFP) was implanted into the window chamber.[11] Using this approach, we examined tumor development in the same living mouse over time using fluorescent microscopy coupled with a digital camera. We were able to visualize tumor formation from a single tumor cell (Fig. 2). Several intriguing observations were obtained from this study.

1. *Tumor cells alter the host "normal vasculature" morphology.* Implantation of only 50 tumor cells in the window chamber caused dramatic vascular morphologic changes in just a few days (Fig. 2A and B). The surrounding host "normal" vessels became dilated and torturous.

[10] D. Vordermark, T. Shibata, and J. M. Brown, *Neoplasia* **3**, 527 (2001).
[11] C. Y. Li, S. Shan, Q. Huang, R. D. Braun, J. Lanzen, K. Hu, P. Lin, and M. W. Dewhirst, *J. Natl. Cancer Inst.* **92**, 143 (2000).

2. *Host environment alters tumor cell behavior.* 4T1 is an epithelial tumor line, which has a typical cobblestone-like morphology when cultured *in vitro*. However, a few days after implantation *in vivo* the cells became enlongated or polarized. The tumor cells underwent epithelial-to-mesenchymal transformation (EMT) indicative of cell motility, and they migrated toward nearby blood vessels (Fig. 2D and F). When the cells reached to the blood vessels, they grew around the vessel and formed a cuff, and then grew along the vessels. This suggests that tumor cells not only actively recruit blood vessels, but they also actively search for blood vessels for nutrients and oxygen. It is a mutual recruitment involving tumor–host interaction.

3. *Tumor angiogenesis starts at a much earlier time than textbooks suggest.* The traditional view is that tumor angiogenesis does not start until a tumor reaches 1–2 mm^3 in size. However, we observed that the earliest tumor angiogenesis started when there were only 300–400 cells in the tumor model (Fig. 2).

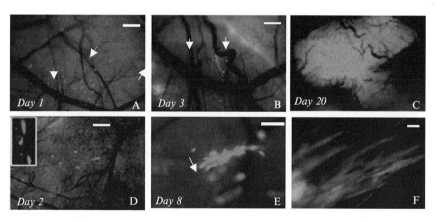

Fig. 2. Optical imaging of tumor–host interaction and tumor angiogenesis using the window chamber model. A small window chamber is installed on the back skinfold of a Balb/c mouse. Approximately 20 cells of a murine mammary tumor line transfected with EGFP (4T1–GFP) are implanted onto the fascial plane in the window chamber, and their growth is followed serially after the initial implantation. A vascularized "green tumor" forms from a single cell (green dot in Panel A) over the period of 20 days after tumor cell implantation (A–C). Interactions of tumor cells and host vasculature can be seen directly. Tumor cells change the local environment, and the surrounding normal blood vessels become dilated and tortuous, as indicated by arrows, in a few days after tumor cell implantation (A–B). Reciprocally, host environment affects tumor cell behavior. 4T1 is an epithelium tumor line and displays cobblestone-like morphology *in vitro*. However, the cells become spindle-shaped, with a fibroblast-like morphology after implantation *in vivo* (D–F). The cells migrate toward surrounding blood vessels. Size bars represent 20 μm. (See color insert.)

Imaging Tumor Vascular Response to Therapy Using the Window Model

Development of imaging parameters that reflect the efficacy of angiogenesis inhibitors has great potential. Notably, antiangiogenic therapy is intended to target the angiogenic vessels. It causes fewer side effects because normal vessels are typically quiescent. Traditional clinical trial design looking for the maximum toxicity dose would not apply to this new type of therapy. Identifying surrogate markers for antiangiogenic therapy is imminent. Optical imaging in tumor window models can be used to study the tumor vascular responses to therapy, as well as evaluate the efficacy of antiangiogenic reagents. Using the window model, we observed the heterogeneous response of tumor blood vessels to irradiation that exists among different types of tumors grown in the same host (C57–BL6). Two different types of tumors were implanted into the tumor windows. Vascularized tumors developed in 1 week, at which time they were treated with local irradiation. Irradiation treatment induced a dose- and time-dependent destruction to tumor blood vessels within the window. Radiation-sensitive tumor vessels in melanoma B16F0 showed a rapid and marked regression following 3 Gy of irradiation. Radiation-resistant tumor vessels in glioma GL261 did not respond to 3 Gy treatment and showed little response even at 6 Gy (Fig. 3). This finding indicates that tumors change the local environment and affect vascular response to therapy.

In addition, we evaluated the combination therapy of anti-vascular endothelial growth model (VEGF) with irradiation on tumor vasculature using the tumor window model. First we established a vascularized tumor in the window model, then we treated the tumors with local irradiation in combination with two different VEGF inhibitors. We observed that blocking VEGF activation by using either a soluble VEGF receptor or a small VEGF receptor kinase inhibitor enhanced irradiation-induced tumor vascular destruction.[12]

Multiphoton Laser-Scanning Microscopy in Tumor Window Model

The multiphoton laser-scanning microscope (MPLSM) offers significant advancements in depth of tissue penetration into intact samples, improved signal to noise ratio, and longer sample lifetimes over other optical sectioning techniques. Rakesh Jain's group has used MPLSM combined with EGFP and chronic animal window models to generate high-resolution, three-dimensional imaging in deeper regions of tumors, which provide powerful insight into gene expression, angiogenesis, physiologic

---

[12] L. Geng, E. Donnelly, G. McMahon, P. C. Lin, E. Sierra-Rivera, H. Oshinka, and D. E. Hallahan, *Cancer Res.* **61,** 2413 (2001).

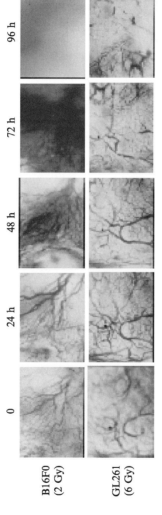

Fig. 3. Optical imaging of tumor vascular response to therapy using the window model. The dorsal window frame is applied to the back skinfold of C57–BL6 mice after removal of the epidermis. Tumor cell lines GL261 and B16F0 are implanted into the window skinfold. Angiogenesis in the implanted tumor occurs over the course of 7 days. At that time, the tumor is treated with local irradiation with 3 Gy for B16F0 melanoma and 6 Gy for GL261 glioma. The tumor vascular responses are photographed daily by microscopy. Magnification is 40×. (See color insert.)

function, and drug delivery in tumors.[9] They used transgenic mice that express EGFP under the control of the VEGF promoter. A tumor window model was installed on the back skinfold of the mice, and VEGF expression during tumor development was evaluated.

A few intriguing observations were obtained. (1) Tumor cells induce VEGF promoter activity in host cells (Fig. 4). Implantation of tumor cells in the window chamber activates VEGF expression in host stromal cells, and these cells exhibit fibroblast-like morphology. (2) Optical cross-sectioning by MPLSM shows that tumors induce VEGF promoter activity in a 35–50 μm-thick layer of host stromal cells interfacing the tumor (Fig. 4A). Below that layer, EGFP-positive stromal cells are seen at least 200 μm into the tumor (Fig. 4B). (3) EGFP-positive host cells inside the tumor frequently are parallel with angiogenic vessels (Fig. 4C). A thin (1 μm) nonfluorescent sheath often lies between the blood vessels and fluorescent cells (Fig. 4D), which presumably represents endothelial cells, pericytes, or basement membrane. Clearly, this type of study can generate novel information about the mechanisms of tumor angiogenesis, which probably cannot be achieved with other imaging technologies.

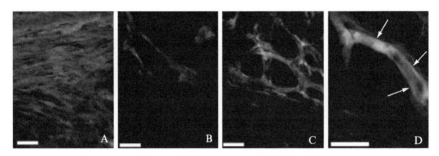

FIG. 4. Multiphoton laser-scanning microscopy imaging of tumor angiogenesis. This figure is adapted from Rekesh Jain's publication, with permission.[9] A MCaIV tumor is grown in transgenic mice that express EGFP under the control of the VEGF promoter (green). The tumor vasculature is highlighted by TMR-BSA injected systemically (red). A–B. Two optical sections of the same region within the tumor, at different depths, showing the three-dimensional resolution and depth of penetration of MPLSM. (A) A highly fluorescent (GFP-expressing) layer is seen within the first 35–50 m of the tumor–host interface. (B) 200 m deeper inside the tumor, EGFP-expressing host cells have successfully migrated into the tumor and tend to colocalize with angiogenic blood vessels. (C) A maximum intensity projection of 27 optical sections, spaced 5 m apart and beginning 45 m from the tumor surface. Colocalization of VEGF-expressing host cells with angiogenic vessels is readily apparent. (D) A single optical section of a vessel from (C) taken at twice the magnification. A thin, nonfluorescent layer is apparent at many locations (arrows). Scale bars are 50 μm. (See color insert.)

Vascular Window Model for Arthritis Angiogenesis Studies

Inflammation and angiogenesis are two fundamental processes that underlie pathologic disorders. Tissue injury induces inflammation, and inflammation triggers angiogenesis, which in turn initiates tissue repair and tissue growth. Rheumatoid arthritis is an inflammatory disease, as well as an angiogenic disease. We established an *in vivo* synovium vascular window model to study angiogenesis in arthritis. This model is adopted from the tumor window model. We installed a window frame on the back skinfold on syngeneic DBA/1 J mice. A 0.1-mm³ piece of rheumatoid arthritis synovium isolated from a mouse paw joint with collagen-induced arthritis (in DBA/J background) was then placed onto the fascial plane, saline solution was added, and the chamber was then sealed with a glass cover slip to form a semitransparent chamber. Rheumatoid arthritis synovium in window chambers were photographed using a microscope for vascular length density measurement. As expected, arthritis synovium is highly angiogenic because it produced high levels of angiogenic factors. Implantation of synovium isolated from an arthritic mouse paw into the window chamber induced dramatic angiogenesis within 8 days, and the synovium survived. In contrast, implanted normal joint tissue failed to induce angiogenesis, and the tissue died within days (Fig. 5). This model can be used to study angiogenesis in arthritis synovium, as well as evaluate angiogenic inhibitors for arthritis treatment.

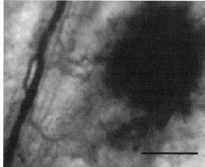

Normal joint   Arthritis joint

FIG. 5. Arthritis synovium window chamber model. A novel vascular window model is established on the back skinfold of a DBA mouse to study arthritis-induced angiogenesis. Synovium samples were isolated from a donor mouse paw with collagen-induced arthritis (CIA). A small piece of CIA-synovium (right panel) or a normal joint tissue (left panel) was implanted in the vascular window. The photos were live pictures taken from mouse synovium windows. Magnification is 40×. Bar represents 1 mm. (See color insert.)

Vascular Reporter Transgenic Mouse Model

Vascular-Specific Reporter Transgenic Mouse Model

Two families of receptor tyrosine kinase, the vascular endothelial growth factor receptor family (VEGF-R) and the Tie family (including Tie1 and Tie2), are expressed mainly or predominantly on vascular endothelial cells. Therefore transgenic mice harboring the promoters of the vascular-specific genes to drive a reporter gene can be valuable tools for imaging vasculature in living animals. Indeed, several lines of transgenic mice have been developed to display vascular-specific reporter gene expression. Here we will use Tie2–GFP mice as an example.[13,14] Because Tie2 is an endothelium-specific receptor, transgenic mice containing the Tie2 promoter to drive GFP expression exhibit vascular-specific GFP expression and display "green blood vessels" (Fig. 6). This model allows direct visualization of blood vessels *in vivo* without any treatment. Here we present a few examples of using the mice for imaging angiogenesis.

Laser-Scanning Confocal Imaging of the Vascular System in Tie2–GFP Mice. We have established a noninvasive, high-resolution imaging system to examine the microcirculation in the ear dermis and conjunctiva in living

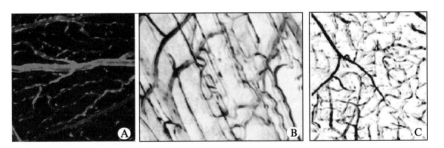

FIG. 6. (A) Conjunctiva capillaries in a Tie2–GFP mouse. Often all elements of the microcirculatory system show uniformly strong fluorescence. This image was obtained noninvasively in a lightly anesthetized Tie2–GFP mouse using a Zeiss 510 confocal microscope. (B) Anaglyph showing 3D structure of microcirculatory network in skeletal muscle of Tie2–GFP mouse. The Z-series stack of images was obtained from the medial quadriceps muscle of an anesthetized mouse through the incised skin. Arterioles (A) and venules (V) can be distinguished from capillaries. (C) Anaglyph showing 3D structure of subpial microvessel network in fixed mouse brain. (See color insert.)

[13] T. Motoike, S. Loughna, E. Perens, B. L. Roman, W. Liao, T. C. Chau, C. D. Richardson, T. Kawate, J. Kuno, B. M. Weinstein, D. Y. Stainier, and T. N. Sato, *Genesis* **28,** 75 (2000).
[14] C. L. Phillips, L. J. Arend, A. J. Filson, D. J. Kojetin, J. L. Clendenon, S. Fang, and K. W. Dunn, *Am. J. Pathol.* **158,** 49 (2001).

Tie2–GFP mice. Using laser-scanning confocal microscopy (LSCM) to image tissues from 8–10-week-old Tie2–GFP mice, we have observed that in the capillary endothelial cells the most intense fluorescence is localized around the nucleus (Fig. 6A and B), and this is consistent with the stereotypical structure of the capillary endothelial cell and its highly attenuated nature. The cytoplasm of endothelial cells is highly attenuated, often measuring 50–100 nm in thickness, thus providing a short distance required for the diffusion of nutrients from the blood to the tissues. Electron micrographs of capillary endothelial cells have shown that the major fraction of the cytoplasm is located in the perinuclear region of the cell, where the cytoplasm is expanded to accommodate the nucleus. Furthermore, in the Tie2–GFP mouse, GFP is expressed as a cytosolic protein that is not restricted from the nucleus. Noninvasive fluorescence imaging of the conjunctival microvasculature and the ear dermal microcirculation in living Tie2–GFP mice showed some variation in fluorescence intensity in different segments of the vascular tree, with relatively stronger staining in small arterioles and venules compared with the fluorescence intensity of the capillary endothelial cells.

IMAGING PROCEDURES. Live Tie2–GFP mice were imaged directly by LSCM. The mice were lightly anesthetized with ketamine and then placed on the stage of a microscope. The tissue was examined in a Zeiss 510 LSCM with either a 10 × 0.5 NA air lens, a 20 × 0.75 NA air lens, or a 40 × 1.3 NA oil immersion lens. Working distance is a limiting factor when using high NA oil immersion lenses, and the use of tissue sections is useful to circumvent these potential problems. Z-sectioning and 3D reconstruction was all done with Zeiss 510 hardware and software (Carl Zeiss, Inc., Oberkochen, Germany).

Corneal Angiogenesis Assay. The corneal micropocket assay is a simple, clean, and reliable assay to evaluate angiogenesis.[6] Capillary vessels sprout from the limbus into the avascular cornea upon stimulation with an angiogenic pellet implanted in a mouse cornea. Typically, when a corneal assay is established in a normal mouse, we cannot see any early events until blood flow has been established at approximately 4–6 days after pellet implantation. However, Tie2–GFP mice present a unique opportunity. Because vascular endothelium is labeled with GFP, one can visualize and record this dynamic process from the very beginning without relying on the establishment of circulation in matured vessels (Fig. 7). The corneas obtained from these animals showed a complex proliferation of capillaries that uniformly expressed high levels of GFP. This result confirms that Tie2 promoter activity remains activated at high levels in growth factor–induced corneal neovascularization and that Tie2–GFP is a suitable probe for noninvasive LSCM imaging in living mice.

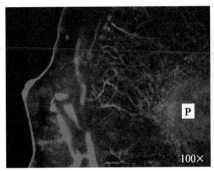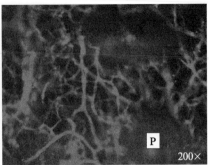

Fig. 7. Corneal micropocket assay in Tie2–GFP mice. A small surgical micropocket is created in the Tie2–GFP mouse cornea 1 mm from the limbus, and a small pellet containing bFGF (25 ng/pellet) is implanted in the pocket afterwards. Mouse cornea is harvested 5 days after pellet implantation and flatly mounted on a glass cover slip. Images are taken from fluorescent microscopy. P, Pellet. (See color insert.)

ESTABLISHMENT OF MICE CORNEAL ASSAY. The corneal micropocket assay is relatively simple and easy to perform. Briefly, angiogenic factors such as basic fibroblast growth factor (bFGF) or VEGF are combined with sterile Hydron casting solution (Interferon Sciences, New Brunswick, NJ), and the solution is pipetted onto the surface of 1.5-mm-diameter Teflon rods (Dupont Co, Wilmington, DE). The pellets, containing approximately 25 ng/pellet of angiogenic factors, are air-dried in a laminar hood for 1 hour and refrigerated overnight. The following day, pellets are rehydrated with a drop of phosphate buffered saline (PBS) buffer and then placed in a surgically created pocket within the cornea stromal, 1 mm from the limbus laterally of Tie2–GFP mice. At day 5 postimplantation, mice are anesthetized and corneas are removed and imaged by confocal microscopy. Measurements of the neovascular area, microvessel diameter, and assessment of length and branching complexity are made using Metamorph software (distributed by Universal Imaging Corp., a subsidiary of Molecular Devices, Downingtown, PA). Alternatively, one can directly image corneal angiogenesis in living animals.

Intravascular Fluorescent Probes as Markers of Capillary Perfusion and Permeability. We have used fluorescent tracer to image capillary perfusion and permeability. Intravenous injection of 0.1 mg rhodamine B dextran in 50 μl PBS in Tie2–GFP mice showed that in both mouse ear skin microvascular elements and mouse conjunctival capillary beds, rhodamine B dextran fills all fluorescent capillaries and only rarely flows through capillary segments that are negative for the GFP signal. This result reflects, in part, the extreme sensitivity of the confocal microscope and indicates the feasibility of this technique to monitor the patents of the endothelial cell lumen. It is generally possible to visualize blood cells within capillary lumen based on their negative staining characteristics against the background of fluorescence in the capillary. Also, it is normally possible to discern the movement of blood cells through the capillaries, and this provides information concerning rate of blood flow through specific capillary segments in a microcirculatory bed. We have found that in both conjunctival and dermal capillaries, flow rates are not uniform across all parts of a microvascular bed, although how the rate of flow is regulated remains unclear.

Dual fluorescence microangiography can measure capillary permeability using two fluorescent probes, and this method appears to be more sensitive to subtle changes in capillary permeability than single probe methods.[15] This system has been used in studies of the retinal vasculature and utilizes simultaneous imaging of two fluorescent tracers with distinct excitation and emission properties, and the two probes differ in molecular size. The relative molecular size of the two probes is important for determining capillary permeability. One of the tracers must be large (>70 kD) in order for the probe to remain confined within the vasculature, whereas the other is small and is free to cross the vascular barrier. The fluorescence intensity of the two tracers in the interstitial tissues is monitored following injection and directly reflects permeability, whereas comparison of the fluorescence intensity between the large feeding arteries and the large veins in the field is used to calculate the extraction of diffusing tracer by the microcirculation.

METHODS. For capillary perfusion measurement, fluorescein dextran (70 kD) or rhodamine B dextran (70 kD) are injected into the tail veins of anesthetized mice and the mice are mounted for imaging by confocal microscopy. In the seconds immediately following injection, the entire vascular system is defined by the intravascular contrast. However, transcapillary transport of tracer from the blood to the tissues occurs at different rates in different tissues, presumably due to permeability characteristics of the endothelial barrier in the different tissues. Because of this, over time, the microvascular anatomy becomes less distinct. For a detailed protocol of dual fluorescence permeability imaging, see Russ et al.[15] Briefly, a mixture of sodium fluorescein (376 kD) to indicate vessel leak and Texas red-dextran (70 kD) to indicate vessel filling are injected intravenously into a mouse prepped for imaging by either wide-field epifluorescence as described[15] or by confocal imaging. For confocal imaging, immediately following the dye injection, the gain and offset functions of the red and green channels are adjusted to maximize the dynamic range in the microvascular bed of interest. Another mouse is then prepped and mounted on the microscope and injected with dye as before, and a time series of data points are then collected and analyzed as described.[15]

[15] P. K. Russ, G. M. Gaylord, and F. R. Haselton, *Ann. Biomed. Eng.* **29,** 638 (2001).

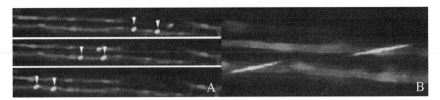

FIG. 8. Blood velocity measurements by LSCM. (A) Skeletal muscle capillaries from Tie2–GFP mice are depicted and show three successive frames from a time-lapse sequence, and it is possible to track a pair of fluorescent RBCs in each of the three frames. (B) The same capillaries imaged with slow scan speed show streaked RBCs. In A, the direct calculation of $\Delta d/\Delta t$ (comparing top to bottom) measurement of the distance traveled reveals the particle velocity at 0.26 pixel per millisecond. Velocity calculated from $\Delta d/\Delta t$ in streaks in B is 0.27 pixel per millisecond. (See color insert.)

Blood Flow Velocity Measurement in Living Animals. For these studies, isolated mouse RBCs are labeled with fluorescein isothiocyanate (FITC) or Texas red-dextran by standard techniques[16] and the fluorescent cells are injected intravenously into the mouse. LSCM has confirmed uniform labeling of all RBCs by this technique. We have found that when time-lapse recordings are made using rapid laser scan speeds (scan time 20–100 μsec), μsec), we can easily track the particle in capillaries, venules, and arterioles in multiple successive frames (Fig. 8A). Because the field size is constant for a given objective lens and the frame acquisition rate is constant and known, the velocity of fluorescent particles in individual microvessels up to 200 μm beneath the muscle surface can easily be determined using the formula:

$$\text{Velocity} = \Delta d/\Delta t$$

where Δd is the distance the cell traveled and Δt is the time interval.

Alternatively, we have found that by using slower scan rates, LSCM is ideal for automatically determining the instantaneous flow velocity of labeled RBCs in the microvasculature in a single scan.[16,17] In laser-scanning images, moving objects are often diagonally streaked (Fig. 8B) and the slope of the streak is related to the velocity of the moving particle. This method of determining particle velocity takes advantage of the precise control of the laser-scanning mechanism that scans the field in horizontal lines, pixel by pixel, over time. The streak results because time is required

[16] J. Seylaz, R. Charbonne, K. Nanri, D. Von Euw, J. Borredon, K. Kacem, P. Meric, and E. Pinard, *Cereb. Blood Flow Metab.* **19,** 863 (1999).

[17] D. Kleinfeld, P. P. Mitra, F. Helmchen, and W. Denk, *Proc. Natl. Acad. Sci. USA* **95,** 15741 (1998).

for the laser-scanning mechanism to scan a horizontal line of pixels. The labeled cell moves during this time interval and occupies a different position in the subsequent horizontal line of pixels. Thus the position of moving objects is different in each successive horizontal line scan. Because the pixel dimensions are constant and the time per pixel is constant and known, the slope of the streak gives the change in position over the change in time, from which the instantaneous particle velocity can be calculated by using Velocity $= \Delta d/\Delta t$, where Δd is the distance the cell traveled and Δt is the time interval.

In addition, Z-series stacks of images of the microvasculature, including capillaries, arterioles, and venules, have been obtained. These stacks are rendered off-line into 3D images and/or are digitized to measure microvessel diameter. In addition, recordings obtained in living, anesthetized mice have demonstrated that at rest there is tremendous variability in the flow rates between adjacent capillaries, and even in the same capillary over time. However, in studies in which we analyzed blood flow velocity in the same capillary bed over time intervals up to 20 minutes, we have found very consistent flow over the 20-minute period in individual capillaries. Tie2–GFP mice show relatively uniform fluorescence in all the elements of the skeletal muscle microvascular system, including arterioles, venules, and capillaries (see Fig. 6B). Use of these animals will confirm that unperfused capillaries are included in the analysis, which cannot be done by other methods.

METHODS. Red blood cells (RBCs) are fluorescently labeled by two different protocols, including reaction of the cells with FITC and hypotonic loading of the cells with sulforhodamine 101. Mouse RBCs are obtained by drawing cells from the inferior vena cava in a heparinized syringe. The cells are separated from serum by low-speed centrifugation in a table-top centrifuge and rinsed two times in heparinized saline and one time in 100 mM sodium bicarbonate (pH 8.0). The cells are resuspended in 1 ml of 100 mM sodium bicarbonate and incubated with 250 μg of FITC dissolved in 50 μl dimethylsulfoxide (DMSO) at room temperature for 30 minutes. Then the cells are washed three times and resuspended in 200 μl of heparinized saline. RBCs have also been loaded with sulforhodamine 101 by hypotonic shock treatment. Sulforhodamine 101 (0.2 mg) is added to 0.9 ml of distilled water, followed by the addition of 0.5 ml whole mouse blood for 20 minutes at room temperature. One hundred microliters of 10× PBS are then added, and the cells are centrifuged at low speed and rinsed three times in heparinized saline. For *in vivo* imaging, labeled RBCs are intravenously injected into the tail veins of anesthetized Tie2–GFP mice that are mounted on the confocal microscope and imaged by standard methods. For experiments where tracking of individual particles is desired, rapid scan rates are used. Alternatively, in experiments in which instantaneous velocity of populations of RBCs is desired, slow scanning speeds are used.

Conclusions

Advances in the biomedical sciences have been accelerated by the development and utilization of new imaging technologies. With animal models widely used in the biologic sciences, finding ways to conduct *in vivo* experiments more accurately and efficiently becomes a key factor in the success of medical research. Optical imaging is inexpensive and high resolution (up to a single cell level or even subcellular level), and it allows real-time imaging. Optical imaging not only provides powerful tools with which to study the molecular mechanisms of tumor angiogenesis, identify therapeutic targets, and validate the targets in animal models, but it will also be a valuable measurement for monitoring drug response in patients. In addition, optical imaging can easily combine with other imaging technologies. The combination of different imaging modalities will provide a detailed story about the molecular and cellular abnormalities of tumor blood vessels and will generate valuable information about the action of angiogenesis inhibitors in cancer patients.

Acknowledgments

We thank Laura DeBusk at Vanderbilt University Medical Center for her critical reading of the manuscript. This work was supported in part by a grant (CA87756) from the National Cancer Institute (P. C. L.), by the Vanderbilt-Ingram Cancer Center (CA68485), the Vanderbilt In Vivo Imaging Center (CA 86283), and the Vanderbilt Diabetes Center (DK20593).

[5] Cardiovascular Magnetic Resonance Imaging: Current Applications and Future Directions

By Milind Y. Desai, João A. Lima, and David A. Bluemke

Introduction

Magnetic resonance imaging (MRI) is of increasing importance in cardiovascular applications. It is widely recognized to provide an accurate and reliable means of assessing function and anatomy of the heart and great vessels (Table I). Previously, however, the means to obtain images of the cardiovascular system had been compromised by extremely long

Copyright 2004, Elsevier Inc.
All rights reserved.
0076-6879/04 $35.00

TABLE I
ADVANTAGES AND DISADVANTAGES OF MRI

Advantages
Unparalleled resolution
Three-dimensional imaging capacity
Noninvasive
Nontoxic contrast agents
Ability to depict soft tissues

Disadvantages
Long scan times
Artifacts, including motion, respiratory, and cardiac motion
Incompatibility with pacemakers, defibrillators, and aneurysm clips

examination times and software that was available only at specialized centers. It appears that with the recent development of specialized cardiovascular MR scanners, the applications of this technique will become more widely utilized on a routine basis. In this chapter, technical aspects of MR scanning of the cardiovascular system are outlined, followed by applications for the heart and aorta (Table II).

Patient Preparation

Often, the most critical portion of the cardiac examination is patient preparation. After myocardial infarction (MI), for example, it has been our experience that claustrophobia is accentuated for patients. Therefore explanation of the procedure to the patient is essential, and in some patients, conscious sedation anxiolytics reduce claustrophobia. The newest cardiovascular MR scanners are being designed as a "short-bore" system, so that the head of the patient is outside of the magnet during imaging, vastly reducing instances of claustrophobia. Finally, patients should be asked to empty their bladder before beginning the examination.

The second preparatory factor in scanning patients is electrocardiographic (ECG) lead placement. The ECGs of patients in the bore of the MR scanner are degraded by superimposed electrical potential from flowing blood in the magnetic field. Electrical potentials are generated when the flow is perpendicular to magnetic field lines; frequently the ECG T-wave increases in amplitude in the MR scanner, potentially confusing the scanner detection of the R-wave. Therefore lead placement is *not* the same as for obtaining an ECG outside of the magnet; for example, no limb leads are used. Instead, typically four chest leads are used.

The third factor is the MR coil. A surface coil is usually placed over the patient's heart. Common terminology used are "cardiac coils" or "torso

TABLE II
APPLICATIONS OF MRI IN CARDIOVASCULAR MEDICINE

Current clinical applications
Assessment of right and left ventricular function, volume, and mass
Assessment of myocardial viability
Assessment of different cardiomyopathies
  Dilated cardiomyopathy
  Hypertrophic cardiomyopathy
  Arrythmogenic right ventricular dysplasia
  Restrictive cardiomyopathy
    Sarcoidosis
    Amyloidosis
    Hemochromatosis
    Endomyocardial fibrosis
Evaluation of pericardial diseases
Evaluation of cardiac and paracardiac masses
  Benign
  Malignant
Evaluation of congenital heart disease, including shunts
Evaluation of valvular heart diseases
Evaluation of aortic diseases
  Aortic dissection
  Aortic aneurysm
  Congenital disorders (Coarctation of aorta and Marfan's syndrome)
Connective tissue disease (Takayasu's arteritis)
Evaluation of pulmonary veins

Emerging applications
Assessment of myocardial perfusion and ischemia
Atherosclerotic plaque imaging
Coronary artery angiography
Therapeutic MRI for electrophysiology and interventional procedures

coils." "Phased array" cardiac or torso coils are preferred, and consist of multiple individual coils electrically coupled in order to increase the signal obtained. Phased array capability is an expensive MR scanner option, but is standard on dedicated cardiovascular MR scanners and is less common on multipurpose MR scanners. If phased array capability is not available, single element "surface coils" can be placed over the chest. Finally, scout images and images of the entire chest are usually obtained using the "body coil" (i.e., the coil built into the MRI bore).

Understanding the Basic Pulse Sequence

It is extremely helpful to understand the strategy of the MR pulse sequence being used. This allows the user to determine the length of time for the acquisition and the corresponding imaging strategy to be employed. A fundamental characteristic of all pulse sequences is that the image is generated in the frequency domain, also referred to as "k-space." Fourier reconstruction is then employed to generate the image. The "k-space" representation of the image is a pixel array; this array is generated one line at a time, and typically 128 to 256 lines must be generated. Traditionally, each line of the image required one heartbeat to acquire, thus imaging times were relatively long. For example, if 256 lines were generated, 256 heartbeats at 60 beats per minute required 256/60 = 4.3 minutes. An improvement in imaging speed by a factor of 4–16 or more is now commonplace. The most useful pulse sequences are briefly explained as follows.

Spin–Echo Imaging

Spin–echo images are the fundamental MR pulse sequence. In all other parts of the body, a complete examination consists of T_1- and T_2-weighted images. T_1 images typically display anatomy more clearly; T_2 images accentual pathologic changes, such as edema from infarction, or relationship of a tumor to the pericardium. Advantages are excellent signal to noise ratio (SNR), and excellent contrast between the heart, epicardial fat, and adjacent structures. The primary disadvantage is that these are very time-consuming sequences. As in the example discussed earlier, they require between 128 to 256 heartbeats or more to acquire, depending on the resolution that is required. Blood is typically made to be dark—"black blood" images. Typically, no cine information is obtained. Despite these drawbacks, spin–echo images are important for: (1) congenital anomalies, (2) pericardium evaluation, (3) right ventricular dysplasia, and (4) cardiac tumors.

Fast Spin–Echo Imaging

Also called turbo spin–echo, the goal is simply to reduce the time required, while obtaining similar SNR and image quality. Although considerations here are complex, imaging times are reduced by an acceleration factor typically ranging from 2–16. Low acceleration factors (from 4–8) have less image degradation and blurring.

Cine Gradient–Echo Imaging

Cine gradient–echo images are used to evaluate motion gated to the cardiac cycle. On older MR scanning systems, these also require 128 to 256 heartbeats to obtain, and typically 2–3 individual slices or cross-sectional levels are obtained during this time, with 12–15 cardiac phases. Gradient moment nulling, or flow compensation, is used to make blood bright—"bright blood images." SNR is moderate, but image acquisition is very time consuming.

Segmented k-space cine MRI allows gradient–echo images to be obtained in one breath-hold. Acceleration factors range from 4–16 times that of conventional cine images. Steady-state free precession images, in particular, provide very high SNR with high imaging speed.

The different clinical and emerging applications of cardiovascular MRI are described as follows.

Assessment of Ventricular Function

MRI is a very accurate and highly reproducible technique for measuring ejection fraction and ventricular volumes noninvasively in three dimensions, and for this reason, it has become a "gold standard" to which other modalities are compared.[1] Simpson's rule is applied to determine ejection fraction and volumes. For assessment of volume and mass, usually bright blood gradient–echo sequences are used, with 10–30 phases per cardiac cycle. Breath-holding techniques with acquisition times of approximately 15–20 seconds are preferred to reduce blurring of the endocardial border. Generally, for accurate measurement of volume and mass, entire coverage of the left ventricle (LV) with short-axis views from the mitral plane are recommended. Slice thickness should not exceed 10 mm, and in cases of subtle changes, the thickness should be reduced appropriately. Automatic edge detection can improve the reliability of the technique. MRI is rapidly becoming the method of choice for longitudinal follow-up in patients who are undergoing therapeutic interventions.[2] From a research perspective, the sample size needed to detect LV parameter changes in a clinical trial is far less, in the range of one order of magnitude, with MRI than with 2D echocardiography, which markedly reduces the time and cost of patient care and pharmaceutical trials.[3]

A unique MR technique called myocardial tagging (SPAMM technique) has been developed, which labels the heart muscle with a dark grid and enables 3D analysis of cardiac rotation, strain, displacement, and deformation of different myocardial layers during the cardiac cycle (Fig. 1). This helps in assessing regional wall motion of the myocardium.[4,5]

[1] Task Force of the European Society of Cardiology, in collaboration with the Association of European Pediatric Cardiologists, *Eur. Heart J.* **19,** 19 (1998).

[2] N. E. Doherty, 3rd, K. C. Seelos, J. Suzuki, G. R. Caputo, M. O'Sullivan, S. M. Sobol, P. Cavero, K. Chatterjee, W. W. Parmley, and C. B. Higgins, *J. Am. Coll. Cardiol.* **19,** 1294 (1992).

[3] R. C. Semelka, E. Tomei, S. Wagner, J. Mayo, C. Kondo, J. Suzuki, G. R. Caputo, and C. B. Higgins, *Radiology* **174,** 763 (1990).

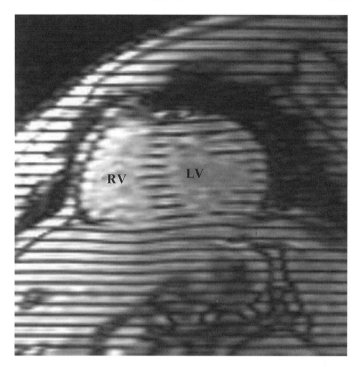

FIG. 1. SPAMM-tagged MR images of the heart in short axis showing deformation of the tagging grid that is produced by the motion of the underlying heart muscle. LV, Left ventricle; RV, right ventricle.

Assessment of Different Cardiomyopathies

MRI is a noninvasive tool that has a high degree of accuracy and reproducibility in visualization of left and right ventricular morphology and function.[6] It is also superior to echocardiography in determination of ventricular mass and volume[3,7] and is fast becoming the "gold standard" for *in vivo* identification of the phenotype of cardiomyopathies.[1]

[4] N. Reichek, *J. Magn. Reson. Imaging* **10,** 609 (1999).

[5] J. Bogaert, H. Bosmans, A. Maes, P. Suetens, G. Marchal, and F. E. Rademakers, *J. Am. Coll. Cardiol.* **35,** 1525 (2000).

[6] H. Benjelloun, G. B. Cranney, K. A. Kirk, G. G. Blackwell, C. S. Lotan, and G. M. Pohost, *Am. J. Cardiol.* **67,** 1413 (1991).

[7] P. B. Bottini, A. A. Carr, L. M. Prisant, F. W. Flickinger, J. D. Allison, and J. S. Gottdiener, *Am. J. Hypertens.* **8,** 221 (1995).

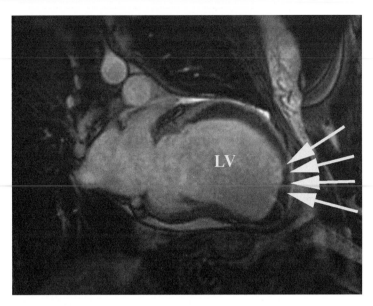

FIG. 2. Bright blood images showing a dilated, thinned-out left ventricle (LV) due to a large anterior wall infarct with a mural thrombus adjacent to the infarcted myocardium (arrows).

Dilated Cardiomyopathy

In dilated cardiomyopathy (DCM), MRI is useful in the study of ventricular morphology and function (Fig. 2), utilizing gradient–echo sequences with a low inter/intraobserver variability of LV mass and volume.[6] It is also useful to analyze wall thickening,[8] impaired fiber shortening,[9] and end-systolic wall stress (which is a very sensitive parameter of a change in LV systolic function).[10] It can also accurately assess the morphology and function of the right ventricle (RV), which is also frequently affected in DCM.[11] MR spectroscopy has also revealed a change in phosphate metabolism in DCM.[12] A ratio of phosphocreatine to adenosine

[8] E. K. Kasper, W. R. Agema, G. M. Hutchins, J. W. Deckers, J. M. Hare, and K. L. Baughman, *J. Am. Coll. Cardiol.* **23,** 586 (1994).

[9] G. A. MacGowan, E. P. Shapiro, H. Azhari, C. O. Siu, P. S. Hees, G. M. Hutchins, J. L. Weiss, and F. E. Rademakers, *Circulation* **96,** 535 (1997).

[10] N. Fujita, A. J. Duerinekx, and C. B. Higgins, *Am. Heart J.* **125,** 1337 (1993).

[11] U. Sechtem, P. W. Pflugfelder, R. G. Gould, M. M. Cassidy, and C. B. Higgins, *Radiology* **163,** 697 (1987).

[12] S. Neubauer, T. Krahe, R. Schindler, M. Horn, H. Hillenbrand, C. Entzeroth, H. Mader, E. P. Kromer, G. A. Riegger, K. Lackner *et al.*, *Circulation* **86,** 1810 (1992).

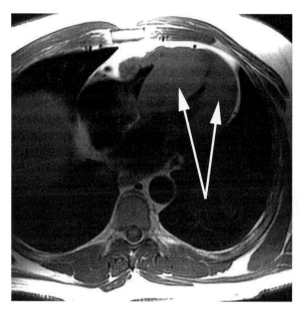

FIG. 3. Long-axis view of the heart (black blood images) revealing grossly hypertrophic left ventricular wall due to hypertrophic cardiomyopathy (arrows).

triphosphate has some prognostic value in DCM. Contrast-enhanced T_1-weighted images are also helpful in detecting changes of acute myocarditis (increased gadolinium accumulation is thought to be due to inflammatory hyperemia-related increased flow, slow wash-in–wash-out kinetics, and diffusion into necrotic cells). There is some evidence of similar changes in chronic DCM.[13] Contrast-enhanced MRI could also increase the sensitivity of endomyocardial biopsy by visualization of inflamed areas, which would aid in determining biopsy site.[14]

Hypertrophic Cardiomyopathy

MRI, due to its high accuracy, is becoming useful to assess morphology, function, tissue characterization, and degree of LV outflow tract (LVOT) obstruction in patients with hypertrophic cardiomyopathy (HCM) (Fig. 3). It is also very accurate in assessing LV mass, regional hypertrophy patterns,

[13] M. G. Friedrich, O. Strohm, J. Schulz-Menger, H. Marciniak, F. C. Luft, and R. Dietz, *Circulation* **97**, 1802 (1998).
[14] G. Bellotti, E. A. Bocchi, A. V. de Moraes, M. L. Higuchi, M. Barbero-Marcial, E. Sosa, A. Esteves-Filho, R. Kalil, R. Weiss, A. Jatene, and F. Pileggi, *Am. Heart J.* **131**, 301 (1996).

and different phenotypes of the disease (e.g., apical HCM).[7] Postsurgical changes after myomectomy can also be reliably monitored.[15] The turbulent jet during systolic LVOT obstruction is also easily detected by using suitable echo times (about 4 ms). MRI can also detect the systolic anterior motion of the mitral valve in the four-chamber view or a short-axis view at the valvular plane.[16] Mitral regurgitation can also be well documented and quantified by MRI.[17] MR spectroscopy also reveals changes in phosphate metabolism in patients with HCM.[18] Analysis of blood flow in the coronary sinus utilizing MRI can be helpful in determining the alterations in coronary flow reserve in patients with HCM.[19] A relatively newer technique is to measure the effective LVOT area by MR planimetry during systole. This method has the potential to overcome the problem of interstudy variability of the LVOT gradient due to its independence from the hemodynamic status.[20] There are preliminary data that assessment of diastolic function utilizing MRI may be superior to conventional parameters utilizing echocardiography. Analysis of early untwisting motion of the myocardium could be helpful in assessing diastolic function.[21] Other functional changes utilizing myocardial tagging include a reduction in posterior rotation, reduced radial displacement of the inferoseptal myocardium, and reduced 3D myocardial shortening and heterogeneity of regional function.[17] MRI is also very useful in the follow-up of patients after surgical or pharmacologic interventions.[16] MRI also easily detects the acute and chronic changes after septal artery ablation.[22]

Arrythmogenic Right Ventricular Dysplasia

Arrythmogenic right ventricular dysplasia (ARVD) is a rare disorder in which there is fibrofatty replacement of the RV free wall. MRI is rapidly becoming the diagnostic technique of choice for ARVD (Fig. 4). It is able

[15] A. Franke, F. A. Schondube, H. P. Kuhl, H. G. Klues, C. Erena, B. J. Messmer, F. A. Flachskampf, and P. Hanrath, *J. Am. Coll. Cardiol.* **31,** 1641 (1998).

[16] R. D. White, N. A. Obuchowski, S. Gunawardena, E. O. Lipchik, H. M. Lever, C. W. Van Dyke, and B. W. Lytle, *Am. J. Card. Imaging* **10,** 1 (1996).

[17] M. G. Friedrich, *J. Cardiovasc. Magn. Reson.* **2,** 67 (2000).

[18] W. I. Jung, L. Sieverding, J. Breuer, T. Hoess, S. Widmaier, O. Schmidt, M. Bunse, F. van Erckelens, J. Apitz, O. Lutz, and G. J. Dietze, *Circulation* **97,** 2536 (1998).

[19] N. Kawada, H. Sakuma, T. Yamakado, K. Takeda, N. Isaka, T. Nakano, and C. B. Higgins, *Radiology* **211,** 129 (1999).

[20] J. Schulz-Menger, O. Strohm, J. Waigand, F. Uhlich, R. Dietz, and M. G. Friedrich, *Circulation* **101,** 1764 (2000).

[21] M. Stuber, M. B. Scheidegger, S. E. Fischer, E. Nagel, F. Steinemann, O. M. Hess, and P. Boesiger, *Circulation* **100,** 361 (1999).

[22] J. Suzuki, R. Shimamoto, J. Nishikawa, T. Yamazaki, T. Tsuji, F. Nakamura, W. S. Shin, T. Nakajima, T. Toyo-Oka, and K. Ohotomo, *J. Am. Coll. Cardiol.* **33,** 146 (1999).

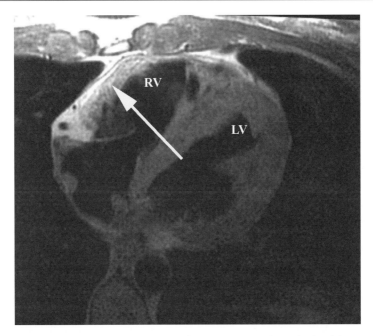

Fɪɢ. 4. Black blood images of the heart (LV is the left and RV is the right ventricle) showing bright signals in the RV free wall (arrow) consistent with fatty deposits suggestive of arrythmogenic RV dysplasia.

to visualize ventricular cavities and walls with an excellent depiction of myocardial anatomy. T_1-weighted spin–echo images reveal fatty infiltration, thinned walls, and dysplastic trabecular structures. Axial and short-axis views are usually recommended for optimal results.[17] Standard gradient–echo images reveal characteristic regional wall motion changes, localized early diastolic bulging, wall thinning, and saccular aneurysmal out-pouchings.[23,24] However, MRI is not as specific as endomyocardial biopsy in detection of myocardial fat,[25] but biopsy may be falsely negative. Currently, the working group classification proposed by McKenna *et al.* is the recognized standard to arrive at the diagnosis.[26]

[23] L. M. Blake, M. M. Scheinman, and C. B. Higgins, *Am. J. Roentgenol.* **162,** 809 (1994).

[24] C. Ricci, R. Longo, L. Pagnan, L. Dalla Palma, B. Pinamonti, F. Camerini, R. Bussani, and F. Silvestri, *Am. J. Cardiol.* **70,** 1589 (1992).

[25] L. Menghetti, C. Basso, A. Nava, A. Angelini, and G. Thiene, *Heart* **76,** 467 (1996).

[26] D. Corrado, G. Fontaine, F. I. Marcus, W. J. McKenna, A. Nava, G. Thiene, and T. Wichter, *Circulation* **101,** 101 (2000).

Restrictive Cardiomyopathy

Primary infiltration of the myocardium by fibrosis or other tissues leads to the development of restrictive cardiomyopathy, which is characterized by normal LV size and systolic function, severe diastolic dysfunction, and biatrial enlargement. It needs to be differentiated from constrictive pericarditis, which is a primary disease of the pericardium rather than the myocardium. LV size and thickness is quantified using gradient–echo sequences. Atrial enlargement is assessed in a four-chamber view. Mitral regurgitation should also be assessed. The following different restrictive diseases can be effectively assessed using MRI.[17]

Sarcoidosis. The incidence of myocardial involvement in systemic sarcoidosis is 20–30%,[27] and up to 50%[28] of the deaths in sarcoidosis may be due to cardiac involvement. MRI is becoming a useful tool in the assessment of sarcoidosis. Sarcoid lesions may lead to different signal intensities, most likely due to different stages of the disease. Some instances have reported high-intensity areas in T_2-weighted MRI, whereas other instances have reported a central low-intensity area on T_1- and T_2-weighted imaging surrounded by a high signal ring.[29] Gadolinium has also been reported to accumulate in the sarcoid lesions and is thought to be due to fibrotic nonactive granulomatous nodules with inflammatory response of the surrounding tissue.[30] Thus T_2-weighted followed by T_1-weighted spin–echo techniques in short and long axis with and without gadolinium could be useful to detect and/or exclude sarcoid granulomas. Occasionally, MRI may be useful in guiding endomyocardial biopsy.[17]

Hemochromatosis. Extensive iron deposits leading to wall thickening, ventricular dilatation, congestive heart failure (CHF), and death characterize cardiac hemochromatosis. Usually the iron deposits are subepicardial, and hence the endomyocardial biopsy may fail to confirm the diagnosis.[17] On the other hand, MRI is used to detect the iron deposits, which is possible due to very strong paramagnetic properties of iron. This leads to an extensive signal loss in native T_1- and T_2-weighted images.[31] The pattern of focal signal loss in a dysfunctional myocardium associated with an abnormally "dark" liver might be sufficient to confirm the diagnosis of systemic hemochromatosis. LV function can also be accurately assessed using MRI. It is also useful in follow-up of these patients undergoing intensified medical therapy.[17]

[27] R. J. Kim, E. L. Chen, J. A. Lima, and R. M. Judd, *Circulation* **94,** 3318 (1996).
[28] G. S. Flora and O. P. Sharma, *Sarcoidosis* **6,** 97 (1989).
[29] S. Otake, T. Banno, S. Ohba, M. Noda, and M. Yamamoto, *Radiology* **176,** 145 (1990).
[30] S. Seltzer, A. S. Mark, and S. W. Atlas, *Am. J. Neuroradiol.* **12,** 1227 (1991).
[31] E. S. Siegelman, D. G. Mitchell, and R. C. Semelka, *Radiology* **199,** 13 (1996).

Amyloidosis. Infiltration of the heart by amyloid deposits is found in almost all cases of primary amyloidosis and in 25% of familial amyloidosis. MRI can be useful in detection of amyloidosis and its differentiation from HCM. A thickness of atrial septum or right atrial posterior wall >6 mm is fairly specific for amyloid infiltration and consistent with previously published echocardiographic data.[32] Tissue characterization in cardiac amyloidosis has not been well studied, and little data are available. In one study, there was a decrease in signal intensity of the amyloid-infiltrated myocardium in comparison to the reference tissue.[32] The mechanism underlying this hypointensity is uncertain, but is likely due to abnormal interstitium and fibrosis.

Endomyocardial Fibrosis. Endomyocardial fibrosis (also termed Loeffler's endocarditis) leads to posterobasal concentric wall thickening, followed by extensive subendocardial fibrosis, apical thrombus formation, progressive diastolic dysfunction, and reduced stroke volume.[17] The morphologic and functional features can be well quantified by MRI. The fibrosis may be visible as a dark apical rim in bright blood prepared gradient–echo sequences. However, there is a lack of sensitivity and specificity in detecting fibrosis.[33]

Assessment of Myocardial Viability

The concept of myocardial viability is of great significance in patients with ischemic heart disease in terms of achieving the maximum benefit from revascularization.[34] Revascularization improves function of the viable myocardium, but not the myocardium that is replaced by fibrous scar, and hence it is imperative to distinguish between scar and viable myocardium. Contrast-enhanced MRI (CE-MRI), is coming of age in terms of its ability to accurately predict myocardial viability.[35–37] Two enhancement patterns have been described. On first-pass perfusion images, there is an area of hypoenhancement within the infarcted region, which has been found to

[32] R. Fattori, G. Rocchi, F. Celletti, P. Bertaccini, C. Rapezzi, and G. Gavelli, *Am. Heart J.* **136,** 824 (1998).

[33] D. L. Huong, B. Wechsler, T. Papo, D. de Zuttere, O. Bletry, A. Hernigou, A. Delcourt, P. Godeau, and J. C. Piette, *Ann. Rheum. Dis.* **56,** 205 (1997).

[34] R. Senior, S. Kaul, and A. Lahiri, *J. Am. Coll. Cardiol.* **33,** 1848 (1999).

[35] R. J. Kim, D. S. Fieno, T. B. Parrish, K. Harris, E. L. Chen, O. Simonetti, J. Bundy, J. P. Finn, F. J. Klocke, and R. M. Judd, *Circulation* **100,** 1992 (1999).

[36] J. A. Lima, R. M. Judd, A. Bazille, S. P. Schulman, E. Atalar, and E. A. Zerhouni, *Circulation* **92,** 1117 (1995).

[37] R. M. Judd, C. H. Lugo-Olivieri, M. Arai, T. Kondo, P. Croisille, J. A. Lima, V. Mohan, L. C. Becker, and E. A. Zerhouni, *Circulation* **92,** 1902 (1995).

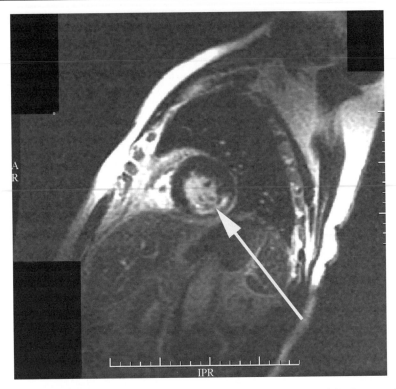

FIG. 5. Short-axis image of the heart obtained 10 minutes after contrast injection revealing a hyperenhanced area (arrow) in the inferolateral left ventricular wall consistent with nonviable infarcted myocardium.

correlate with microvascular obstruction, related to "no reflow" inside the infarct zone.[27,35] The second enhancement pattern is noticed at 10–30 minutes after contrast injection (delayed hyperenhancement [DHE]) and can be depicted with a breath-hold inversion recovery gradient–echo sequence (Fig. 5). This pattern of DHE after both acute and chronic MI reflects nonviable, infarcted tissue and fibrosis, respectively. In patients with a chronic MI, DHE can accurately locate and determine the extent and transmurality of the infarct.[27,35,38] Thus, because of high spatial resolution, high reproducibility, and predictive value, MRI is fast becoming a reference standard for assessment of myocardial viability. Further development of necrosis-specific contrast agents and refinement of MR spectroscopy to assess regional chemistry and metabolism will enhance the utility of MRI.

[38] R. J. Kim, E. Wu, A. Rafael, E. L. Chen, M. A. Parker, O. Simonetti, F. J. Klocke, R. O. Bonow, and R. M. Judd, *N. Engl. J. Med.* **343,** 1445 (2000).

Assessment of Pericardial Disease

MRI is ideally suited to image the pericardium (Fig. 6). T_1-weighted spin–echo imaging demonstrates normal pericardium as a thin band (<2 mm) of low signal, bordered by epicardial and pericardial fat, which has a high signal. Because pericardial thickness varies at different levels, it is generally recommended to measure the thickness in axial images at the level of the right atrium, the right ventricle, and the left ventricle. A thickness of >4 mm is considered abnormal and suggestive of fibrous pericarditis, either acute or chronic (due to surgery, uremia, tumor, infection, or connective tissue disease).[39] Use of CE-MRI may help better delineate the pericardium in cases of effusive–constrictive pericarditis.[40] Breath-hold

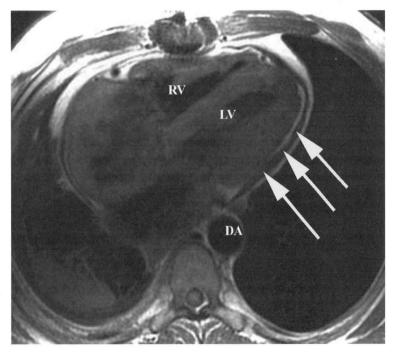

FIG. 6. Black blood images of the heart in long axis revealing thickened pericardium (arrows). LV, Left ventricle; RV, right ventricle; DA, descending aorta.

[39] T. Masui, S. Finck, and C. B. Higgins, *Radiology* **182,** 369 (1992).
[40] A. Watanabe, Y. Hara, M. Hamada, K. Kodama, Y. Shigematsu, S. Sakuragi, K. Kawachi, and K. Hiwada, *Magn. Reson. Imaging* **16,** 347 (1998).

or real-time cine gradient–echo images of the ventricles, and phase velocity mapping of the cardiac valves may be helpful in assessing the significance of pericardial pathology. It is also useful to detect other disorders, such as congenital absence of pericardium, pericardial cysts, or pericardial effusion undetected by other modalities.[41]

Evaluation of Cardiac and Paracardiac Masses

Primary cardiac tumors are rare (0.002—0.3% incidence). The majority (75%) of those are benign.[42,43] Metastatic tumors (Fig. 7) are 20- to 40-fold more common than primary tumors. MRI is excellent for delineating the morphologic details of a mass (including extent, origin, hemorrhage, vascularity, calcification, and effects on adjacent structures). Protocols include the combined use of axial black blood sequences and axial bright blood cine

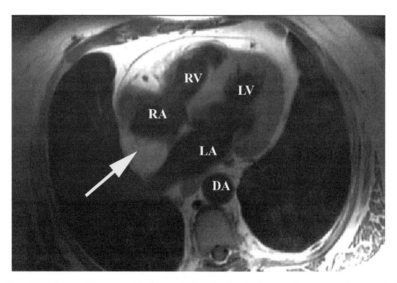

FIG. 7. Black blood images of the heart in longitudinal axis revealing a mass (arrow) in the right atrium (RA), which was subsequently diagnosed as lipoma. LV, Left ventricle; LA, left atrium; RV, right ventricle; DA, descending aorta.

[41] W. H. Smith, D. J. Beacock, A. J. Goddard, T. N. Bloomer, J. P. Ridgway, and U. M. Sivananthan, *Br. J. Radiol.* **74,** 384 (2001).
[42] P. R. Schvartzman and R. D. White, *J. Thorac. Imaging* **15,** 265 (2000).
[43] W. Urba and D. Longo, *in* "Cancer of the Heart" (A. Kapoor, ed.). New York, Springer-Verlag, 1986.

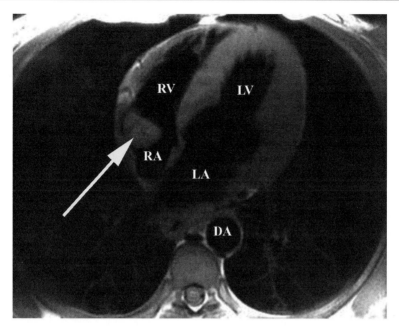

FIG. 8. Black blood images of the heart in longitudinal axis revealing a mass (arrow) in the right atrium (RA), which was diagnosed as an organized thrombus. LV, Left ventricle; LA, left atrium; RV, right ventricle; DA, descending aorta.

images. Functional MR images are useful to study the pathophysiologic consequences of the cardiac mass.[43]

Specifically, benign myxomas (the most common cardiac tumor) appear brighter on T_2 weighting than myocardium, and cine images may reveal the characteristic mobility of the pedunculated tumor. Lipomas (Fig. 8) appear brighter on spin–echo T_1-weighted images, and the diagnosis is verified by a decrease in signal intensity using a fat presaturation technique. An acute thrombus (Fig. 9) appears brighter than myocardium on T_1-weighted images due to an alteration in paramagnetic properties.[43,44]

Evaluation of Congenital Heart Disease

Because of its unparalleled resolution and 3D imaging capacity, MRI is playing an important role in diagnosing and serially following patients with various congenital heart diseases, complementary to echocardiography. It

[44] E. Castillo and D. A. Bluemke, *Radiol. Clin. North Am.* **41,** 17 (2003).

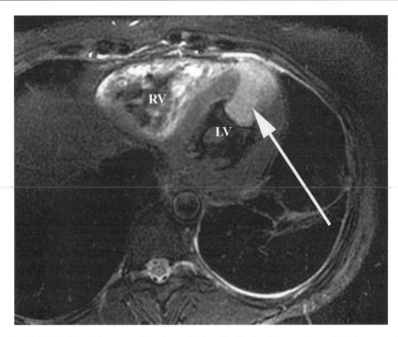

Fig. 9. Black blood images of the heart in longitudinal axis revealing a highly aggressive mass infiltrating the apex (arrow) of the left ventricle (LV). The patient had known leiomyosarcoma, and cardiac involvement was secondary metastasis. RV, Right ventricle.

is a very good tool for identifying and sizing atrial and ventricular septal defects. Phase velocity mapping at the level of the shunt also allows calculation of shunt fraction (Qp/Qs ratio).[45] Using different imaging sequences, including MR angiography (MRA), lesions such as anomalous pulmonary venous return, transposition of great vessels, aortic rings, truncus arteriosus, double outlet right ventricle, tetralogy of Fallot, pulmonary atresia, and pulmonary artery stenosis can be diagnosed (Fig. 10).[46] It can also be used to follow the patients for effects and complications following corrective surgery.[46] MRA can also be useful to diagnose venous anomalies such as persistent superior vena cava and interruption of the inferior vena cava with azygous continuation.[46]

[45] R. H. Mohiaddin and D. J. Pennell, *Cardiol. Clin.* **16,** 161 (1998).
[46] G. M. Pohost, L. Hung, and M. Doyle, *Circulation* **108,** 647 (2003).

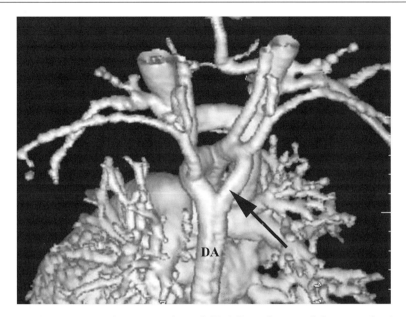

FIG. 10. Reconstructed contrast-enhanced 3D MR angiogram of the aorta showing a congenital aortic ring (arrow). DA, Descending aorta.

Evaluation of Valvular Diseases

Echocardiography with color Doppler is usually the first-line imaging modality for diagnosing valvular diseases. MRI is generally reserved for use when other modalities fail or provide suboptimal information.[44] Double inversion recovery sequences can show valve morphology and show evidence of secondary changes associated with that (chamber dilatation, myocardial hypertrophy, poststenotic changes in the great vessels, or thrombus in any of the chambers).[47] Semiquantitative assessment of valvular stenosis or regurgitation can be obtained by measuring the area of signal void on gradient–echo images. The duration or extent of the signal void correlates with the severity of aortic stenosis, and the total area of signal loss correlates with the severity of mitral regurgitation.[48] It has a very high sensitivity (98%), specificity, (95%) and accuracy (97%) for diagnosing aortic and mitral regurgitation.[46] The signal void, however, is dependent upon certain scan parameters such as echo time, voxel size, and image orientation relative to flow jet. Phase contrast MR allows the assessment of severity of

[47] A. E. Arai, F. H. Epstein, K. E. Bove, and S. D. Wolff, *J. Magn. Reson. Imaging* **10,** 771 (1999).
[48] A. de Roos, N. Reichek, L. Axel, and H. Y. Kressel, *J. Comput. Assist. Tomogr.* **13,** 421 (1989).

valvular stenosis (by measuring the peak jet velocity) by calculating the valve orifice area and transvalvular pressure gradient.[44]

Evaluation of Aortic Diseases

Three-dimensional MRA is performed using a T_1-weighted pulse sequence after a bolus intravenous injection of gadolinium chelate. This method has been firmly established as an accurate noninvasive imaging modality for assessment of the aorta.[49] This technique is able to overcome the limitations of conventional black blood and bright blood sequences, including cardiac and repiratory artifacts, poor SNR, susceptibility effects, and artifacts at air–tissue interfaces, along with significantly shortening acquisition times and not requiring cardiac gating.[49] Although computed tomography (CT) scanning is the modality of choice in acute, life-threatening aortic dissection, CE-MRI is the modality of choice in the more stable patient[49] (Fig. 11). CE-MRI has higher sensitivity and specificity (98% and 98%, respectively), compared with CT (94% and 87%, respectively) and transesophageal echocardiography (98% and 77%, respectively).[46] It is useful to evaluate the extent and localization of the dissection, associated aortic insufficiency, evidence of intrapericardial hemorrhage, relationship with branched vessels, and separation of true and false lumen.[46] In cases where aortic dissection occurs without an intimal flap (intramural hematoma or as a result of aortitis), an additional delayed T_1-weighted image following contrast administration is recommended. This reveals concentric thickening of the aortic wall with increased intramural signal intensity (intramural hematoma) and enhancement of aortic wall and surrounding structures (aortitis).[48] This technique is also useful in the evaluation and follow-up of patients with vasculitis involving the great vessels (e.g., Takayasu's arteritis, Fig. 12).[50]

MRI can also be useful in assessment of thoracic aneurysms (Fig. 13) by demonstrating its length, site, morphology, relationship to branch vessels, presence of a thrombus, or penetrating ulcer. Both CE-MRI and postcontrast T_1-weighted images should be performed to comprehend the full extent of the problem.[49]

CE-MRI is also useful in initial assessment and longitudinal follow-up (both preoperatively and postoperatively) of congenital aortic disorders such as Marfan's syndrome and coarctation of aorta. MRI and velocity mapping can accurately assess the anatomy and severity of the coarctation.[51] In Marfan's syndrome, MRI is useful in assessment and follow-up

[49] F. M. Vogt, M. Goyen, and J. F. Debatin, *Radiol. Clin. North Am.* **41,** 29 (2003).
[50] M. K. Atalay and D. A. Bluemke, *Curr. Opin. Rheumatol.* **13,** 41 (2001).

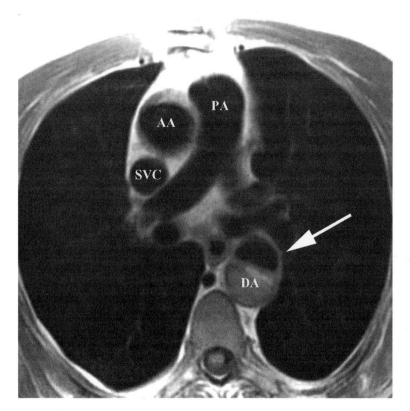

FIG. 11. Black blood axial images of the chest showing a dissection (arrow) in the descending aorta (DA). The false lumen has a higher signal intensity due to organized thrombus. AA, Ascending aorta; PA, pulmonary artery; SVC, superior vena cava.

of the size of the aorta, as well as in revealing the luminal irregularities associated with medial degeneration.

Evaluation of the Pulmonary Veins

CE-MRI is also finding a clinical role in evaluating pulmonary veins (Fig. 14), particularly as an adjunct to the radiofrequency ablation of the pulmonary vein pathways causing atrial fibrillation. MRI is useful to identify and size the pulmonary veins at baseline and serial imaging, and assessment of the size of these veins is very useful in detecting pulmonary vein stenosis, a frequent complication of the ablation procedure.

[51] C. Holmqvist, F. Stahlberg, K. Hanseus, P. Hochbergs, S. Sandstrom, E. M. Larsson, and S. Laurin, *J. Magn. Reson. Imaging* **15,** 39 (2002).

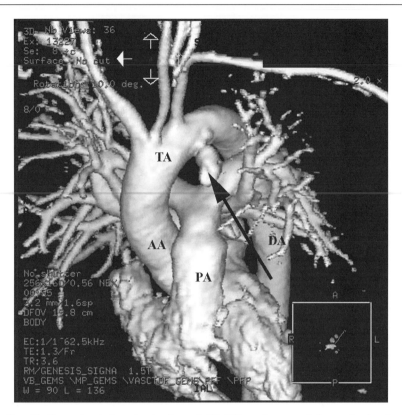

FIG. 12. Reconstructed contrast-enhanced 3D MR angiogram of the aorta showing a mycotic aneurysm (arrow) of the transverse aorta (TA). DA, Descending aorta; AA, ascending aorta; PA, pulmonary artery.

Evaluation of Myocardial Perfusion and Ischemia

MRI is emerging as a reliable and useful tool in assessment of regional left ventricular perfusion. Currently, it relies upon monitoring the first pass of contrast agents (Fig. 15). After a rapid intravenous contrast injection, there is a marked signal enhancement, first in the right ventricular cavity, the left ventricular cavity, and then in the left ventricular myocardium. This is completed within 20–30 seconds and involves a prolonged breath-hold. Gradient–echo sequences have been used for single-slice (midcavity, short axis) perfusion MRI in humans.[52,53] Recently, echo planar techniques have

[52] D. J. Atkinson, D. Burstein, and R. R. Edelman, *Radiology* **174,** 757 (1990).

[53] W. J. Manning, D. J. Atkinson, W. Grossman, S. Paulin, and R. R. Edelman, *J. Am. Coll. Cardiol.* **18,** 959 (1991).

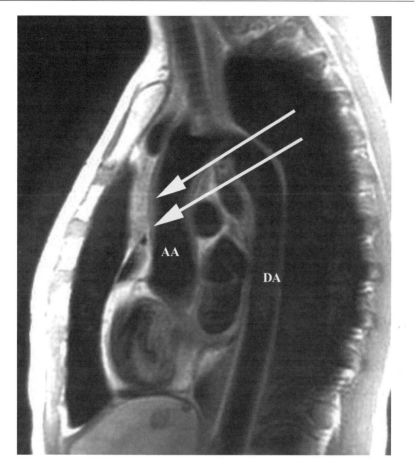

FIG. 13. Black blood oblique sagittal images of the thoracic aorta (candy cane view) showing thickened wall in the aorta consistent with Takayasu's arteritis (arrow). AA, Ascending aorta; DA, descending aorta.

been employed for ultrafast multislice MRI.[54,55] The peak signal intensity is related to the concentration of the contrast agent in the local tissue and is directly proportional to the coronary blood flow. Perfusion MR at rest and after infusion of pharmacologic agents (adenosine and persantine) have been compared with standard methods (angiography or radionuclide scintigraphy) and demonstrated decent sensitivity (67–83%) and specificity

[54] R. R. Edelman and W. Li, *Radiology* **190,** 771 (1994).
[55] M. F. Wendland, M. Saeed, T. Masui, N. Derugin, M. E. Moseley, and C. B. Higgins, *Radiology* **186,** 535 (1993).

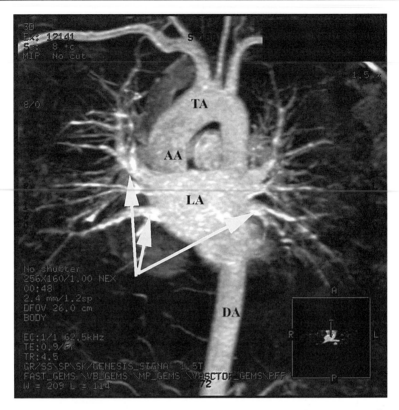

FIG. 14. Reconstructed contrast-enhanced 3D MR angiogram of the thorax revealing the left atrium (LA) and the pulmonary veins (arrows). AA, Ascending aorta; TA, transverse aorta; DA, descending aorta.

(75–100%).[56,57] Additionally, assessing wall motion assessment further improved the performance of MRI.

Dobutamine stress MR has also been studied as a stress-testing modality. Stress-induced wall motion abnormality is an early and reliable sign of myocardial ischemia. Quantitative assessment of systolic wall thickening provides better results than does the visual assessment of response to dobutamine (sensitivity of up to 91%, specificity of around 80%).[44,58,59]

[56] A. C. Eichenberger, E. Schuiki, V. D. Kochli, F. W. Amann, G. C. McKinnon, and G. K. von Schulthess, *J. Magn. Reson. Imaging* **4**, 425 (1994).

[57] M. A. Klein, B. D. Collier, R. S. Hellman, and V. S. Bamrah, *Am. J. Roentgenol.* **161**, 257 (1993).

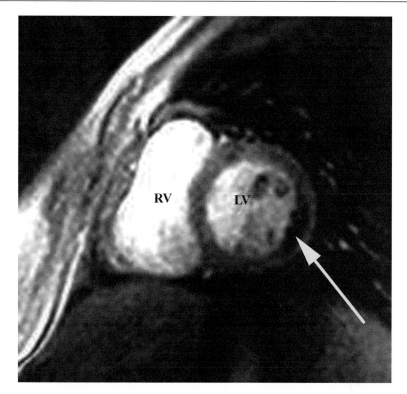

FIG. 15. First-pass, short-axis image of the heart obtained after injection of contrast agent revealing a hypoenhanced area in the lateral wall (arrow) consistent with myocardial ischemia. LV, Left ventricle; RV, right ventricle.

Coronary Artery Imaging

Imaging of the coronary arteries using MR (Fig. 16) still remains a very challenging proposition and hence is predominantly relegated to research centers. In patients with left main or three vessel disease, MR has a sensitivity of 100%, specificity of 85%, and accuracy of 87% for the diagnosis of coronary artery disease.[60] The drawback of current MR technology

[58] E. Nagel, H. B. Lehmkuhl, W. Bocksch, C. Klein, U. Vogel, E. Frantz, A. Ellmer, S. Dreysse, and E. Fleck, *Circulation* **99,** 763 (1999).

[59] D. J. Pennell, S. R. Underwood, C. C. Manzara, R. H. Swanton, J. M. Walker, P. J. Ell, and D. B. Longmore, *Am. J. Cardiol.* **70,** 34 (1992).

[60] W. Y. Kim, P. G. Danias, M. Stuber, S. D. Flamm, S. Plein, E. Nagel, S. E. Langerak, O. M. Weber, E. M. Pedersen, M. Schmidt, R. M. Botnar, and W. J. Manning, *N. Engl. J. Med.* **345,** 1863 (2001).

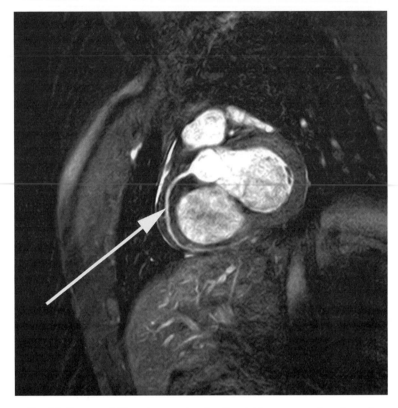

Fig. 16. Coronary MR angiogram showing the right coronary artery (arrow).

is its inability to visualize smaller distal vessels.[60,61] However, with the development of more sophisticated sequences, better contrast agents, routine noninvasive MR coronary angiography might soon become a reality.

However, the role of MRI in assessing for anomalous origin and course of coronary artery is well established. MR techniques have shown excellent results for definition and identification (93–100% cases) of anomalous coronary arteries. Also, MRI can aid in classifying cases that were not classified or were misclassified by conventional angiography.[62,63]

[61] J. C. Post, A. C. van Rossum, M. B. Hofman, J. Valk, and C. A. Visser, *Am. J. Roentgenol.* **166,** 1399 (1996).
[62] M. V. McConnell, P. Ganz, A. P. Selwyn, W. Li, R. R. Edelman, and W. J. Manning, *Circulation* **92,** 3158 (1995).

Unlike native coronary arteries, MR has a definite current role in the assessment of saphenous vein and internal mammary bypass grafts, with a sensitivity, specificity, and accuracy in the 90% range. It is also useful in detecting saphenous vein graft aneurysms that require surgical repair. However, metallic clips, graft markers, and sternal wires may cause local artifacts.[64]

MR Imaging of the Atherosclerotic Plaque

Atherosclerotic plaque stability is inherently dependent upon lipid core, thickness of the fibrous cap, and inflammation within the cap.[65] MRI can provide information regarding all these factors in different vascular territories, including the aorta, carotid arteries, and now, coronary arteries. Atherosclerotic plaque in the aorta (Fig. 17) can be assessed volumetrically and characterized using double inversion recovery fast-echo sequences to suppress the signal from flowing blood.[66,67] Studies have also been performed using transesophageal MRI to improve the signal intensity.[67] Similarly, carotid arteries have been imaged and attempts at plaque characterization have been performed.[68] More recently, coronary vessel wall imaging has been attempted using black blood MRI.[69] This will serve as an important tool to assess regression of the coronary plaque in the future, particularly with the advent of intravascular MRI techniques.

Newer Applications

As the MR technology evolves, newer applications are being developed. In the horizon of cardiac MR technology is therapeutic MR. Preliminary work is already underway in developing MR-compatible catheters for electrophysiology studies–ablation, as well as interventional cardiology-related procedures.

[63] J. C. Post, A. C. van Rossum, J. G. Bronzwaer, C. C. de Cock, M. B. Hofman, J. Valk, and C. A. Visser, *Circulation* **92,** 3163 (1995).

[64] A. C. van Rossum, W. L. Bedaux, and M. B. Hofman, *J. Magn. Reson. Imaging* **10,** 734 (1999).

[65] C. M. Kramer, *Am. J. Cardiol.* **90,** 15L (2002).

[66] K. A. Shunk, J. Garot, E. Atalar, and J. A. Lima, *J. Am. Coll. Cardiol.* **37,** 2031 (2001).

[67] R. Corti, V. Fuster, Z. A. Fayad, S. G. Worthley, G. Helft, D. Smith, J. Weinberger, J. Wentzel, G. Mizsei, M. Mercuri, and J. J. Badimon, *Circulation* **106,** 2884 (2002).

[68] C. Yuan, K. W. Beach, L. H. Smith, Jr., and T. S. Hatsukami, *Circulation* **98,** 2666 (1998).

[69] W. Y. Kim, M. Stuber, P. Bornert, K. V. Kissinger, W. J. Manning, and R. M. Botnar, *Circulation* **106,** 296 (2002).

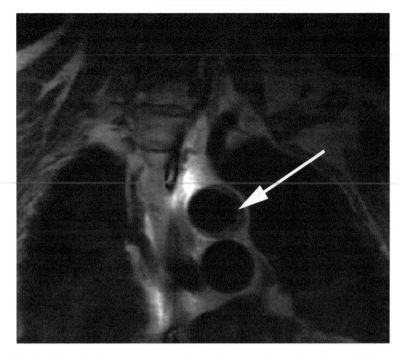

Fɪɢ. 17. Black blood coronal image of the chest revealing atherosclerosis in the aorta (arrow).

Conclusion

MRI is the newest, most complex, and most rapidly emerging noninvasive test of choice for patients with a multitude of cardiovascular problems. Its emerging role in being the dominant imaging modality in every facet of cardiology cannot be understated. It is entering an important phase in its evolution, with an anticipated exponential growth in its current applications and development of new applications.

[6]   MRI of Animal Models of Brain Disease

By LOUISE VAN DER WEERD, DAVID L. THOMAS, JOHN S. THORNTON, and
MARK F. LYTHGOE

Introduction

This chapter will be restricted to the most common application of magnetic resonance imaging (MRI) in the brain, namely, the depiction of hydrogen nuclei (protons) of mobile water molecules. The utility of MRI with respect to animal models of disease lies in the sensitivity of the technique to both microscopic and macroscopic molecular motions. Nervous tissue consists of 70–80% water by weight, these water molecules being distributed through a variety of microscopic environments and physiologic compartments. It is possible to generate MR images whose contrast reflects, among other factors, random molecular rotational motions (T_1 and T_2), random translational motion (diffusion), exchange with macromolecular protons (MTC), and blood flow (perfusion). The concentration and mobility of water molecules is modified in many pathologies, and this is the basis of the high sensitivity of MRI to cerebral disease processes.

Biophysical Background and Methods

Some important principles are discussed here to provide a background for the rest of this chapter. For a more detailed introduction to MRI, we refer to various excellent textbooks.[1,2] Protons possess a nuclear magnetic moment (or "spin"). In the absence of an external magnetic field, these magnetic moments are randomly distributed in every direction. In the presence of a magnetic field, however, a thermal equilibrium is achieved between spins oriented parallel and antiparallel to the magnetic field. The result is a net macroscopic magnetic moment, the bulk magnetization (M_0), orientated in the direction of the external field (conventionally taken to be the z-axis). The individual spins precess around the z-axis at the Larmor frequency (ω, rad s^{-1}), which is proportional to the external magnetic field (B_0, Tesla):

[1] D. G. Gadian, "NMR and Its Applications to Living Systems." Oxford University Press, Oxford, UK, 1995.
[2] R. Smith and R. Lange, "Understanding Magnetic Resonance Imaging." CRC Press, Boca Raton, FL, 1999.

METHODS IN ENZYMOLOGY, VOL. 386

Copyright 2004, Elsevier Inc.
All rights reserved.
0076-6879/04 $35.00

$$\omega = \gamma B_0 \qquad (1)$$

where γ is a constant called the gyromagnetic ratio (26.751 10^7 rad T^{-1} s^{-1} for protons).

Relaxation

M_0 is proportional to the total number of protons present in the sample, and hence is also called the proton density. To be able to distinguish the magnetization M_0 from the external magnetic field, M_0 is rotated by 90° into the transverse (xy) plane using a radiofrequency (RF) pulse at the Larmor frequency. Immediately following this 90° pulse, the initial magnetization level can be detected. In time, the thermal equilibrium is restored, and the magnetization vector returns to the z-axis. The characteristic times involved in this process are the spin relaxation times: longitudinal relaxation time (T_1) for the restoration of the magnetization along the z-axis, and the transverse relaxation time (T_2) for the decay in the xy-plane (Fig. 1).

T_2 and T_2*. The transverse relaxation time T_2 is also called the spin–spin relaxation time, referring to the molecular interactions behind the transverse relaxation process. The protons experience intramolecular dipolar interactions between two protons within the same molecule, as well as intermolecular interactions with protons of neighboring molecules. This interaction becomes more efficient when the contact time between protons is relatively long (e.g., in viscous media). When the rotational correlation time of the molecules is short, as is the case for free water molecules, T_2 is relatively long ($\sim$3 s). Water molecules interacting with macromolecules or solid surfaces generally have slower tumbling rates, which leads to a reduction in the relaxation time. Because water mobility often varies substantially between tissue types, and changes in situations of cellular stress, T_2-dependent contrast is very commonly used in MRI studies.

In addition to these spin–spin interactions, the transverse magnetization is also perturbed by small local magnetic field differences. This results in different (local) Larmor precession frequencies of the spins under observation and thus in a loss of phase coherence, causing a faster decay of the magnetization in the xy-plane. The corresponding apparent relaxation time is called T_2* to distinguish it from the intrinsic transverse relaxation time T_2.

This field-disturbing effect can be exploited as a source of contrast in tissue, since such magnetic field inhomogeneities typically occur at interfaces of structures with differing magnetic susceptibilities, such as soft tissue and bone, or tissue and blood. T_2* contrast is of specific importance in blood oxygenation level dependent (BOLD) imaging, which is widely used in functional MRI investigations.[3] The paramagnetic nature of

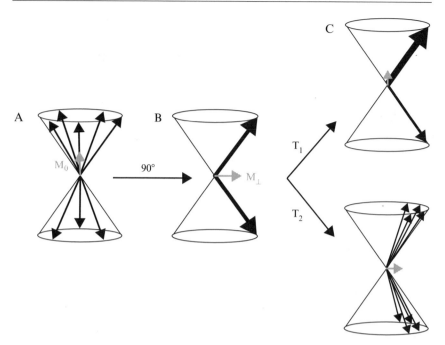

FIG. 1. Schematic representation of the nuclear magnetic resonance principle. (A) The sample magnetization M_0 arises from the uneven distribution of the spins (black arrows) between two different states, either parallel or antiparallel to the main magnetic field B_0. The spins precess around the main magnetic field direction with the Larmor frequency ω. (B) After the application of a 90° pulse, the original distribution is shifted into the horizontal plane and phase coherence is established (the spins are all aligned along the same axis). The result is a sample magnetization $M_\perp$. (C) The spins return to the original distribution through T_1 relaxation. The loss of phase coherence is called T_2 relaxation. Both processes occur simultaneously but are depicted separately in the figure.

deoxygenated blood generates magnetic field gradients in blood vessels and surrounding tissues, leading to signal loss in T_2^*-weighted images. Fast T_2^*-weighted imaging is performed continuously to track transient changes in the magnetic field disturbances associated with the balance between oxyhemoglobin and deoxyhemoglobin in the blood, thus providing information on local neuronal activity. The second important application of T_2^* contrast is its use in contrast-enhanced MRI.[4] Specific exogenous paramagnetic contrast agents, analogous to the tracers used in nuclear

[3] S. Ogawa, R. S. Menon, S. G. Kim, and K. Ugurbil, *Annu. Rev. Biophys. Biomol. Struct.* **27,** 447 (1998).

[4] M. Modo and S. C. R. Williams, "Biomedical Imaging in Experimental Neuroscience" (N. Van Bruggen and T. P. L. Roberts, eds.). p. 29. CRC Press, Boca Raton, FL, 2002.

medicine, are being developed continuously in order to target specific areas or molecules, thus providing a means to map molecular events *in vivo*.

T_2 AND T_2* MEASUREMENTS. As already described, for detection the bulk magnetization M_0 is rotated by $90°$ into the xy-plane. The ensuing loss of phase coherence due to T_2* effects can be reversed by the application of a series of $180°$ RF pulses following the initial $90°$ RF pulse, forming the so-called spin–echo (SE) sequence. The restoration of coherent magnetization between the $180°$ pulses is called an echo. The amplitude of this echo is only attenuated by T_2 relaxation (Fig. 2).

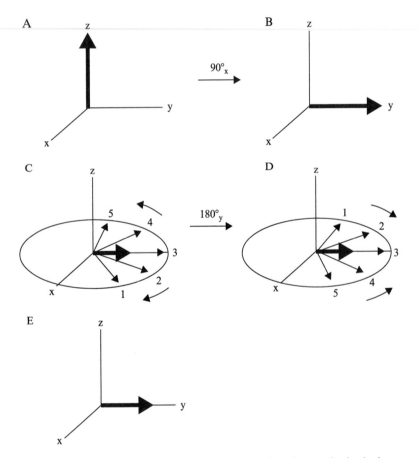

FIG. 2. (A) Diagram showing the fanning out and refocusing of magnetization in the course of a spin–echo sequence. (B) After the application of a $90°$ pulse, the original distribution is shifted into the horizontal plane. (C) The loss of phase coherence is primarily due to T_2* effects. (D) The $180°$ pulse flips the spins in the xy-plane, and the magnetization refocuses along the y-axis. The attenuation of the net magnetization vector is due to T_2 relaxation.

T_2^* can be measured by means of gradient echoes. In the spin–echo sequence the 180° pulse reverses the effects of local field inhomogeneities, whereas in a gradient–echo sequence the echo is generated by reversing a magnetic field gradient. The main difference between a spin echo and a gradient echo is that the gradient echo does not refocus the dephasing due to field inhomogeneities, and therefore the echo is weighted according to T_2^* rather than T_2. In addition to its use in T_2^* imaging, gradient echoes are commonly used in rapid imaging sequences, as the echo time can be made much shorter than in spin–echo sequences.

T_1. The other relaxation time that must be considered is the longitudinal relaxation time T_1, also referred to as the spin–lattice relaxation time. The mechanisms behind this relaxation are complex, but it is facilitated by the presence of microstructures (macromolecules, membranes, etc.), also called the lattice, that via dipolar interactions can absorb the energy of the excited protons. This energy transfer is most efficient when the rotational correlation rate of the molecules is in the same range as the Larmor frequency. In practice this means that T_1 becomes shorter as the molecular mobility decreases, but increases again for very slow molecular motion, as in solids. Both T_1 and T_2 reflect the properties of the physical microenvironment of water in tissue, albeit not in exactly the same way.

T_1 MEASUREMENTS. The most well-known sequence to measure T_1 is the inversion recovery sequence. This sequence starts with a 180° RF pulse causing inversion of M_0, which then gradually recovers to its equilibrium. To detect the amount of magnetization left, a 90° pulse is applied after a range of delay times. This rotates the magnetization into the xy-plane, where it can be detected (Fig. 3). From the different delay times and the corresponding residual magnetization levels, T_1 can be calculated.

Diffusion

Up to now, the translational motion of individual water molecules has not been considered. However, all molecules in a fluid are subject to Brownian movements, the extent of this motion depending on the temperature and the viscosity of the fluid. When an ensemble of molecules is followed in time, the root mean square displacement (x, m) shows a $\sqrt{t}$ dependence:

$$x = \sqrt{2dDt} \tag{2}$$

where D is the bulk diffusion coefficient of the fluid ($m^2 \ s^{-1}$), t is the displacement time (s), and d ($= 1, 2,$ or 3) is the dimensionality of the diffusion displacement. Normally, the displacement distribution of all molecules is Gaussian, where the mean displacement distance increases with increasing displacement times. However, if the molecules encounter barriers to

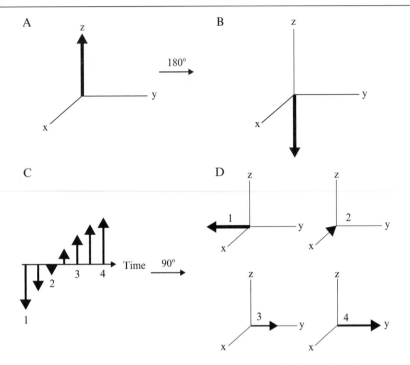

Fig. 3. (A) Diagram showing the magnetization changes during an inversion recovery sequence. (B) The $180°$ pulse inverts the magnetization M_0. (C) In time, the magnetization returns to equilibrium due to T_1 relaxation. (D) $90°$ pulses are applied to detect the residual magnetization at a number of time points. After each $90°$ pulse, a waiting time is introduced to let the magnetization return to equilibrium, after which the next cycle of 180–$90°$ pulses is performed.

diffusion (e.g., cell membranes) these determine the maximum displacement. These boundary restrictions imply that the displacement distribution is no longer Gaussian and is going to depend on the diffusion time. As a result, the measured apparent diffusion coefficient (ADC) is smaller than the intrinsic D. This ADC value is sensitive to the number of barriers, their geometry, and their permeability: in other words, to the tissue microstructure.

The previous is true if isotropic diffusion can be assumed (i.e., diffusion that exhibits no directionality). Many biologic tissues have a microstructure that favors molecular motion in a certain direction. In the brain, diffusional anisotropy occurs primarily in white matter tracts, caused by the myelin sheaths and other structures surrounding the nerve fibers, which restrict diffusion perpendicular to the axonal length. The anisotropic

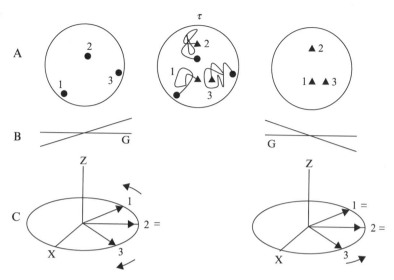

Fig. 4. (A) Water diffusion for three different spins within a sample. (B) Two diffusion gradients are applied with opposite sign and a delay time τ between them. (C) Diffusing spins experience different phase shifts (dependent on their position in the direction of the applied diffusion gradient) and are incompletely refocused, leading to a net loss in signal intensity.

diffusion that arises when displacement along one direction occurs more readily than along another is defined by a diffusion tensor, and diffusion tensor imaging (DTI) is the term used for measurements of the full diffusional properties of the sample in all three dimensions.[5]

Diffusion Measurements. The ADC can be measured using a pulsed field gradient (PFG) experiment. In this experiment a sequence of two pulsed magnetic field gradients of equal magnitude but opposite sign and separated by an interval Δ temporarily change the resonance frequency of the observed spins (Eq. 1) as a function of their position. If the spins remain at exactly the same position, the effects of the opposing gradient pulses compensate each other. However, as soon as translational motion occurs, the gradient-induced frequency shifts do not exactly compensate each other anymore and a phase shift occurs. Because diffusion is random in all directions, no net phase shift results but phase *coherence* is partially lost, resulting in attenuation of the echo amplitude (Fig. 4). The amount of this attenuation is determined by the length, amplitude, and separation of the gradient pulses, summarized in the so-called b factor, and by the mean translational distance traveled during the interval Δ, which depends

[5] M. Moseley, J. Kucharczyk, and H. Asgari, *Magn. Reson. Med.* **19,** 321 (1991).

on the ADC. The signal intensity in a diffusion-weighted image (DWI) can therefore be described as:

$$S_b = S_0 e^{-b \cdot ADC} \tag{3}$$

where S_b is the DWI signal intensity and S_0 is the signal without any diffusion gradients applied.

Perfusion

Perfusion (or cerebral blood flow [CBF]) is the amount of blood delivered to the capillary bed of a block of tissue in a certain period of time (units ml blood/100 g/min), and is closely related to the delivery of oxygen and other nutrients to the tissue. It is this quantity that determines the energy status of the tissue, and for this reason much effort has been put into its measurement. Two main MRI approaches have been developed for CBF measurement: bolus tracking and arterial spin labeling.

Bolus Tracking. A bolus of paramagnetic contrast agent is injected intravenously into the subject. Soon after, the contrast agent passes through the vasculature of the brain and causes the signal of a T_2- or T_2*-weighted image to change due to a difference in magnetic susceptibility between the intravascular and extravascular compartments. This reduction is transient because the bolus of contrast agent washes through the vasculature of the tissue. The passage of the bolus is relatively fast (of the order of seconds), particularly in small animals. It is therefore important to acquire images as quickly as possible, in order to characterize the signal time course accurately, which is crucial for the quantification of CBF. The rate at which successive bolus tracking experiments can be performed is limited by the clearance rate of the contrast agent from the vasculature. Residual levels of contrast agent will reduce the signal change induced by subsequent boluses and therefore reduce the precision of the CBF measurement, and repeated injections of contrast agents such as gadolinium are eventually limited by their toxicity.[6]

To calculate CBF values from the bolus passage time course, the arterial input function (AIF) (i.e., the concentration of contrast agent entering the voxel of interest at a certain time point) needs to be known. The AIF is usually estimated from the signal time course within a major artery, such as the middle cerebral artery. Although not impossible in small animals,[7,8]

[6] F. G. Shellock and E. Kanal, *J. Magn. Reson. Imaging* **10**, 477 (1999).

[7] W. H. Perman, M. H. Gado, K. B. Larson, and J. S. Perlmutter, *Magn. Reson. Med.* **28**, 74 (1992).

[8] L. Porkka, M. Neuder, G. Hunter, R. M. Weisskoff, J. W. Belliveau, and B. R. Rosen, "Arterial Input Function Measurement with MRI." Proceedings of SMRM 10th Annual Meeting, p. 120, 1991.

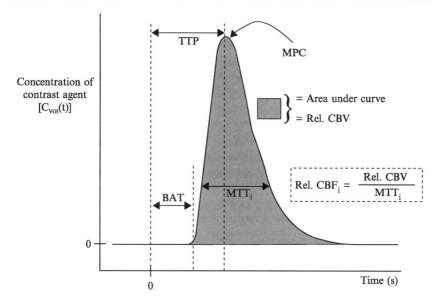

FIG. 5. Schematic representation of the concentration time curve observed in brain tissue following intravenous injection of a paramagnetic contrast agent. The summary parameters illustrated are: BAT, bolus arrival time; TTP, time to peak; MPC, mean peak concentration; MTT_i, mean transit time index; rel. CBV, relative cerebral blood volume; rel. CBF_i, relative cerebral blood flow index.

reliable characterization is not straightforward due to the small size of arteries. Additionally, the AIF should strictly define the entry of contrast agent into the voxel of interest, and this may be different to the passage of the bolus through a major artery, due to delay and dispersion effects between the two locations.[9,10] As a result, reliable quantification of CBF can be problematic, and so summary parameters have been widely used as an alternative. Their advantage is that they are quick and easy to calculate, with minimal assumptions required; their disadvantage is that they do not give values for the real physiologic variables CBF and cerebral blood volume (CBV), and depend on experimental conditions (e.g., injection rate, vascular geometry, cardiac output); Fig. 5 illustrates the most commonly used summary parameters: bolus arrival time (BAT), time to peak (TTP), maximum peak concentration (MPC), mean transit time index (MTT_i), relative CBV (rel. CBV), and relative CBF index (rel. CBF_i).

[9] F. Calamante, D. G. Gadian, and A. Connelly, *Magn. Reson. Med.* **44,** 466 (2000).
[10] L. Ostergaard, R. M. Weisskoff, D. A. Chesler, C. Glydensted, and B. R. Rosen, *Magn. Reson. Med.* **36,** 715 (1996).

These parameters are related to different aspects of the cerebral hemodynamics: BAT, the time after injection at which contrast agent begins to
arrive in the voxel of interest, can provide information about the role of
collateral blood supply when CBF is compromised; MTT_i is an approximation of the average time for any given particle of tracer to pass through the
tissue, and it gives an indication of different vascular territories. The CBF
value obtained from the relative CBV ("relative" because it does not take
the AIF into account) and MTT_i is dubbed relative CBF_i. Although the use
of summary parameters does not provide absolute values for CBF and
CBV, it provides a useful tool for the investigation of perfusion during
normal and ischemic conditions.

Arterial Spin Labeling. Arterial spin labeling (ASL) is a CBF measurement method that uses magnetically labeled blood water as an endogenous
tracer. It is appealing for imaging animal models of brain disease because it
does not require injection of exogenous tracers and places no restriction of
the number of repeat measurements that can be made in a single study.
This makes ASL well suited to the monitoring of CBF during experimental
time courses in animal models. The images obtained with ASL can be converted into CBF maps (with units of ml/100 g/min) as long as certain other
MR parameters (such as T_1 and labeling efficiency; see the following) are
also measured.

In ASL, perfusion-weighted images are generated by labeling inflowing
blood water by inversion of its longitudinal magnetization. Two images
must be acquired: one in which spin labeling is performed (the labeled
image) and one in which spin labeling is not performed (the control image).
The difference between the two images is proportional to the amount of
labeled blood that enters the imaging slice during the time between labeling and image acquisition, and therefore is also proportional to CBF. Spin
labeling can be achieved using several alternative approaches, which has
given rise to a family of ASL techniques that will now be discussed in turn.

CASL MEASUREMENTS. The original ASL approach[11] is now generally
known as *continuous* ASL (or CASL), because inflowing blood magnetization is inverted continuously over a period of several seconds prior to
image acquisition. This allows an accumulation of spin-labeled water as it
is deposited in the perfused cerebral tissue. The measurement approach
is illustrated in Fig. 6. A magnetic field gradient is applied to give the
proton magnetization a position-dependent frequency spread. An RF pulse
of 1–3 seconds is applied with a frequency offset corresponding to water
protons flowing through the carotid and vertebral arteries in the neck. This

[11] J. A. Detre, J. S. Leigh, D. S. Williams, and A. P. Koretsky, *Magn. Reson. Med.* **23,** 37
(1992).

causes inversion of flowing spins via a process known as flow-induced adiabatic inversion.[12] Following the labeling RF pulse, an image is acquired using a rapid imaging technique (e.g., EPI). A problem associated with the labeling RF pulse is chemical exchange of protons between water and macromolecules, which causes the water magnetization to decrease (the MT effect; see also MTC imaging in this chapter). For this reason, an RF pulse with the reverse frequency offset is applied for the control image, which results in the same MT effect in the labeled and control image, and the only difference between the images is due to the perfusion-weighting.

PASL MEASUREMENTS. Soon after the introduction of CASL, alternative approaches using shorter adiabatic inversion pulses were proposed, known as *pulsed* ASL (PASL). They have the advantage of a lower RF energy deposition than CASL due to the short ($\sim$10 ms) inversion pulses, but also have a lower intrinsic signal to noise ratio (SNR) for the perfusion signal. One of the PASL approaches is known as flow-sensitive alternating inversion recovery, or *FAIR* (Fig. 6). This technique uses a pair of inversion recovery (IR) images, one following a slice-selective inversion pulse and the other following a nonselective (global) inversion pulse. The slice-selective IR image is flow-sensitive because the inflowing blood water magnetization is fully relaxed, and so accelerates the apparent T_1 relaxation of the tissue that it flows into. In the nonselective IR case, both tissue and inflowing blood magnetization are inverted, greatly reducing the flow sensitivity. Because the same inversion time is used in both the slice-selective and nonselective acquisitions, the signal from the static tissue is the same in both, but a difference between the images is observed in the presence of flow.

Perfusion can be quantified using ASL images and some additional information such as T_1; for a description of the principles see Refs. 13 and 14. An important factor that affects the accuracy of CBF quantification is a parameter known as the transit time. Following spin inversion, blood water travels through the vascular tree until it reaches the capillary bed and exchanges into the cerebral tissue. During this time (the transit time), the spin label decays at a rate determined by the T_1 of arterial blood (T_{1a}). This must be accounted for in order to achieve accurate CBF quantification. Transit time effects are particularly problematic in CASL, where the labeling plane must be a minimum distance from the imaging slice to ensure that it coincides

[12] W. T. Dixon, L. N. Du, D. D. Faul, M. Gado, and S. Rossnick, *Magn. Reson. Med.* **3**, 454 (1986).

[13] F. Calamante, D. L. Thomas, G. S. Pell, J. Wiersma, and R. Turner, *J. Cereb. Blood Flow Metab.* **19**, 701 (1999).

[14] D. L. Thomas, M. F. Lythgoe, G. S. Pell, F. Calamante, and R. J. Ordidge, *Phys. Med. Biol.* **45**, R97 (2000).

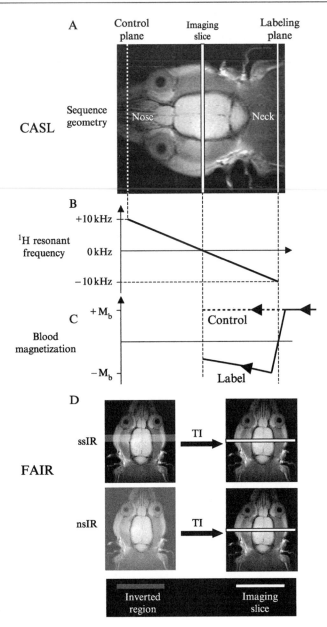

FIG. 6. (A) Continuous arterial spin labeling (CASL). Axial MR image of the rat brain showing the relative positions of the imaging slice (coronal) and the labeling and control planes. (B) A magnetic field gradient causes water protons to resonate over a range of frequencies dependent on their position, so that the labeling plane corresponds to −10 kHz

with a major artery. With PASL, transit times tend to be less because the labeling region can be placed immediately adjacent to the imaging slice. To decrease the sensitivity to transit time effects, the sequences can be modified by inserting a postlabeling delay, which allows time for all the blood to travel from where it was labeled into the imaging region.[15–17]

Magnetization Transfer Contrast

A certain fraction of protons in tissue exist in a so-called bound state (i.e., their motion is restricted because they are part either of macromolecules, or of water molecules within the hydration layers around macromolecules). These protons possess very short T_2 relaxation times (<100 μs) and hence are not accessible by standard spin–echo MRI methods. (With currently available small-bore technology, minimum echo times are of the order of several ms, and hence the signal from these protons has already decayed before signal acquisition.) Although simple models for proton relaxation behavior predict that such a proton population will have some influence upon T_1 and T_2, Wolff and Balaban[18] proposed the method of magnetization transfer contrast (MTC) imaging,[19] as a more direct means of probing these bound protons.

MTC imaging is based on an inverse Fourier relationship between T_2 and the range of frequencies over which protons respond to RF excitation: protons in the bound fraction possess a very short T_2 and hence exhibit a broad resonance width ($\sim$20 kHz) in the frequency domain. Conversely, bulk water protons have a long T_2 with a correspondingly narrow ($\sim$10 kHz) frequency domain resonance (Fig. 7A). This difference can be used to excite the bound fraction independently from the bulk protons. If RF energy is supplied at a frequency offset from the central Larmor frequency (typically 1–5 kHz) the magnetization in the bound fraction will

[15] D. C. Alsop and J. A. Detre, *J. Cereb. Blood Flow Metab.* **16**, 1236 (1996).

[16] W. M. Luh, E. C. Wong, P. A. Bandettini, and J. S. Hyde, *Magn. Reson. Med.* **41**, 1246 (1999).

[17] E. C. Wong, R. B. Buxton, and L. R. Frank, *Magn. Reson. Med.* **39**, 702 (1998).

[18] S. D. Wolff and R. S. Balaban, *Magn. Reson. Med.* **10**, 135 (1989).

[19] R. M. Henkelman, G. J. Stanisz, and S. J. Graham, *NMR Biomed.* **14**, 57 (2001).

and the control plane to +10 kHz. (C) By simultaneously applying a −10 kHz off-resonance RF pulse, blood water ^{1}H spins are inverted as they flow through the inversion plane (spin labeled). The control plane does not intersect any major arteries, and so application of a +10 kHz RF pulse causes MT effects but not spin labeling. (D) Pulsed arterial spin labeling (PASL). In FAIR, a pair of images is also acquired, one following a slice-selective inversion (upper) and the other following a nonselective inversion (lower). The difference between the images is caused by the difference in the magnetization state of the inflowing blood water.

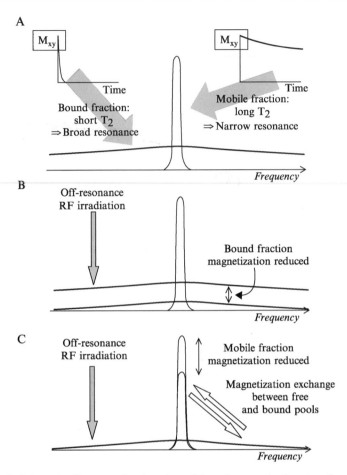

Fig. 7. Schematic diagrams showing the origins of magnetization transfer contrast. (A) Macromolecular-bound protons possess a very short T_2 and hence a broad resonance response in the frequency domain. Mobile water protons conversely exhibit a narrow frequency domain linewidth. (B) In the absence of exchange, the application of RF energy at a frequency away from resonance perturbs only the bound proton fraction, causing the magnetization of this pool to reduce toward zero. (C) Exchange of magnetization between free and bound protons causes a reduction in the magnetization of the mobile pool, and hence an observable reduction in MR image intensity.

be reduced by saturation, whereas in the absence of exchange, the bulk water pool would remain unaffected (Fig. 7B). However, on the time scale of a typical MRI experiment there is a significant exchange of magnetization between the two proton fractions, either by chemical exchange or by magnetic interactions. Because the bound proton magnetization has been reduced by the off-resonance irradiation, such exchange also leads to a reduction in both the magnitude of the observable bulk water magnetization and its associated T_1 (Fig. 7C). The degree of reduction depends upon both the relative sizes of the two fractions and upon the rate of magnetization exchange between them; both of these factors may be influenced by tissue pathology.

MTC Measurements. In its most simple form,[20] the MTC imaging experiment involves collecting an image (S_s) that is preceded by a long (~3 s) saturating off-resonance RF pulse, followed by a second image (S_0) without a presaturation pulse. The magnetization transfer ratio (MTR) is then quantified as:

$$MTR = (S_0 - S_s)/S_0 \qquad (4)$$

A high MTR signifies the presence of a significant proton pool associated with macromolecules or cellular microstructure *and* a significant exchange of magnetization between these protons and those of the bulk water. Reduction of the MTR suggests a disruption of tissue microstructure, the most successful application of MTC in experimental neuroscience being the investigation of pathologic disruption of white matter due to demyelination.

To reduce imaging time, selective saturation of the bound fraction may also be achieved by pulsed methods whereby gradient–echo images with short TR are obtained with a low-angle excitation pulse preceded by a short (~10 ms) off-resonance pulse.[21] If the repetition time is sufficiently short (~100 ms) compared to the T_1 relaxation time of the bound pool, an equilibrium is established after a number of cycles, resulting in substantial saturation of the bound fraction.

It should be noted that unless total selective saturation of the bound pool is achieved, a situation impossible to achieve in practice, the magnitude of the MTC effect is dependent upon the duration, intensity, and frequency offset of the off-resonance pulses. Caution is therefore required in the quantitative comparison of MTC imaging results obtained using differing experimental schemes.

Applications of MRI to Experimental Neuropathology

The following section contains an overview of some the possibilities for MRI investigations of cerebral pathology in experimental animal models (for an extensive review of applications, see Ref. 22).

[20] R. J. Ordidge, J. A. Helpern, R. A. Knight, Z. X. Qing, and K. M. A. Welch, *Magn. Reson. Imaging* **9,** 895 (1991).
[21] V. Dousset, R. I. Grossman, and K. N. Ramer, *Radiology* **182,** 483 (1992).
[22] N. Van Bruggen and T. P. L. Roberts, "Biomedical Imaging in Experimental Neuroscience." CRC Press, Boca Raton, FL, 2002.

Cerebral Ischemia

The use of MRI in experimental neuroimaging stroke research has focused on a number of areas that include the use of MRI both as a diagnostic and as a predictive tool.[14] Some of the most important topics are the investigation of the underlying processes that lead to cell death,[23] defining and understanding the concept of the penumbra[24] as a possible salvageable region, and related to that, possible therapies for the treatment of stroke.[25,26]

ADC Changes Following a Stroke. It is now well established that following a stroke, the ADC of water decreases minutes after the reduction in blood supply (e.g., Fig. 8).[27,28] Yet it was first demonstrated, by the use of DWI, that following occlusion of the middle cerebral artery (MCAO), the ADC decreases and the corresponding ischemic lesion in both rat and cat is visible within minutes and expands primarily during the first 2 hours.[29] Although the lesion size has nearly fully evolved at 2 hours, the ADC value continues to decrease for a period of up to 4–6 hours following MCAO (40% of control),[30] with some studies reporting the lowest ADC values at 24 to 48 hours (40–50% of control).[31] In the chronic stages of cerebral ischemia, the ADC of water exhibits a different pattern. Approximately 24–48 hours after a vessel occlusion, the ADC starts to rise again and slowly returns to a normal value at 3 days.[32] Following this, a subsequent increase in the diffusion of water above that of the ischemic control can be observed after 1 week.[30] The elevated ADC of tissue water (or pseudo-normalization) above control values is associated with cellular lysis or the loss of cellular barriers, combined with an excessive accumulation of edematous water.[33]

[23] M. Hoehn-Berlage, *NMR Biomed.* **8,** 345 (1995).

[24] K. A. Hossmann, *Ann. Neurol.* **36,** 557 (1994).

[25] M. Fischer and T. G. Brott, *Stroke* **34,** 359 (2003).

[26] M. Rudin, N. Beckmann, R. Porszasz, T. Reese, D. Bochelen, and A. Sauter, *NMR Biomed.* **12,** 69 (1999).

[27] M. F. Lythgoe, S. R. Williams, A. L. Busza, L. I. Wiebe, A. J. B. McEwan, D. G. Gadian, and I. Gordon, *Magn. Reson. Med.* **41,** 706 (1999).

[28] M. E. Moseley, Y. Cohen, J. Mintorovitch, L. Chileuitt, H. Shimizu, J. Kucharczyk, M. F. Wendland, and P. R. Weinstein, *Magn. Reson. Med.* **14,** 330 (1990).

[29] S. A. Roussel, N. Van Bruggen, M. D. King, J. Houseman, S. R. Williams, and D. G. Gadian, *NMR Biomed.* **7,** 21 (1994).

[30] R. A. Knight, M. O. Dereski, J. A. Helpern, R. J. Ordidge, and M. Chopp, *Stroke* **25,** 1252 (1994).

[31] M. Hoehn-Berlage, M. Eis, T. Back, K. Kohno, and K. Yamashita, *Magn. Reson. Med.* **34,** 824 (1995).

[32] R. A. Knight, R. J. Ordidge, J. A. Helpern, M. Chopp, L. C. Rodolosi, and D. Peck, *Stroke* **22,** 802 (1991).

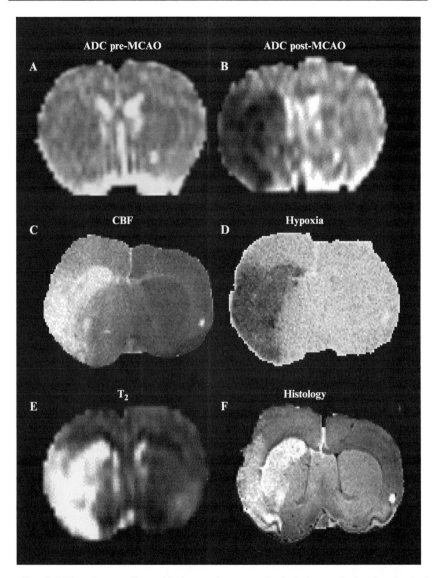

FIG. 8. MR and autoradiographic images from a rat brain before and after intraluminal suture occlusion of the middle cerebral artery: (A) ADC image pre-MCAO; images of (B) ADC, (C) CBF, and (D) hypoxia 2 hours post-MCAO; and (E) MR T_2 image and (F) histology 7 hours post-MCAO. After permanent MCAO in the rat, the ADC decreases and the lesion area at 2 hours is comparable with the area of hypoxia. The area of blood flow decrease is larger than the area of hypoxia or ADC decrease, indicating the so-called diffusion–perfusion mismatch region. The area of infarction, as indicated by the H + E stain, was similar to the area of T_2 increase at 7 hours, which matched the area of ADC change at 2 hours.[27]

Cerebral Blood Flow Following a Stroke. Although DWI may provide unique information about the effect of an ischemic insult as early as a few minutes after occlusion, it is clearly desirable to obtain information regarding the integrity of the vascular bed. MRI experiments in rat and cat models of focal cerebral ischemia demonstrate an absence of contrast agent in the ischemic core of the lesion.[34] Interestingly, in the periphery of the lesion, a diffusion–perfusion mismatch can be observed; that is, perfusion deficits not great enough to cause energy failure may go unnoticed on DWI. A cat model of hypoperfusion shows the same pattern; that is, the diffusion-weighted images show no evidence of abnormality in the territory of the occluded MCA, whereas the dynamic susceptibility contrast (DSC)-MRI blood flow measurements show decreased CBF throughout the MCA region.[35] This early work using MRI has been followed by studies using noninvasive ASL techniques for the quantitation of CBF in animal models.[36] By combining DWI and CBF data, three different tissue areas can be assigned following an ischemic insult: areas with normal perfusion and diffusion, denoting normal tissue; tissue that has a CBF decrease without diffusion changes, corresponding to a region of hypoperfusion; and a region in which there is a concomitant diffusion and perfusion decrease, corresponding to a severely compromised region.[37]

The Penumbra. The term *penumbra* was introduced to designate a zone of brain tissue with moderate ischemia and impaired neuronal function, with the neuronal paralysis being fully reversed upon reperfusion.[38] Currently the term has a wider definition and is used to describe a region of ischemic tissue peripheral to the core where viable neurons may be found, and thus may be potentially salvageable with suitable intervention or therapy.[24,39] Penumbral zones can be delineated using several different imaging modalities, each of which have definitions that rely on the underlying mechanisms of that technique.[24] Moseley *et al.*[28] observed an ADC decrease

---

[33] C. Pierpaoli, A. Righini, I. Linfante, J. H. Tao-Cheng, J. R. Alger, and G. Di Chiro, *Radiology* **189,** 439 (1993).

[34] M. J. Quast, N. C. Huang, G. R. Hillman, and T. A. Kent, *Magn. Reson. Imaging* **11,** 465 (1993).

[35] N. Derugin and T. P. L. Roberts, *Microsurgery* **15,** 70 (1994).

[36] M. F. Lythgoe, D. L. Thomas, and F. Calamante, "Biomedical Imaging in Experimental Neuroscience" (N. Van Bruggen and T. P. L. Roberts, eds.), pp. 21–54. CRC Press, Boca Raton, FL, 2002.

[37] F. Calamante, M. F. Lythgoe, G. S. Pell, D. L. Thomas, M. D. King, C. H. Sotak, A. L. Busza, S. R. Williams, R. J. Ordidge, and D. G. Gadian, *Magn. Reson. Med.* **41,** 479 (1999).

[38] J. Astrup, B. K. Siesjo, and L. Symon, *Stroke* **12,** 723 (1981).

[39] H. Kinouchi, F. R. Sharp, J. Koistinaho, K. Hicks, H. Kamii, and P. H. Chan, *Brain Res.* **619,** 334 (1993).

in an animal model of cerebral ischemia; this reduction of tissue water diffusion is attributed to an osmotically obliged shift of extracellular water to intracellular compartments, as a result of a disruption of ion homeostasis and formation of cytotoxic edema. The region of signal intensity change in DWI corresponds closely to the region of peri-infarct acidosis, but also encompasses the area of ATP depletion (infarct core). Therefore it was postulated that the outer margin of the DWI visible lesion corresponds with that of the penumbra.[40] However, this is only part of the picture, as it is now acknowledged that regions of DWI change that normalize on reperfusion may proceed on to infarction at a later time point[41] and that the areas of so-called diffusion–perfusion mismatch need to be considered as well.[42]

Spreading Depression. The occurrence of transient waves of membrane depolarization known as *spreading depression* (SD) emanating from an infarct core has been suggested as one mechanism for the expansion of tissue injury into the penumbral zone.[43] During SD the metabolic rate of the tissue increases in response to the greatly enhanced energy demands of the activated ion exchange pumps.[44] In the penumbra, flows are suppressed, and as a result, the increased metabolic demand is not compensated by an increase in oxygen and glucose.[45] Eventually adenosine triphosphate (ATP) stores will be depleted, followed by the cascade of pathophysiologic events leading to tissue infarction. Using DWI and ADC maps to monitor cell volume change,[46] and gradient–echo MRI to follow apparent changes in blood flow, both of these consequences can be imaged (Fig. 9).[47,48] The change in ADC associated with SD has been used to study the pathologic basis by which SD leads to infarct growth and the mechanism whereby neuroprotective drugs may have their therapeutic effect.[49,50]

[40] K. Kohno, M. Hoehn-Berlage, G. Mies, T. Back, and K. A. Hossmann, *Magn. Reson. Imaging* **13**, 73 (1995).

[41] M. van Lookeren Campagne, G. R. Thomas, H. Thibodeaux, J. T. Palmer, S. P. Williams, D. G. Lowe, and N. Van Bruggen, *J. Cereb. Blood Flow Metab.* **19**, 1354 (1999).

[42] A. E. Baird and S. Warach, *J. Cereb. Blood Flow Metab.* **18**, 583 (1998).

[43] M. Nedergaard and J. Astrup, *J. Cereb. Blood Flow Metab.* **6**, 607 (1986).

[44] M. Kocher, *J. Cereb. Blood Flow Metab.* **10**, 564 (1990).

[45] T. Back, K. Kohno, and K. A. Hossmann, *J. Cereb. Blood Flow Metab.* **14**, 12 (1994).

[46] L. L. Latour, Y. Hasegawa, J. E. Formato, M. Fisher, and C. H. Sotak, *Magn. Reson. Med.* **32**, 189 (1994).

[47] A. R. Gardner Medwin, N. Van Bruggen, S. R. Williams, and R. G. Ahier, *J. Cereb. Blood Flow Metab.* **14**, 7 (1994).

[48] M. L. Gyngell, T. Back, M. Hoehn-Berlage, K. Kohno, and K. A. Hossmann, *Magn. Reson. Med.* **31**, 337 (1994).

[49] J. Rother, A. J. de Crespigny, H. E. D'Arceuil, K. Iwai, and M. E. Moseley, *Stroke* **27**, 980 (1996).

[50] K. Takano, L. L. Latour, J. E. Formato, R. A. D. Carano, K. G. Helmer, Y. Hasegawa, C. H. Sotak, and M. Fisher, *Ann. Neurol.* **39**, 308 (1996).

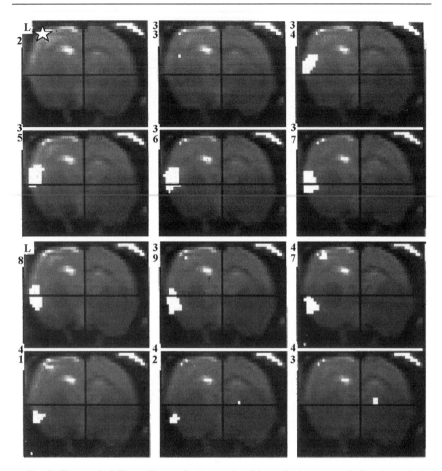

FIG. 9. Changes in MR gradient–echo image signal intensity in a coronal slice through the brain of an anesthetized rat. Spreading depression was initiated with 150 mM KCl applied to the left parietal cortex through a 4.5-mm craniotomy at the point indicated by the white star. Gradient–echo images were obtained every 13 seconds following the induction of spreading depression. White zones indicate where the signal intensity has increased more than 5.5% of the control image (No. 2) and have been superimposed upon conventional (spin–echo) images to reveal the relation to brain structure. Changes in signal intensity as observed propagating through the cortex. (Courtesy of van Bruggen.)

Brain Reorganization Following Stroke. Stroke patients show some functional recovery over time, which is commonly thought to be associated with brain plasticity. Recent BOLD MRI studies have demonstrated changes in activation patterns following a stroke.[51,52] With the use of behavioral tests and MRI in a rat stroke model, Dijkhuizen *et al.*[51] assessed the correlation between temporal changes in sensorimotor function, brain activation patterns, and cerebral ischemic damage. Unilateral stroke induced acute dysfunction of the contralateral forelimb, which significantly recovered at later stages. Forelimb impairment was accompanied by loss of functional MRI (fMRI)-activation in the lesioned cortex. At 3 days after stroke, extensive fMRI responses were detected in the contralesional hemisphere. After 14 days, they found reduced involvement of the contralesional hemisphere, and significant responses in the lesion periphery. A further study suggests that the degree of shift of activation balance toward the contralesional hemisphere early after stroke increases with the extent of tissue injury and that functional recovery is associated mainly with preservation or restoration of activation in the ipsilesional hemisphere.[53]

Other MRI Contrast Mechanisms for Imaging Stroke. During the last few years, animal studies have demonstrated that several MRI parameters provide additional information to the now clinically implemented T_2, diffusion, and perfusion measurements.

Because BOLD MRI is sensitive to changes in regional tissue oxygenation status,[3] it can be used to monitor acute deoxygenation following induction of ischemia, as well as reoxygenation after reperfusion.[54] BOLD MR signal intensity, measured by T_2^*-weighted MRI, drops immediately upon the onset of ischemia, and rises when reflow occurs. A transient overshoot in signal intensity during reperfusion has been described and may reflect postreperfusion hyperemia.[55] These hemodynamic responses indirectly report on local changes in CBF, CBV, and oxygen extraction fraction, and their individual contributions cannot easily be distinguished. More recently, early changes in T_2 values have been reported in conditions of ischemia[37,56] and oligemia.[57] In the latter studies of cerebral ischemia, two patterns were observed following an initial T_2 decrease: (1) T_2 values

[51] R. M. Dijkhuizen, J. Ren, J. B. Mandeville, O. Wu, F. M. Ozdag, M. A. Moskowitz, B. R. Rosen, and S. P. Finklestein, *Proc. Natl. Acad. Sci. USA* **98,** 12766 (2001).

[52] M. Hoehn, K. Nicolay, C. Franke, and B. van der Sanden, *J. Magn. Reson. Imaging* **14,** 491 (2001).

[53] R. M. Dijkhuizen, A. B. Singhal, J. B. Mandeville, O. Wu, E. F. Halpern, S. P. Finklestein, B. R. Rosen, and E. H. Lo, *J. Neurosci.* **23,** 510 (2003).

[54] S. A. Roussel, N. Van Bruggen, M. D. King, and D. G. Gadian, *J. Cereb. Blood Flow Metab.* **15,** 578 (1995).

[55] A. J. de Crespigny, M. F. Wendland, N. Derugin, E. Kozniewska, and M. E. Moseley, *Magn. Reson. Med.* **27,** 391 (1992).

[56] O. H. J. Gröhn, J. A. Lukkarinen, J. M. E. Oja, P. C. van Zijl, J. A. Ulatowski, R. J. Traystman, and R. A. Kauppinen, *J. Cereb. Blood Flow Metab.* **18,** 911 (1998).

[57] M. F. Lythgoe, D. L. Thomas, F. Calamante, G. S. Pell, A. L. Busza, M. D. King, C. H. Sotak, S. R. Williams, R. J. Ordidge, and D. G. Gadian, *Magn. Reson. Med.* **44,** 706 (2000).

remained depressed throughout the study without an ADC change, indicating a mild hypoperfusion condition[37,56]; (2) T_2 and ADC values were decreased throughout the study, indicating a severe hypoperfusion condition.[37,54] Furthermore, early detection (within 15 minutes) of ischemia[37] and oligemia[57] are possible using the MRI parameter T_1. The underlying mechanism for this change is still unclear, but may depend on tissue oxygenation status. These data highlight that changes in T_2 and T_1 are not always related to vasogenic edema, and early changes in these parameters may provide information as to the pathophysiologic nature of ischemia or oligemia. In deciding whether tissue is suitable for treatment, it is necessary to distinguish compromised yet recoverable tissue from permanently damaged tissue. To this end, Gröhn et al.[58] have demonstrated early detection of irreversible cerebral ischemia in the rat using a parameter known as $T_{1\rho}$ (T_1 in the rotating frame), which probes specific spin populations taking place in the macromolecules or in the macromolecular–water interface. Thus various combinations of MRI parameters may provide more reliable MRI tissue signatures to clarify type and timing of drug therapy.

Epilepsy

Although the majority of experimental epilepsy research has been performed outside the neuroimaging field, there has been increasing interest following some early nuclear MR spectroscopy investigations and, more recently, imaging studies using MRI. The present goal of imaging is to characterize both the metabolic derangement and brain injury associated with seizures, using functional (diffusion and perfusion imaging) and structural (T_1 and T_2) neuroimaging.

Short-Term Changes. Following kinate injection in the rat brain, tissue damage is detectable by MRI, in which the T_1-weighted images provide better lesion contrast than the T_2 approach, and DWI shows improved contrast for edematous tissue.[59] BOLD MRI shows increases in blood flow that were associated with minor behavioral seizure signs in a similar model, but as seizure activity progresses, the signal intensity remains near control values, possibly due to the increased oxygen extraction of the tissue.[60] Zhong et al.[61] used diffusion-weighted MRI to investigate changes associated with epilepsy. Following intraperitoneal injection of bicuculline in

[58] O. J. Gröhn, M. I. Kettunen, H. I. Makela, M. Penttonen, A. Pitkanen, J. A. Lukkarinen, and R. A. Kauppinen, *J. Cereb. Blood Flow Metab.* **20,** 1457 (2000).

[59] M. D. King, N. van-Bruggen, R. G. Ahier, J. E. Cremer, J. V. Hajnal, M. Doran, and S. R. Williams, *Magn. Reson. Med.* **20,** 158 (1991).

[60] S. Ogawa and T. Lee, *Proc. Soc. Magn. Reson. Med.* **1,** 501 (1992).

[61] J. Zhong, O. A. C. Petroff, J. W. Pritchard, and J. C. Gore, *Magn. Reson. Med.* **30,** 241 (1993).

the rat, the ADC in the brain decreases 14–18% during seizures. No changes occur in T_1 or T_2. This result demonstrates that during a seizure the ADC changes in a similar fashion to that reported in ischemia,[14] but under different circumstances as the blood flow is increased and the ATP stores are only modestly reduced. This ADC decrease during seizure activity leads to the provisional hypothesis that the ADC change may be due to perturbations in intracellular cytosolic streaming.[61] Although this study has been supported by more recent work,[62] there is still no complete explanation for the ADC change in stroke or epilepsy.[14]

Long-Term Changes. The long-term temporal evolution of lesion development, investigation of epileptic activity, and tissue damage are well suited to the noninvasive nature of MRI. Several studies have now monitored T_2, ADC, and blood–brain barrier (BBB) breakdown following epilepsy.[63,64] In parts of the cortex, and in the amygdala, the T_2 was increased by 24 hours, then progressively normalized by 5–9 days, and finally increased again in the chronic phase (9 weeks). The chronic T_2 increase corresponds to gliosis and characterizes the initial step leading to development of epilepsy, which could result from spontaneous seizures.[63]

Neurodegenerative Disorders

Neurodegenerative disorders such as Alzheimer's disease (AD) or Parkinson's disease (PD) are devastating progressive illnesses. A combination of MRI approaches such as pharmacologic MRI, DTI, and anatomic studies of brain atrophy is particularly useful to understand the etiology and progression of these diseases.[22,65]

Parkinson's Disease. The disease is characterized by striatal dopamine depletion, resulting in progressive loss of motor control and the characteristic symptoms of the disease. Experimental models of Parkinson's disease fall into two major categories: pharmacologic, for example, reserpine or amphetamine administration to deplete dopamine, a largely reversible treatment, and lesioning using neurotoxins, which is permanent. Imaging experimental models has largely been in the domain of positron-emission tomography (PET) imaging with the availability of radioligands to monitor

[62] T. Q. Duong, J. J. H. Ackerman, H. S. Ying, and J. J. Neil, *Magn. Reson. Med.* **40**, 1 (1998).
[63] C. Roch, C. Leroy, A. Nehlig, and I. J. Namer, *Epilepsia* **43**, 325 (2002).
[64] T. Tokumitsu, A. Mancuso, P. R. Weinstein, M. W. Weiner, S. Naruse, and A. A. Maudsley, *Brain Res.* **744**, 57 (1997).
[65] I. Y. Choi, S. P. Lee, D. N. Guilfoyle, and J. A. Helpern, *Neurochem. Res.* **28**, 987 (2003).

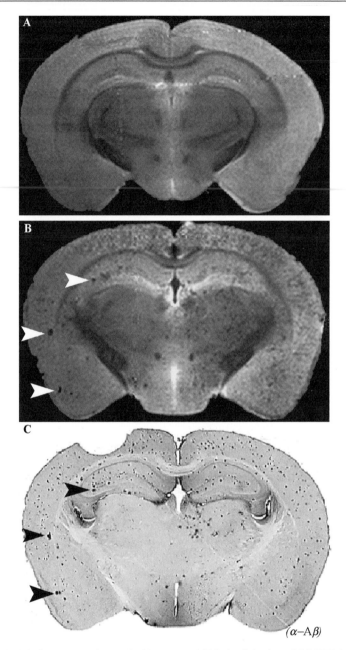

FIG. 10. Aβ plaques were detected with *ex vivo* μMRI after injection of Gd-DTPA–Aβ1-40 with mannitol. *Ex vivo* T$_2$-weighted SE coronal μMR images show (A) 6-month-old control and (B) APP/PS1-transgenic mouse brains. Both brains were extracted and prepared for

both presynaptic and postsynaptic function.[66] This methodology has now been largely supplanted by fMRI[67] or pharmacologic MRI (a term used to describe drug-induced fMRI changes) with activity induction by levodopa,[68] D1 and D2 receptor agonists,[69] amphetamine,[70] or dopamine transporter agonists.[71] Using more classical anatomic MRI sequences, temporal signal changes in T_1 and T_2 have been used to map regions of degeneration after lesioning in nonhuman primates,[72] and attempts have been made to correlate T_1 relaxation times with the abnormal accumulation of iron in degenerating dopaminergic neuron in a neurotoxin rat model.[73]

Alzheimer's Disease. Alzheimer's disease is the most common dementia in Western societies. Pathohistologic findings include widespread neuronal degeneration and neuritic plaques containing β-amyloid (Aβ). Although only limited studies have been performed in this area,[65] a recent study has demonstrated that using magnetically labeled Aβ1–40 peptide in an Alzheimer's transgenic mouse (Fig. 10), which has a high affinity to Aβ, enabled the detection of Aβ plaques.[74] This may suggest a diagnostic approach to detect Aβ in patients.

Huntington's Disease. This inherited disease is characterized by degeneration of GABAergic neurons localized mainly within the deep gray matter of the basal ganglia. Experimental models are based on acute lesioning using GABAergic antagonists or chronic lesioning using systemic administration of 3-nitroproprionic acid.[75,76] Since the discovery of the

[66] A. L. Brownell, E. Livni, W. Galpern, and O. Isacson, *Ann. Neurol.* **43**, 387 (1998).
[67] G. Pelled, H. Bergman, and G. Goelman, *Eur. J. Neurosci.* **15**, 389 (2002).
[68] B. G. Jenkins, Y. I. Chen, and J. B. Mandeville, "Biomedical Imaging in Experimental Neuroscience" (N. Van Bruggen and T. P. L. Roberts, eds.), pp. 155–209. CRC Press, Boca Raton, FL, 2002.
[69] Z. Zhang, A. Andersen, R. Grondin, T. Barber, R. Avison, G. Gerhardt, and D. Gash, *Neuroimage* **14**, 1159 (2001).
[70] Y. I. Chen, A. L. Brownell, W. Galpern, O. Isacson, M. Bogdanov, M. F. Beal, E. Livni, B. R. Rosen, and B. G. Jenkins, *Neuroreport* **10**, 2881 (1999).
[71] Y. C. Chen, W. R. Galpern, A. L. Brownell, R. T. Matthews, M. Bogdanov, O. Isacson, J. R. Keltner, M. F. Beal, B. R. Rosen, and B. G. Jenkins, *Magn. Reson. Med.* **38**, 389 (1997).
[72] R. S. Miletich, K. S. Bankiewicz, M. Quarantelli, R. J. Plunkett, J. Frank, I. J. Kopin, and G. Di Chiro, *Ann. Neurol.* **35**, 689 (1994).
[73] S. Hall, J. N. Rutledge, and T. Schallert, *J. Neurol. Sci.* **113**, 198 (1992).
[74] Y. Z. Wadghiri, E. M. Sigurdsson, M. Sadowski, J. I. Elliot, Y. Li, H. Scholtzova, C. Tang, G. Aguinaldo, M. Pappolla, K. Duff, T. Wisniewski, and D. H. Turnbull, *Magn. Reson. Med.* **50**, 293 (2003).

imaging 6 h after carotid injection of Gd-DTPA–Aβ1-40 with 15% mannitol. Note the obvious matching of many larger plaques (arrowheads) between μMRI (B) and immunohistochemistry (C) (Courtesy Wadghiri and co-workers.)

gene mutation for the disease, new transgenic mouse models are being developed.[77] Imaging data are largely PET-based, mapping regions of reduced metabolism indicative of cell loss.[78] Striatal lesions are visible with T_2-weighted and DWI protocols,[79] although at acute time points only diffusion-weighted imaging provides sufficient sensitivity.[80]

CNS Inflammation

The inflammatory response is part of the hosts' defense to injury and infection, but can exacerbate tissue injury when excessive or inappropriate. It has become clear that inflammation contributes not only to the archetypal central nervous system (CNS) inflammatory disease, multiple sclerosis, but also to a wide variety of acute neurologic and chronic neurodegenerative diseases, such as stroke, head trauma, Alzheimer's disease, prion disease, and human immunodeficiency virus (HIV)-related dementia. Despite this, little is known of the effects of inflammatory processes within the CNS, or their contribution to MR images of human neuropathologies.

Multiple Sclerosis. Despite a wealth of studies, the pathogenesis of multiple sclerosis (MS) is still not fully understood. A primary goal of animal studies is to determine the relationship between the histopathology of a disease and the MRI signal changes. Experimental allergic encephalopathy (EAE) is an autoimmune CNS disorder that can be induced in susceptible species, such as mice, rats, guinea pigs, and nonhuman primates.[81,82] However, few of these models adequately represent all of the features of human MS, and the lesions evolve spontaneously at any site within the brain and often exhibit varying temporal progressions. An alternative model is the delayed-type hypersensitivity (DTH) model in the rat,[83] which involves sensitization of the immune system to a non-CNS antigen previously deposited in the brain. This model exhibits all the primary

---

[75] C. V. Borlongan, T. K. Koutouzis, and P. R. Sanberg, *Neurosci. Biobehav. Rev.* **21**, 289 (1997).

[76] P. Hantraye, *Nucl. Med. Biol.* **25**, 721 (1998).

[77] L. B. Menalled and M. F. Chesselet, *Trends Pharmacol. Sci.* **23**, 32 (2002).

[78] D. M. Araujo, S. R. Cherry, K. J. Tatsukawa, T. Toyokuni, and H. I. Kornblum, *Exp. Neurol.* **166**, 287 (2000).

[79] P. Hantraye, A. Leroy-Willig, A. Denys, D. Riche, O. Isacson, M. Maziere, and A. Syrota, *Exp. Neurol.* **118**, 18 (1992).

[80] H. B. Verheul, R. Balazs, J. W. Berkelbach van der Sprenkel, C. A. Tulleken, K. Nicolay, and C. M. Van Lookeren, *Brain Res.* **618**, 203 (1993).

[81] R. Gold, H. P. Hartung, and K. V. Toyka, *Mol. Med. Today* **6**, 88 (2000).

[82] H. Brok, J. Bauer, M. Jonker, E. Blezer, S. Amor, R. E. Bontrop, J. D. Laman, and B. A. Hart, *Immunol. Rev.* **183**, 173 (2001).

[83] M. K. Matyszak and V. H. Perry, *Neuroscience* **64**, 967 (1995).

features of MS lesions: T-cell and macrophage infiltration, BBB breakdown, edema and tissue damage, and primary demyelination. A major advantage of this model for longitudinal MRI studies is that the site of the lesion is precisely dictated by the location of the intracerebral antigen injection.

MRI findings in the EAE models include increased T_1, increased T_2, and contrast enhancement,[84,85] thus corresponding broadly to those found most commonly in MS patients. However, only a relatively small number of studies have investigated correlations between MRI and histopathology obtained at the same time point. An increase in T_2 has been found to correspond to regions of macrophage recruitment and edema in both guinea pig[86] and rat EAE models.[87] However, T_2 changes were also associated with demyelination in the rat,[87] but not in the guinea pig.[86] In the *Callithrix jacchus* marmoset model of EAE, increased proton density and T_2 appear to be associated with regions of either perivascular cuffing, demyelination, or perivascular gliosis.[88] The variability of these findings suggests that T_2-weighted MRI alone is not a reliable method of distinguishing purely inflammatory lesions from either demyelinating or remyelinating lesions.[88,89] In MS patients, areas that are enhanced after an injection of an MRI contrast agent are generally considered to be a sensitive indicator for disease activity. In both guinea pig[90] and rat[91] models of EAE, BBB breakdown has been found to correlate with macrophage recruitment. In contrast, in the *C. jacchus* marmoset EAE model, which arguably provides the most accurate representation of the relapsing–remitting form of MS, contrast-enhancing areas correlated solely with acute, actively demyelinating lesions.[89] However, recent work in the rat DTH model[92] has demonstrated

[84] T. L. Richards, E.-C. J. Alvord, J. Peterson, S. Cosgrove, R. Petersen, K. Petersen, A. C. Heide, J. Cluff, and L. M. Rose, *NMR Biomed.* **8,** 49 (1995).

[85] M. R. Verhoye, E. J. Gravenmade, E. R. Raman, J. Van-Reempts, and A. Van der Linden, *Magn. Reson. Imaging* **14,** 521 (1996).

[86] R. I. Grossman, R. P. Lisak, P. J. Macchi, and P. M. Joseph, *Am. J. Neuroradiol.* **8,** 1045 (1987).

[87] H. J. Duckers, H. J. Muller, J. Verhaagen, K. Nicolay, and W. H. Gispen, *Neuroscience* **77,** 1163 (1997).

[88] E. K. Jordan, H. I. McFarland, B. K. Lewis, N. Tresser, M. A. Gates, M. Johnson, M. Lenardo, L. A. Matis, H. F. McFarland, and J. A. Frank, *Am. J. Neuroradiol.* **20,** 965 (1999).

[89] B. A. Hart, J. Bauer, H. J. Muller, B. Melchers, K. Nicolay, H. Brok, R. E. Bontrop, H. Lassmann, and L. Massacesi, *Am. J. Pathol.* **153,** 649 (1998).

[90] C. P. Hawkins, P. M. Munro, F. MacKenzie, J. Kesselring, P. S. Tofts, E. P. du-Boulay, D. N. Landon, and W. I. McDonald, *Brain* **113,** 365 (1990).

[91] S. P. Morrissey, H. Stodal, U. Zettl, C. Simonis, S. Jung, R. Kiefer, H. Lassmann, H. P. Hartung, A. Haase, and K. V. Toyka, *Brain* **119,** 239 (1996).

[92] T. A. Newman, S. T. Woolley, P. M. Hughes, N. R. Sibson, D. C. Anthony, and V. H. Perry, *Brain* **124,** 2203 (2001).

that axonal injury and inflammatory events occurring within a lesion are not restricted to the period of BBB breakdown and contrast enhancement. These findings suggest that MS disease progression may persist despite an intact BBB, and, consequently, contrast enhancement may not be an accurate marker of disease activity. Recently it has been suggested that a more appropriate method may be to monitor monocyte infiltration ("the major source of demyelination in EAE") via cells labeled with iron oxide particles that may be detected by MRI, thereby allowing assessment of the inflammatory activity induced by EAE.[93]

Other imaging modalities have been used less frequently, but have yielded some interesting findings. It has been suggested that DTI may distinguish between acute and chronic EAE lesions,[94] such that in acute lesions diffusion increases in all directions, probably as a consequence of edema, whereas in chronic lesions diffusion only increases perpendicular to the main axon axis, possibly reflecting demyelination. Recent measurements of MTR in EAE[95] have suggested that decreases in MTR, which are frequently assumed to reflect demyelination in human MS, may in fact result from inflammatory-related changes to white matter structure rather than myelin loss per se. In contrast, the short component of tissue water T_2 may more accurately reflect myelin content.[95] New contrast agents have recently provided an alternative approach to imaging EAE. A recent study has shown that the use of a superparamagnetic iron oxide contrast agent enables macrophage recruitment to the CNS to be followed *in vivo*.[96]

Conclusion

The combination of appropriate imaging techniques and suitable animal models of disease can greatly elucidate our understanding of human brain pathologies. This experimental imaging partnership has contributed to the development of novel imaging techniques, to the promotion of better diagnostic and prognostic measures, and to elucidation of the basic mechanisms of cellular injury, leading to improved therapies.

[93] M. Rudin, M. Rausch, P. Hiestand, D. Baumann, and C. Cannet, *Magn. Reson. Med.* **50,** 309 (2003).

[94] T. L. Richards, E.-C. J. Alvord, Y. He, K. Petersen, J. Peterson, S. Cosgrove, A. C. Heide, K. Marro, and L. M. Rose, *Mult. Scler.* **1,** 109 (1995).

[95] P. J. Gareau, B. K. Rutt, S. J. Karlik, and J. R. Mitchell, *J. Magn. Reson. Imaging* **11,** 586 (2000).

[96] V. Dousset, L. Ballarino, C. Delalande, M. Coussemacq, P. Canioni, K. G. Petry, and J. M. Caille, *Am. J. Neuroradiol.* **20,** 223 (1999).

Acknowledgments

The authors would like to thank Dr N. R. Sibson and Dr N. G. Harris for their contribution to the manuscript. We also acknowledge the Wellcome Trust and the BBSRC for their support of the work carried out at Radiology and Physics Unit of the Institute of Child Health, Great Ormond Street Hospital, and the Wellcome Trust High Field MR Research Laboratory.

[7] Magnetic Resonance Imaging in Animal Models of Pathologies

By PASQUINA MARZOLA and ANDREA SBARBATI

Introduction

Animal models of human pathologies are useful at a fundamental level to understand the dynamics of a disease and its underlying mechanisms. They also play a fundamental role in pharmaceutical research. Animal models of diseases are used in several phases of pharmaceutical research, from the drug target identification to the efficacy tests of the experimental drug. Magnetic resonance imaging (MRI) is a well recognized tool for *in vivo* characterization of animal models of human pathologies in preclinical research. Because of its noninvasiveness, high soft contrast, and high space resolution, which can reach 50–100 μm in small laboratory animals, MRI constitutes a powerful morphologic technique. MR signal is sensitive to a number of parameters, including proton density, water proton relaxation times (T_1, T_2, and T_2^*), microscopic water self-diffusion, and macroscopic flow (i.e., blood flow in arteries). Thanks to its multiparameter dependence, MRI is not a pure morphologic technique, but it also provides physiologic, functional, and metabolic information. Examples range from neuronal activation in brain (functional MRI, fMRI) induced by stimuli or drugs,[1,2] assessment of organ viability after transplantation,[3] assessment of blood volume and flow in brain,[4] and characterization of neovasculature in tumors.[5,6]

[1] J. J. Marota, C. Ayata, M. A. Moskowitz, R. M. Weisskoff, B. R. Rosen, and J. B. Mandeville, *Magn. Reson. Med.* **41,** 247 (1999).

[2] J. J. Marota, J. B. Mandeville, R. M. Weisskoff, M. A. Moskowitz, B. R. Rosen, and B. E. Kosofsky, *Neuroimage* **11,** 13 (2000).

[3] N. Beckmann, R. P. Hof, and M. Rudin, *NMR Biomed.* **13,** 329 (2000).

[4] F. Calamante, D. L. Thomas, G. S. Pell, J. Wiersma, and R. Turner, *J. Cereb. Blood Flow Metab.* **19,** 701 (1999).

Copyright 2004, Elsevier Inc.
All rights reserved.
0076-6879/04 $35.00

The availability of contrast agents with different physical, chemical, and biologic properties further enhances the usefulness of the technique. Standard low molecular weight ($\approx$500 D) contrast agents that are widely used in clinics are characterized by a rapid equilibrium between vascular and extracellular space. At the equilibrium distribution such agents are markers of tissue integrity (e.g., blood–brain barrier defects), whereas their dynamic penetration in tissues can be used as a marker of tissue microcirculation.[7] However, a number of applications require the use of contrast agents with different properties. For example, high molecular weight contrast agents have been used in preclinical research for characterization of tumor blood volume and microvessel permeability.[8] Iron-based nanoparticles that are characterized by a long blood half-time have been used as markers of cerebral blood volume alteration induced by somatosensory stimulation or drug administration.

Several pathologies are now recognized to have a genetic origin, and consequently, genetically modified animals play an important role in this field.[9] Compared with the traditional invasive techniques, MRI allows repeated observations of the same subject during the time evolution of the pathology or therapy follow-up. The possibility of repeated observations on the same subject drastically decreases the number of experiments needed to reach a statistical significance, with a consequent reduction in the number of animals and in the quantity of drug used, providing economic and ethical benefits.

In this chapter, we describe representative experimental approaches used in our laboratory in studying and characterizing animal models of representative pathologies, namely, models of experimental infectious, ischemic, neoplastic, and metabolic–degenerative diseases.

Bacterial Infections

In antibacterial research, animal models of bacterial infection are used to test the efficacy of novel compounds. Systemic or localized microbial infections are traditionally monitored by measuring the bacterial load

---

[5] A. Gossmann, T. H. Helbich, N. Kuriyama, M. D. Ostrowitzki, T. P. L. Roberts, D. M. Shames, N. Van Bruggen, M. F. Wendland, M. A. Israel, and R. C. Brasch, *J. Magn. Reson. Imaging* **15,** 233 (2002).

[6] Z. M. Bhujwalla, D. Artemov, and J. Glockner, *Top. Magn. Reson. Imaging* **10,** 92 (1999).

[7] W. T. C. Yuh, *J. Magn. Res. Imaging* **10,** 221 (1999).

[8] R. C. Brasch, K. C. Li, J. E. Husband, M. T. Keogan, M. Neeman, A. R. Padhani, D. Shames, and T. Turetschek, *Acad. Radiol.* **7,** 812 (2000).

[9] N. Beckmann, T. Mueggler, P. R. Allegrini, D. Laurent, and M. Rudin, *Anat. Rec.* **265,** 85 (2001).

(i.e., the number of bacterial colonies per organ, and/or the survival of the infected animals). The assessment of the chronologic evolution of infection, and drug efficacy, consequently requires the sacrifice of a large number of animals and suffers from the intrinsic interindividual variability of data. With the aim to establish if and at what extent *in vivo* MRI can substitute or complement traditional assays, we have investigated, using MRI, two experimental models of bacterial infections, namely, thigh infection[10] and pneumonia.[11] To this purpose, thigh infection was induced in three groups of animals: the first group was used for histologic examination, the second group was used for bacterial viable count examination, and the third group was used for MRI examination. Experiments on lung infection were conducted on a single group of animals that were sacrificed, after MRI, for histology. Although the presence of infection in the thigh muscle was detectable by standard T_2-weighted (T_2W) spin–echo (SE) images, detection of pneumonia was more critical. In general, lung imaging is particularly challenging for standard MRI, being the lung parenchyma characterized by a very low ^1H MRI signal intensity. Low signal is due to the low proton density of the lung tissue (roughly 20% less than that of other tissues[12]) and to the peculiar morphologic architecture constitued by microscopic arrangement of air–tissue interfaces that produces relevant susceptibility artifact and strongly decreases the T_2^* relaxation time. New experimental approaches based on the use of hyperpolarized Xe and He gases have been proposed[13,14] for lung imaging; although extremely promising, these techniques are not widely available at present and are not usable with standard instrumentation. Several ^1H MRI techniques have been also proposed and are aimed at increasing the signal to noise ratio (SNR) of the lung parenchyma, including projection reconstruction techniques,[15] SE or gradient–echo (GRE) sequences with short echo time,[16] or inversion recovery

[10] P. Marzola, E. Nicolato, E. Di Modugno, P. Cristofori, A. Lanzoni, C. H. Ladel, and A. Sbarbati, *Magn. Reson. Mat. Physics Biol. Med. (MAG*MA)* **9,** 21 (1999).

[11] P. Marzola, E. Nicolato, A. Sbarbati, L. Piccoli, E. Di Modugno, P. Cristofori, and A. Lanzoni, "^1H MRI Investigation of Pneumococcal Pneumonia in a Murine Model," abstract presented at the 17th Annual Meeting of European Society for Magnetic Resonance in Medicine and Biology, Paris, September 14–17, 2000.

[12] G. A. Johnson and L. W. Hedlund, *Nat. Med.* **2,** 1192 (1996).

[13] M. S. Albert, G. D. Cates, B. Driehuys, W. Happer, B. Saam, C. S. Springer, Jr., and A. Wishnia, *Nature* **370,** 199 (1994).

[14] H. Middleton, R. D. Black, B. Saam, G. D. Cates, G. P. Cofer, R. Guenther, W. Happer, L. W. Hedlund, G. A. Johnson, K. Juvan *et al.*, *Magn. Reson. Med.* **33,** 271 (1995).

[15] S. L. Gewalt, G. H. Glover, L. W. Hedlund, G. P. Cofer, J. R. MacFall, and G. A. Johnson, *Magn. Reson. Med.* **29,** 99 (1993).

[16] H. Hatabu, J. Gaa, D. Kim, W. Li, P. V. Prasad, and R. R. Edelman, *Magn. Reson. Med.* **36,** 503 (1996).

preparation to the acquisition sequence with suppression of the muscle and/ or the fat signal.[17] The previously mentioned techniques are based on the idea that the signal from the lung parenchyma has to be increased. In the detection of pneumonia, we have exploited the high contrast between normal lung tissue and inflamed regions. In inflamed regions of the lung, where the functionality is impaired, the air content of pulmonary tissue is reduced, with a consequent increase in T_2^* and in the SNR compared with the normal parenchyma.

Bacterial Thigh Infection

Thigh infection was induced in male Cr1:CD-1(ICR)BR mice by intramuscular (IM) injection of 0.125 ml of an overnight Mueller-Hinton broth culture of *Staphylococcus aureus* ($\approx 3 \times 10^7$ bacteria per mouse) in the left hind leg. To investigate the sensitivity of MRI to different therapeutic treatments, three groups of animals were observed: control group ($n = 5$), a group treated with imipenem–cilastatin ($n = 5$), and a group treated with vancomycin ($n = 5$). Although imipenem–cilastatin has marked bactericidal effect against *S. aureus*, vancomycin has only a bacteriostatic effect.[10] Animals were anesthetized by intraperitoneal (IP) injection of chloral hydrate (400 mg/kg^{-1}) and placed in a homemade 5-cm internal diameter (i.d.) coil. SE T_1 and T_2-weighted (T_1W, T_2W) images were acquired at 4.7 T with the following parameters: repetition time (TR) and echo time (TE) amounted to 800 ms and 20 ms, respectively, for T_1W images and to 2000 ms and 70 ms, respectively for T_2W images. Other parameters were: slice thickness = 2 mm, field-of-view (FOV) = 8×4 cm^2, matrix size = 256×128 (corresponding to a in-plane space resolution of 312×312 μm^2). Images were acquired in both coronal and transversal orientation; 11 and 7 contiguous slices were acquired in the two orientations, respectively. Two additional groups of animals were used for microbiologic investigations ($n = 35$) and for pathology ($n = 14$). Each group was further subdivided into three groups according to the therapeutic dosing regimen. Details about experimental protocols used for bacterial counts (BC) assessment and pathology examination can be found in Marzola *et al.*[10] Briefly, for BC, at the designed time points, mice were sacrificed by cervical dislocation, and the infected leg was excised and homogenized in physiologic saline. The homogenized solution was serially diluted, and the quantitative BC were performed by plating 10 μl of each dilution on agar-containing plates, which were incubated at 37° for 24 h. The resulting count was expressed as CFU ml^{-1} (CFU = colony-forming units per

[17] V. M. Mai, J. Knight-Scott, and S. S. Berr, *Magn. Reson. Med.* **41,** 866 (1999).

milliliter). For pathology, animals were sacrificed by blood withdrawal under deep carbon dioxide (CO_2) anesthesia. After skin removal, the whole body of the animals was retained in 10% buffered formalin for fixation (1-week fixation period). Afterward, the left leg sample was removed from the coxofemoral joint, decalcified in Christenson's fluid, and cut in three transversal slices; these were then processed, included in paraffin blocks and cut to obtain 5-μm-thick sections, and stained with hematoxylin eosin. Two blind investigators evaluated and graded the following parameters: fascial inflammation, inflammation of the muscle tissue (myositis), and abscess of flemmon.

In Fig. 1 (A,B), we show images obtained 48 h after infection for one animal belonging to the control group (no antibiotic treatment). No abnormality in the infection-bearing leg is apparent in T_1W images, apart from an increased dimension of this leg. It is interesting to note that one traditional assessment of the entity of bacterial infection was a measurement of the leg diameter performed by calliper.[18] In T_2W images the infected leg appears strongly hyperintense compared with the normal leg; a subcutaneous edematous region is easily visible (arrow). The efficacy of the drug is clearly reflected in MR images: Fig. 1 (C and D) shows T_1W and T_2W images acquired for one animal belonging to the group treated with imipenem–cilastatin; no hyperintensity in the infected leg muscle is detectable compared with the normal leg. In the group treated with vancomycin, the hyperintensity of the muscle in the infected leg was comparable to that observed in the control group (images not shown).

The ratio between the signal intensity (SI) of the muscle in the infected and in the normal leg obtained from T_2W images was used to quantify the lesion:

$$Ri/n = (SI_{infected\ leg})/(SI_{normal\ leg}) \tag{1}$$

The signal intensities were obtained by the region-of-interest (ROI) analysis. Five transversal contiguous slices were selected across the lesion with the criterion to cover the whole thigh. An operator-defined ROI was manually traced on the muscle of both the infected and the normal leg, avoiding the bone, as well as the subcutaneous edema, which appears strongly hyperintense. Figure 2A reports the time dependence of the ratio Ri/n for the three groups. $Ri/n = 1$ indicates absence of abnormality, whereas higher values of Ri/n indicate more marked abnormality. Figure 2B reports the time dependence of BC. In the untreated group of animals, Ri/n increases with time, indicating a progressive worsening

[18] P. Acred, "The Selbie or Thigh Lesion Test, Experimental Models in Antimicrobial Chemotherapy" Vol. I, p. 109. Academic Press, London, 1986.

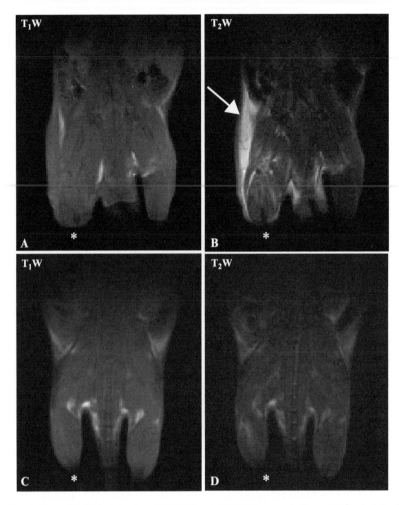

FIG. 1. Bacterial thigh infection. (A) T_1W and (B) T_2W coronal slices obtained 48 h after infection from an animal belonging to the control group. The dimension of the lesioned leg (*) is clearly increased, and strong alteration in the signal intensity of T_2W image is detected. (C) T_1W and (D) T_2W coronal slices obtained 48 h after infection from an animal belonging to the group treated with imipenem–cilastatin. No hyperintensity in the infected leg muscle (*) is detectable. Adopted from Marzola et al.,[10] with permission.

of the infection; a qualitative similar trend has been obtained for BC. In the group treated with vancomycin, an antibiotic characterized by a bacteriostatic effect against *S. aureus,* the increase in Ri/n is less pronounced than in the control group and the value of Ri/n remains substantially constant

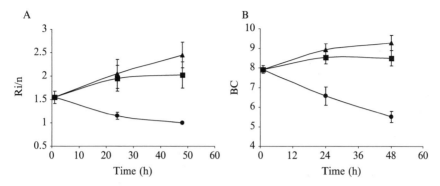

FIG. 2. Bacterial thigh infection. (A) Dependence of Ri/n and (B) BC on time elapsed after infection induction. (triangles) Control group; (squares) group treated with vancomicin; (circles) group treated with imipenem–cilastatin. Adapted from Marzola *et al.*,[10] with permission.

between 24 and 48 h; this trend is similar to that observed for BC. In the group treated with imipenem–cilastatin, an antibiotic characterized by a marked bactericidal effect, Ri/n continuously decreases, reaching a value close to 1, 48 h after infection induction. A similar decreasing trend was observed for BC in this treatment group. A good correlation was also found with histologic data.

Pneumonia

Infection was induced in a group of mice ($n = 5$) weighing 18–22 g by intranasal administration of about 1×10^6 CFU/mouse of *Streptococcus pneumoniae*. A group of noninfected animals ($n = 5$) was used as the control group. Animals were anesthetized by IP injection of pentobarbital (at a 60 mg/kg dose) and placed in the prone position in a 35-mm i.d.: birdcage coil. Three copper electrocardiogram (ECG) electrodes were subcutaneously implanted in the chest of the animals for ECG monitoring. The ECG signal was filtered and transformed in a square wave that was used as triggering for the acquisition sequence. Axial, ECG-gated, spoiled GRE images were acquired at each heartbeat in the systolic phase with the following parameters: TE = 5 ms, slice thickness = 1.2 mm, FOV = 6×6 cm^2, and matrix size = 256×256 (corresponding to in-plane space resolution of 234×234 μm^2). To obtain a reasonable SNR, four averages were used, resulting in an acquisition time of about 200 s. Images were acquired 48 h after infection induction. Immediately after imaging, animals were sacrificed by anesthetic overdose and the lungs were excised, inflated with 10% buffered formalin, and prepared for histology. For each animal, transversal sections (5 μm thick), 1.2 mm apart, were obtained and stained with hematoxylin and eosin. In each slice the area of the lesion was obtained by standard image analysis software. Quantification of lesion volume from MR images was obtained by manually delineating an ROI to cover the whole lung surface in each acquired slice. A second ROI was taken on the background noise. The number of pixels belonging to the lung surface and having signal intensities greater than five times the standard deviation of the noise were counted for each slice containing lung tissue. This analysis was performed in pathologic and normal animals; the average value obtained for normal animals was subtracted as a baseline.

Bacterial lesions in lung were evident in MR images as relatively high-intensity pixels compared with the normal lung parenchyma, whose signal was at the noise level. Figure 3 shows two images acquired for mice belonging to the infected (Fig. 3A) and control (Fig. 3B) group at approximately the same anatomic position. In the control mouse, lungs are characterized by SI at the level of the background noise, whereas vessels and heart ventricles show very high SI; heart and skeletal muscle have similar SIs. In Fig. 3A, relevant alterations of the SI in localized areas of the lung parenchyma can be observed compared with the normal lung parenchyma (see arrows). In this model of infection, lesions are generally localized in the perivascular and pericardiac regions. In Fig. 4 we compare MRI with histologic slices obtained at a similar anatomic position in the same animal. At histology, foci of pneumonia, characterized by perivascular–peribronchiolar and alveolar inflammatory cell infiltrates, and pleuritis are observed. Figure 4 indicates that the regions that appear lesioned at

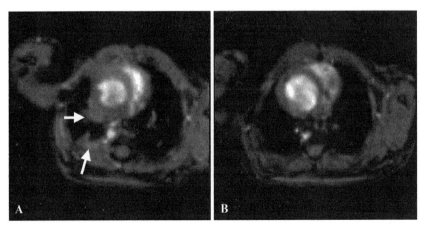

Fig. 3. Pneumonia. Comparison between transversal images obtained at similar anatomic position for mice belonging to the infected (A) and control (B) group. Arrows indicate regions of infection.

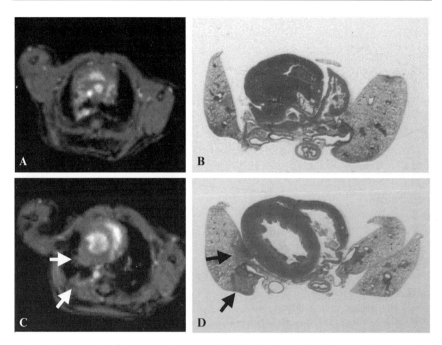

FIG. 4. Pneumonia. Comparison between (A, C) MRI and (B, D) histologic slices obtained at a similar anatomic position in the same animal. Arrows indicate regions of infection. (See color insert.)

histology (arrows), specifically, the apical and pericardial regions of the lung, roughly correspond to regions of high SI in the MRI slices. However, a quantitative comparison between MRI and histology slices cannot be obtained, since the well-known alteration in tissue shape and morphology is caused by histologic processing of animal tissues. A quantitative comparison between histology and MRI could be obtained by comparing the volume of the lesion with the two techniques. The results showed that there is a good correlation ($r^2 = 0.8478$) between MRI and histologic determined volumes.

Ischemic Pathologies

Although useful in different anatomic districts, MRI has found major applications in brain imaging. The development of diffusion-weighted MRI (DW-MRI) has revolutionized research in cerebral ischemia because it is sensitive to the hyperacute phase.[19] After the pioneer work by Mosely

[19] M. Hoehn, N. Nicolay, C. Franke, and B. Van der Sanden, *J. Magn. Res. Imaging* **14,** 491 (2001).

et al.,[20] several papers have shown that a decrease in water apparent diffusion coefficient (ADC) occurs within minutes from the occlusion of the middle cerebral artery (see Ref. 19 for a review), whereas abnormalities in T_2 occur only after hours. The increase in ADC in the early phase of ischemic insult is mainly due to cytotoxic cell swelling that causes a massive transfer of water from extracellular to intracellular space. The assessment of regional impairment in cerebral blood perfusion also plays an important role in ischemic lesion characterization. Cerebral blood perfusion can be assessed by MRI by acquiring fast images (weighted by the T_2^* relaxation time) during the first passage of a bolus of contrast agent, typically gadolinium (Gd)-DTPA[21] or iron oxide particles.[22] Recently, iron oxide particles characterized by long blood half-time (2–3 h in rats) also have been made available for experimental studies; these contrast agents reach a steady-state concentration in blood and produce a decrease in SI proportional to the local blood volume.[23] This experimental approach has been recently used by our group to characterize the blood volume alterations occurring in an experimental model of epilepsy.[24] Other methods are based on the arterial spin labeling technique, which consists in acquiring images made sensitive to the inflowing spins.[19]

We have implemented experimental protocols aimed at studying cerebral ischemia, its time evolution, and also the effect of an experimental drug [25] by using DW-MRI on the experimental model of permanent middle cerebral artery occlusion (MCAO). An experimental protocol also has been developed for assessing regional cerebral blood volume (rCBV) deficits in the same model.

Permanent MCAO

Permanent MCAO was performed in male Sprague-Dawley rats according to the method published by Tamura *et al.*[26] with minor modifications.[25,27] A group of $n = 24$ rats weighing 300–400 g was used for a

[20] M. E. Moseley, Y. Cohen, J. Mintorovitch, L. Chileuitt, H. Shimizu, J. Kucharczyk, M. F. Wendland, and P. R. Weinstein, *Magn. Reson. Med.* **14,** 330 (1990).

[21] B. R. Rosen, J. W. Belliveau, J. M. Vevea, and T. J. Brady, *Magn. Reson. Med.* **14,** 249 (1990).

[22] P. Loubeyre, T. De Jaegere, Y. Miao, W. Landuyt, and G. Marchal, *Magn. Reson. Imaging* **17,** 627 (1999).

[23] L. M. Hamberg, P. Boccalini, G. Stranjalis, G. J. Hunter, Z. Huang, E. Halpern, R. M. Weisskoff, M. A. Moskowitz, and B. R. Rosen, *Magn. Reson. Med.* **35,** 168 (1996).

[24] P. F. Fabene, P. Marzola, A. Sbarbati, and M. Bentivoglio, *Neuroimage* **18,** 375 (2003).

[25] A. Reggiani, C. Pietra, R. Arban, P. Marzola, U. Guerrini, L. Ziviani, A. Boicelli, A. Sbarbati, and F. Osculati, *Eur. J. Pharmacol.* **419,** 147 (2001).

[26] A. Tamura, D. H. Graham, J. McCulloch, and G. M. Teasdale, *J. Cereb. Blood Flow Metab.* **1,** 53 (1981).

DW-MRI study, whereas a group of $n = 18$ rats ($n = 9$ ischemic, $n = 6$ normal, and $n = 3$ sham operated) weighing about 200 g was used in the vascular deficit study. After completion of MRI experiments, animals were sacrificed for histology and electron microscopy. The first study was based on T_2W and DW images, acquired 6, 24, and 144 h after MCAO. Animals were anesthetized with fentanyldroperidol at 2.5 ± 0.5 mg/kg IP (Leptofen, Farmitalia, Carlo Erba, Italy) and placed in a homemade 5-cm i.d. radiofrequency (RF) coil. Transversal images were acquired at 4.7 T with the following parameters: TR = 1000 ms; TE = 70 ms; b-factor = 500 s/mm^2, diffusion-sensitive gradient pulses applied along the z direction (slice direction). Eleven transversal slices with 0.5-mm slice thickness and 1-mm slice separation were acquired. T_2W images were acquired with the same parameters as DW images, but without diffusion-sensitive gradient pulses. The drug, GV150526, a selective glycine receptor antagonist of the N-methyl-D-aspartate (NMDA) receptor[25,28] was administered through the tail vein at a dose of 3 mg/kg 5 min before MCAO.

The vascular deficit study was based on acquisition of T_2W images before and after administration of a iron-based contrast agent, AMI-25 (Endorem, Guerbet, France). Animals were anesthetized by inhalation of a mixture of air and oxygen (O_2)-containing 1–2% of halothane and placed in the supine position in a 72-mm i.d. transmitter–receiver birdcage coil. Images were acquired at 4.7 T 24 h after ischemia induction and at longer time points (15 days–3 months). A T_2W Rapid Acquisition with Relaxation Enhancement (RARE) sequence was used to delineate the ischemic lesion; afterward, T_2W images were acquired using SE sequences with the following parameters: TR = 1200 ms, TE = 60 ms, FOV = 8×8 cm^2, in-plane space resolution 625×625 μm^2. AMI-25 (Endorem) is a commercially available suspension of superparamagnetic iron particles that is characterized by a blood half-time of 15 min in rats. Considering that the acquisition time of the SE sequence was about 2 min and that only the central lines of the k-space contribute significantly to the signal, we can hypothesize that AMI-25 concentration in blood is steady-state during the acquisition of the relevant k-space lines. *In vivo* relative blood volume maps were obtained according to the relationship[23]:

$$rCBV \propto K \times \log(SI_{pre}/SI_{post}) \tag{2}$$

where SI_{pre} and SI_{post} are the signal intensities before and after contrast medium administration, respectively; the constant K is a function of TE

[27] A. Sbarbati, A. Reggiani, E. Lunati, R. Arban, E. Nicolato, P. Marzola, R. M. Asperio, P. Bernardi, and F. Osculati, *NeuroImage* **12,** 418 (2000).
[28] E. Ratti, M. Corsi, C. Carignani, A. Cugola, M. Mugnaini, G. Gaviraghi, D. Trist, and A. Reggiani, *Eur. J. Neurol.* **2,** 57 (1995).

and other experimental conditions, but is constant over the different pixels of the image. rCBV maps can be calculated from the previously written equation without considering the factor K. The rCBV values are not absolute and allow comparison only between different regions of the same image. The use of iron-based contrast agents characterized by longer blood half-time as, for example, AMI-227 (Sinerem, Guerbet, France) that has a blood half-time of 2–3 h would present several advantages over AMI-25: its concentration remains approximately constant for longer time intervals, allowing acquiring images with longer acquisition times (improved SNR and/or space resolution), and also acquiring repeated images during the previously mentioned time interval.

T_2W and DW images are characterized by different sensitivity windows during the evolution of ischemic lesion. Table I reports ischemic lesion volumes as determined by either T_2W or DW images. Six hours after MCAO, the lesion volumes were greatly underestimated by T_2W; this was observed in both vehicle and drug-treated animals. On the contrary, at later time points (144 h) no lesion was detectable in DW images while still clearly detected in T_2W images. The effect of treatment with GV150526 was clearly assessed by MRI. At each time point investigated, the infarct volume was smaller in treated than in control animals ($p < 0.001$). Results reported here are referred to an experimental protocol in which the drug is administered before MCAO. The complete study showed that MRI is sensitive to the effect of the treatment also when administered after MCAO.[25]

Parametric maps of rCBV were obtained at different time points after MCAO. The technique allowed to detect substantial reduction in blood volume (from 22–77%) in the ischemic region relative to controlateral regions, 24 h after MCAO. Images taken at longer time intervals (15 days or longer) after MCAO showed that a residual vascular deficit is still

TABLE I
PERMANENT MCAO. INFARCT VOLUMES (EXPRESSED IN MM^3) MEASURED FROM
MRI (T_2W AND DW)

Time elapsed after MCAO	Drug (preischemia administration)		Vehicle	
	T_2W	DW	T_2W	DW
6 h	1.1 ± 0.5	8.4 ± 3.7	10.3 ± 5.1	43.5 ± 14.5
24 h	14.7 ± 5.9	28.2 ± 12.2	93.1 ± 18.5	100.8 ± 20.3
144 h	16.4 ± 6.5	n.d.[a]	87.6 ± 19.6	n.d

[a] No data available.

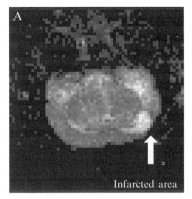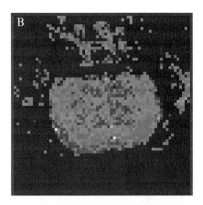

0.3   0.4   0.5   0.6

Fig. 5. Permanent MCAO. (A) T_2W transversal image acquired 30 days after MCAO and (B) rCBV map obtained for the same slice. A region with reduced blood volume is observed in the cerebral cortex. This region corresponds to an hyperintensity area in T_2W. (See color insert.)

present in about half of the animals. The presence of vascular deficit correlates with the initial (24 h) size of the lesion. Figure 5 shows representative T_2W image and rCBV map obtained for one animal 30 days after MCAO. A region with reduced blood volume is detected in rCBV map in correspondence with a hyperintense region in T_2W image.

Neoplastic Pathologies

MRI plays a crucial role both in diagnosis of tumors and in monitoring of their response to therapies in both clinical and preclinical environments. Experimental tumor tissues are generally characterized by a long T_2 relaxation time (compared with neighboring healthy tissues) and consequently are easily detected and delineated by T_2W images. Standard T_2W images allow for noninvasive *in vivo* measurement of tumor volume and its time evolution, although in the presence of irregular or infiltrating lesions, such evaluation can be difficult. Tumor size has been used for a long time as a standard endpoint for experimental drug screening, and as a prospective endpoint for planning clinical trials, and it is widely used in clinical practice for making decisions in tumor treatments.[29] However, there are

[29] A. R. Padhani, *Eur. J. Cancer* **38**, 2116 (2002).

several recognized limitations to the use of tumor size changes as a marker of therapeutic efficacy. Such limitations are particularly evident when evaluating the effect of antiangiogenic drugs, which selectively inhibit tumor vasculature before any effect on tumor size can be visible.[30] Functional techniques sensitive to tumor vascularization have been developed based on dynamic acquisition of images at high time resolution during the arrival of contrast agents in the tumor tissue.[6] These techniques, known as dynamic contrast-enhanced MRI (DCE-MRI) can be divided into (1) techniques that use low molecular weight contrast agents and that rapidly distribute in the extracellular space and (2) techniques that use high molecular weight contrast agents designed for prolonged intravascular retention. Preclinical studies have indicated that DCE-MRI with macromolecular contrast agents is more specific (compared with low molecular weight contrast agents) for detection of tumor microvessel features. This is due to the peculiar characteristic of tumor vasculature of being hyperpermeable to macromolecules. Gd-DTPA–albumin is a prototype contrast agent obtained by covalent binding of several Gd-DTPA moieties to albumin,[31] resulting in a molecule with 90 kD molecular weight. We have implemented an experimental protocol exploiting DCE-MRI with Gd-DTPA–albumin for the characterization of tumor vasculature in a experimental model of colon carcinoma and for the assessment of an antiangiogenic efficacy of an experimental drug.

Colon Carcinoma

HT-29 human colon carcinoma fragments were implanted subcutaneously in the flank of 10 nude mice weighing approximately 25 g. Animals were inserted in the study when the tumors reached a weight of approximately 500 mg. The experimental drug (SU6668) is a small molecule inhibitor of the angiogenic receptor tyrosine kinases (RTKs) Flk-1/KDR (VEGFR-2), PFGFRβ, and FGFR.[32] Animals were divided in two groups; one group received the drug and the other group received the vehicle. Tumor volumes were obtained by caliper measurement of tumor diameters (d and D), according to the formula $d^2 \times D/2$. Gd-DTPA–albumin, synthesized according to Ogan,[31] was obtained from R. Brasch (Contrast Media Laboratory, University of California, San Francisco) and is characterized

[30] K. M. Brindle, *NMR Biomed.* **15,** 87 (2002).
[31] M. D. Ogan, *Invest. Radiol.* **23,** 961 (1988).
[32] A. D. Laird, P. Vajkoczy, L. K. Shawyer, A. Thurnher, C. Liang, M. Mohammadi, J. Schlessinger, A. Ullrich, S. R. Hubbard, R. A. Blake, T. A. T. Fong, L. M. Strawn, L. Sun, C. Tang, R. Hawtin, F. Tang, N. Shenoy, K. P. Hirth, G. McMahon, and J. M. Cherrington, *Cancer Res.* **60,** 4152 (2000).

by an average molecular weight of 94,000 D, corresponding to approximately 45 molecules of Gd-DTPA covalently bound to each albumin molecule. Mice were anesthetized by inhalation of a mixture of air and O_2 containing 0.5–1% halothane and placed in prone position into a 3.5-cm i.d. transmitter–receiver birdcage coil. Images were acquired using a Biospec tomograph (Bruker, Karlsruhe, Germany) equipped with a 4.7-T, 33-cm-bore horizontal magnet (Oxford Ltd., Oxford, UK). Coronal SE and transversal multislice, fast SE T_2W (RARE, TEeff = 70 ms) images were acquired for tumor localization and good visualization of extratumoral tissues. Afterwards, a dynamic series of 3D, transversal spoiled-gradient–echo (SPGR) images were acquired with the following parameters: TR/TE = 50/3.5 ms, flip angle (α) = 90°, matrix size = 128 × 64 × 32, FOV = 5 × 2.5 × 3 cm^3 (corresponding to 0.39 × 0.39 mm^2 in-plane resolution and 0.94 mm slice thickness), number of averages (NEX) = 1. The acquisition time for a single 3D image was 104 s; a dynamic scan of 24 images was acquired with 30-s time intervals between each image (total acquisition time 53 min). A bolus of Gd-DTPA–albumin was injected into the tail vein at a dose of 30 μmol of Gd/kg (26 ml/kg, typically 60 μl for mouse) during the time interval between the first and the second scan. A phantom containing 1 mM Gd-DTPA in saline was inserted in the field of view and used as an external reference standard. The MRI protocol was derived from that reported in Daldrup *et al.*,[33] with some modifications.[34] Briefly, precontrast T_1 values were measured using the IR-Snapshot Flash technique[35] a series of 12 consecutive Snapshot Flash images were acquired after a 5000-ms inversion pulse with the following parameters: FOV = 5 × 2.5 cm^2, slice thickness = 2 mm, matrix size 128 × 64, TR/TE = 10/2.8 ms, α = 7°, and a centric phase-encoding scheme. Some images were obtained at lower space resolution (matrix size 64 × 32) in order to verify that the T_1 values obtained were not influenced by the relatively long acquisition time of the single image. In our experience, when working with mice, it is not possible to measure the signal directly from the blood vessels because the signal of the blood is hyperintense due to flow effects. Consequently, the plasma kinetics of the contrast medium was determined *ex vivo* by withdrawing blood samples from a different group of mice before and 5, 15, 30, and 60 min after Gd-DTPA–albumin administration. A total of 15 animals was used, corresponding to three animals for each time point. The

[33] H. Daldrup, D. M. Shames, M. Wendland, Y. Okuhata, T. M. Link, W. Rosenau, Y. Lu, and R. C. Brasch, *Pediatr. Radiol.* **28**, 67 (1998).
[34] P. Marzola, P. Farace, L. Calderan, C. Crescimanno, E. Lunati, E. Nicolato, D. Benati, A. Degrassi, A. Terron, J. Klapwijk, E. Pesenti, A. Sbarbati, and F. Osculati, *Int. J. Cancer* **104**, 462 (2003).
[35] A. Haase, D. Matthaei, R. Bartkowski, E. Duhmke, and D. Leibfritz, *J. Comput. Assist. Tomogr.* **13**, 1036 (1989).

blood samples were examined using the same sequences as performed in animals (see Ref. 34 for details).

From the SI of MR images acquired with a gradient–echo sequence, it is possible to calculate the longitudinal relaxation time of the different tissues (after T_{1pre} has been independently measured), according to:

$$\frac{SI(t)_{post}}{SI_{pre}} = \frac{1 - \exp[-TR/T_1(t)_{post}]}{1 - \exp(-TR/T_{1pre})} \tag{3}$$

Longitudinal relaxation rates are correlated to contrast agent concentration through:

$$\Delta R1(t) = [R1(t) - R1_0] = \frac{1}{T_1(t)_{post}} - \frac{1}{T_{1pre}} = [Gd(t)] \times r_1 \tag{4}$$

where T_{1pre} and $T_1(t)_{post}$ represent, respectively, the longitudinal relaxation times of the tissue before and at different time points after contrast agent injection. [Gd(t)] indicates the molar concentration and r_1 (expressed in $mM^{-1}s^{-1}$), the relaxivity of Gd ion in the tissue. Assuming that r_1 is constant in the different tissues examined, the parameter $\Delta R1(t)$ is directly proportional to the Gd concentration. This assumption represents an inherent approximation in DCE-MRI techniques. In fact, relaxivity of contrast agents is strictly dependent on the microviscosity of the medium, and can change in different tissues. We have performed *in vitro* measurements of Gd-DTPA–albumin relaxivity in pure water and in a protein-rich aqueous solution,[34] and we have observed a 68% increase of the Gd-DTPA–albumin relaxivity. However, the increase was more pronounced (200%) for Gd-DTPA, thus enforcing the use of macromolecules against small molecules contrast agents in this specific application.

The time dependence of $\Delta R1(t)$ was analyzed in terms of a two-compartment tissue model composed of plasma and interstitial water-equilibrating pools.[33,34] In this model it is assumed that the exchange is due to passive diffusion of the contrast agent. As explained in detail previously,[34] the time dependence of $\Delta R1(t)$ can be expressed by:

$$\Delta R1(t) \propto CT(t) = kPS \int_0^t Coexp - (\vartheta A) \, d\vartheta + fPV \, Coexp(-At) \tag{5}$$

where $CT(t)$ is the total concentration of Gd in both interstitial water and plasma space, $Coexp(-At)$ is the experimentally *ex vivo* measured concentration of Gd in plasma, fPV is the fractional plasma volume of the tissue, and kPS is the transendothelial permeability. This expression was fitted to the $\Delta R1(t)$ values extracted from the experimental signal intensity using a two-parameter best-fit nonlinear algorithm.

After acquisition, data were transferred onto a personal computer (PC) for analysis. The fitting routine was written in Matlab 5.2 (The MathWorks, Inc., Natck, MA). Images were analyzed on a pixel-by-pixel basis to obtain parametric maps of kPS and fPV, or on a ROI basis to obtain the average value of kPS and fPV in the selected ROI; in each animal, the central five slices of the 3D data set were analyzed. For each considered slice, ROIs were manually tracked to cover the tumor rim and the tumor core. A band approximately 2 mm wide at the periphery, on the external side of the tumor, was considered as the rim. The signal in the rim (and in the core) was averaged and analyzed to obtain the mean kPS and fPV values in the rim (or in the core) of the selected slice.

After the last MRI examination, mice were sacrificed and tumoral tissue removed and prepared for histology. The experimental protocol used for histology and immunohistochemistry has been reported in detail.[36] The mean number of vessels and vessel area was obtained from CD31 immunohistochemical staining of tumor tissue.

Animals were examined at MRI before and 24 h after receiving a single administration of SU6668 or vehicle. Figure 6 shows axial T_2W and T_1W precontrast and T_1W postcontrast images obtained for one animal bearing a subcutaneous tumor. The T_2W image clearly delineates the tumor tissue as an area of strongly hyperintense signal. In the precontrast T_1W image, the SI of the tumor, skeletal muscle, blood, and myocardium is very similar. After contrast medium administration (the image shown was acquired about 50 min after injection), the signal from blood and vascularized regions of the tumor is greatly enhanced. As described before, from the time dependence of the signal intensity, it is possible to extract the time dependence of the enhancement in T_1 relaxation rate ($\Delta R1$) of the different tissues that is proportional to the contrast agent concentration. Figure 6B shows the time dependence of $\Delta R1$ (or Gd-concentration) in pixels belonging to the periphery or the core of the tumor; the best fitting to experimental data is also shown. The intercept of the fitting curve with the y axis measures the blood volume, whereas its rate of rising measures the vascular permeability. In this experimental model, and, in general, in xenografts, the periphery of the tumor is well vascularized, whereas the core is mainly necrotic. The fitting of theoretical equation to experimental data can be performed also pixel by pixel, providing a parametric map of fractional plasma volume and transendothelial permeability. The maps

[36] P. Marzola, A. Degrassi, L. Calderan, P. Farace, C. Crescimanno, E. Nicolato, A. Giusti, E. Pesenti, A. Terron, A. Sbarbati, T. Abrams, L. Murray, and F. Osculati, *Clin. Cancer Res.* **10,** 739 (2004).

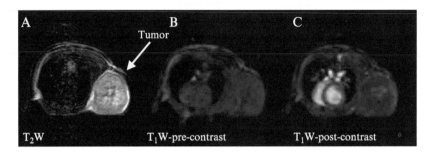

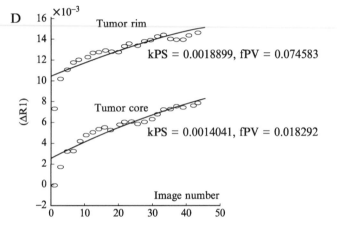

FIG. 6. Colon carcinoma. T_2W, T_1W-pre-contrast, and T_1W-post-contrast images acquired approximately at the tumor center for one animal belonging to the group treated with SU6668, before treatment. (A) Axial T_2W RARE image; (B) axial T_1W GE images acquired before and (C) 50 min after injection of Gd-DTPA–albumin. (D) Time dependence of $\Delta R1$ (or Gd-concentration) in pixels belonging to the periphery or the core of the tumor; the best fitting to experimental data is also shown. fPV and kPS are expressed in ml per cm^3 of tissue and ml per min per cm^3 of tissue, respectively. Adopted from Marzola et al.,[36] with permission.

relative to the slice shown in Fig. 6A are reported in Fig. 7 (upper line). In this particular slice, regions with high plasma volume correspond to regions with high vascular permeability, but this is not always true; in some tumors, we have found areas of high vascular permeability in mostly necrotic regions[34] in agreement with other investigators.[37] Twenty-four hours after receiving a single-dose treatment of SU6668, the fPV and kPS maps were strongly altered with a strong decrease in both parameters, as apparent in Fig. 7. From quantitative analysis on the group of treated animals, we

[37] Z. M. Bhujwalla, D. Artemov, K. Nararajan, M. Solaiyappan, P. Kollars, and P. E. G. Kristjansen, Clin. Cancer Res. 9, 355 (2003).

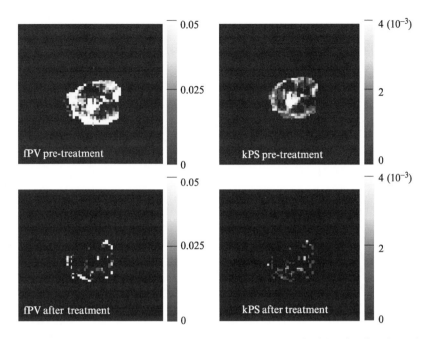

FIG. 7. Colon carcinoma. *Upper line:* fPV and kPS maps obtained for the slice shown in Fig. 6 (i.e., before treatment). *Lower line:* fPV and kPS maps obtained for the same animal but 24 h after treatment. Adopted from Marzola *et al.*,[36] with permission. (See color insert.)

observed a decrease in kPS of 51% ($p < 0.0001$) and 26% ($p < 0.05$) in the tumor rim and core, respectively. The decrease in fPV amounted to 58% and 35% in the tumor rim and core, respectively. No statistically significant alteration was observed in the control group. By DCE-MRI, it was possible to assess the effect of SU6668 on blood vessels, after only 24 h of treatment (single-dose administration), well before any effect on tumor size was detectable.

Lipid Accumulation in Metabolic–Degenerative Disorders

Lipid accumulation represents a common feature of several pathologic or degenerative states (e.g., obesity and age-induced atrophy of organs[38]). Obesity is a health problem widely diffused in industrialized countries that causes complications such as hypertension, diabetes (type 2 diabetes), and

[38] P. Marzola, E. Mocchegiani, E. Nicolato, A. Tibaldi, A. Sbarbati, and F. Osculati, *J. Magn. Reson. Imaging* **10,** 97 (1999).

atherosclerosis. Today at least 30% of adult Americans are obese.[39] The causes of obesity are still under investigation. Overeating and/or scarce physical activity play an important role, but some hereditary factors also seem to be relevant. In 1950 a genetic defect was identified in obese mice[40] the sequencing of the mouse obese gene was later identified and sequenced,[41] and it was demonstrated that the ob gene is expressed in white and brown adipose tissue and encodes a protein called leptin.[39] In the ob–ob mice, two mutations have been demonstrated: extinguished leptin mRNA expression and a nonsense mutation leading to the production of an inactive form of leptin that results in a compensatory 20-fold increase in leptin mRNA.[39] Obesity in the mutant ob–ob mice can be consequently attributed to a deficiency in the active leptin. Ob–ob mutant mice have been used in a number of studies aimed at clarifying the role of leptin in obesity and in development of antiobesity drugs. MRI and localized magnetic resonance spectroscopy (MRS) can be exploited to characterize these mice *in vivo*. In ^{1}H MRI, the signal is derived from water and lipids that occur as fatty droplets in the cytoplasm of adipocytes. Lipid protons have longitudinal relaxation times shorter than those of other nonlipid tissues; a well-known consequence of this is the fact that fat tissues appear as very bright structures disturbing identification of other organs and requiring selective suppression of fat signal. The fat signal originates mainly from the CH2–CH3 groups of the lipidic chain, which resonate 3.3 ppm apart from water. This frequency shift, which is responsible for the chemical shift artifact, allows for selective excitation (or suppression) of fat or water protons. Several techniques have been proposed in biomedical MRI to obtain separate images of water and fat protons in tissues and to evaluate the ratio between these two components.[42] At low magnetic fields, the Dixon method,[43] based upon the dephasing of fat and water signals, may be applied. At high magnetic fields, where the Dixon method is limited by susceptibility effects, the chemical shift between water and fat protons yields frequency separations large enough to permit selective excitation of fat or water protons. The excitation pulse in a single-slice excitation SE experiment can be designed to be frequency selective in such a way that only one of the two signals is excited. We have used this technique in characterization of brown adipose tissue (BAT) and

[39] F. Lonnqvist, L. Nordfors, and M. Schalling, *J. Intern. Med.* **245,** 643 (1999).

[40] A. M. Ingalls, M. M. Dickie, and G. D. Snell, *J. Hered.* **41,** 317 (1950).

[41] Y. Zhang, R. Proenca, M. Maffei, M. Barone, L. Leopold, and J. M. Friedman, *Nature* **372,** 425 (1994).

[42] E. Kaldoudi and S. C. R. Williams, *Concepts Magn. Reson.* **4,** 53 (1992).

[43] W. T. Dixon, *Radiology* **153,** 189 (1984).

thymus.[38,44] Localized spectroscopy techniques are also very useful in characterization of adipose tissues. This family of methods that encompasses single voxel or multivoxel techniques (the last known as chemical shift imaging [CSI]) exploits gradients for obtaining localization while the acquisition of the signal is performed in the absence of gradient; the chemical shift information is consequently retained. We have used localized spectroscopy for characterizing *in vivo* BAT in terms of its content of polyunsaturated fatty acids.[45] Here we report some experimental approaches performed by MRS aimed at phenotyping ob–ob mice.

Ob–Ob Mice

Ob–ob animals and relative controls were obtained by Harlan, Italy. A total of $n = 10$ ($n = 5$ ob–ob and $n = 5$ controls) mice were used. For MRI, mice were anesthetized by inhalation of a mixture of oxygen and air containing 0.5–1% of isofluorane and placed in the supine position in a 3.5-cm ID birdcage coil. Images and localized spectra were acquired using a Biospec Avance Tomograph (Bruker, Germany) operating at 4.7 T. After a pilot acquisition, T_1W transversal and coronal slices were acquired using TR/TE = 1009/18 ms, slice thickness = 2 mm and NEX = 1. Other parameters were FOV = 10×5 cm^2, matrix size = 256×128 and FOV = 4.5×4.5 cm^2, matrix size = 256×256 for sagittal and transversal acquisition, respectively. Animals were positioned in the RF coil in such a way that the kidneys were approximately in the coil (and magnet) isocenter. Unlocalized spectra were acquired using a single RF pulse sequence with RF pulse duration of 50 μs and NEX = 16. Localized spectra were acquired using a stimulated-echo sequence with TR = 2500 ms, TM = 8.9 ms, TE = 22 ms, voxel size = $3 \times 3 \times 3$ mm^3, NEX = 128. The voxel was placed upon the liver.

Figure 8 shows T_1W, transversal images of the abdominal region obtained from an ob–ob animal (Fig. 8A) and normal control (Fig. 8B) at the level of the kidneys. The fat tissue appears bright and represents the most abundant tissue in the ob–ob animal, whereas it is quite scarce in the normal control. In Fig. 8 (C and D) nonlocalized spectra obtained from the abdominal region of the animals are shown; the peaks at 4.7 and 1.3 ppm are due to water and fat protons, respectively. Spectra and images were obtained using the same volume coil without moving the animal. In ob–ob animals, abdominal spectra showed two peaks of similar intensity,

[44] E. Lunati, P. Marzola, E. Nicolato, M. Fedrigo, M. Villa, and A. Sbarbati, *J. Lipid Res.* **40,** 1395 (1999).

[45] E. Lunati, P. Farace, E. Nicolato, C. Righetti, P. Marzola, A. Sbarbati, and F. Osculati, *Magn. Reson. Med.* **46,** 879 (2001).

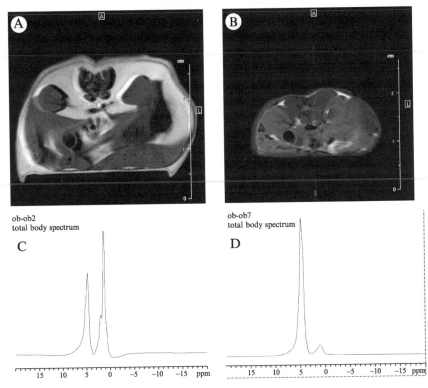

FIG. 8. Ob–ob mice. Representative SE T_1W images of the abdominal region of an ob–ob mouse (A) and control (B). Unlocalized ^1H spectra obtained from the abdominal region of an ob–ob mouse (C) and control (D).

whereas in normal controls the water peak was greatly predominant. By indicating with W the area under the water peak and with F the area under the fat peak, the ratio F/W was calculated for each spectrum. This ratio amounted to 1.46 ± 0.44 and 0.29 ± 0.11 in ob–ob and control mice, respectively. It has been shown that the ratio between water and fat peaks obtained from proton spectra of the abdominal region is well correlated to the total body fat and represents a quantitative measure of it.[46] From the calibration curve reported in Barac-Nieto and Gupta,[46] we can infer that the body fat amounts to 55% in ob–ob animals and 0.5% in control animals. Localized spectroscopy was used to assess the liver content of fat. In Fig. 9 we show localized spectra obtained from the liver of an ob–ob

[46] M. Barac-Nieto and R. K. Gupta, *J. Magn. Reson. Imaging* **1,** 235 (1996).

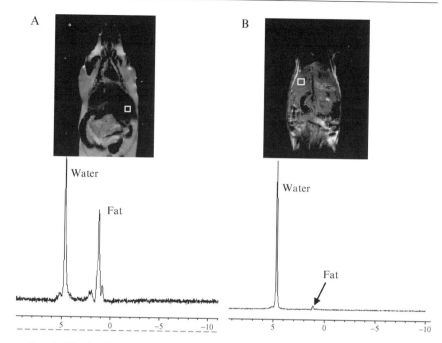

FIG. 9. Ob–ob mice. Localized spectra obtained from a 5-mm³ pixel placed in the liver parenchyma from an ob–ob mouse (A) and a normal control (B).

mouse and normal control. These spectra qualitatively show that the liver of ob–ob animals is rich in fat, different from the liver of normal animals. The signal attributable to fat protons, in fact, accounted for about 50% and less than 1% of the total signal in normal control and ob–ob animals, respectively. Calibration of the sequence with oil phantom is required to obtain a quantification of the fat content. MRI and MRS are valuable tools in characterizing fat tissues in ob–ob animals and in assessing the effect of experimental drugs on fat deposits.

Conclusions

MRI has become an increasingly appreciated diagnostic tool in medicine, and it is probable that in the future, its use in preclinical research will expand. MRI has numerous advantages over conventional techniques of tissue analysis[47]: it is harmless to tissues, the same tissue can be examined

[47] A. Sbarbati and F. Osculati, *Histol. Histopathol.* **11,** 229 (1996).

several times, several organs can be examined at the same time, and sections of relevant structures can be obtained in all planes. *In vivo* examination also results in a significant reduction of artifact due to fixation, embedding, sectioning, and staining. In addition, *in vivo* MRI examination offers the possibility to obtain functional, biochemical, and biophysical data that cannot be obtained by alternative methods. These advantages are, however, accompanied by some important drawbacks. Apart from the cost of the instrument, the main drawback is the spatial resolution, which is relatively low in comparison to conventional microscopes. Finally, the interpretation of the results is often difficult for an incomplete standardization of MRI methods and it needs interdisciplinary competence. For these reasons, to date, preclinical MRI represents a challenge to the scientist and although MRI data must be evaluated with caution, a wide use of MRI-based technologies can permit the execution of experimental paradigms that are not possible using a more traditional approach.

[8] Application of Combined Magnetic Resonance Imaging and Histopathologic and Functional Studies for Evaluation of Aminoguanidine Following Traumatic Brain Injury in Rats

By Jia Lu and Shabbir Moochhala

Introduction

Trauma brain injury (TBI) is the most common cause of brain damage and is one of the leading causes of death. It triggers a large network of morphologic and metabolic changes in the central nervous system. It has been suggested that at least two classes of injury contribute to the overall pathophysiology of TBI. The primary injury results from the initial mechanical event itself, followed by the secondary injury, which involves a cascade of biochemical changes that contribute to delayed tissue damage and neuronal death.[1] Magnetic resonance imaging (MRI) offers a unique window for monitoring pathophysiologic changes associated with brain injury.

---

[1] A. I. Faden, *Pharmacol. Toxicol.* **78**, 12 (1996).

Copyright 2004, Elsevier Inc.
All rights reserved.
0076-6879/04 $35.00

MRI has been used extensively in various kinds of brain lesions, both in humans and in animal models.[2–5] It has been shown to be of great diagnostic value in terms of lesion detectability and depiction of lesion size.[6] Moreover, MRI is also a powerful tool for the investigation of the time course of therapeutical approaches (e.g., MRI has been used to evaluate therapies with neuroprotective agents).[7,8] The major advantage of MRI over histologic examinations is because of its potential to noninvasively assess changes in the brain *in vivo*. It has been widely accepted that T_2-weighted (T_2W) MRI can be used to estimate the extent of cerebral lesion induced because it provides better delineation of the injured area at different time points. The areas with elevated T_2W signal intensity correlate with tissue damage as determined by established histopathologic techniques,[6,9] and hence the extent of cerebral lesion after TBI can be accurately and reproducibly estimated by *in vivo* conventional T_2W imaging. Despite numerous studies using MRI for characterizing stroke or ischemia in animal models,[6,10–13] the utilization of MRI for TBI has been relatively limited.[2,14] Currently, there is only modicum of information regarding potential correlations between MRI changes and histologic or neurologic outcome after TBI.

In this chapter, we study the temporal characterization of a lateral fluid percussion model in rats using T_2W MRI. We also compare MRI data with histologic and neurologic outcome after TBI. In addition, the

[2] B. C. Albensi, S. M. Knoblach, B. G. Chew, M. P. O'Reilly, A. I. Faden, and J. J. Pekar, *Exp. Neurol.* **162,** 61 (2000).

[3] B. L. Bartnik, E. J. Kendall, and A. Obenaus, *Brain Res.* **915,** 133 (2001).

[4] M. F. Lythgoe, N. R. Sibson, and N. G. Harris, *Br. Med. Bull.* **65,** 235 (2003).

[5] M. Cihangiroglu, R. G. Ramsey, and G. J. Dohrmann, *Neurol. Res.* **24,** 7 (2002).

[6] G. C. Palmer, J. Peeling, D. Corbett, M. R. Del Bigio, and T. J. Hudzik, *Ann. NY Acad. Sci.* **939,** 283 (2001).

[7] D. Cash, J. S. Beech, R. C. Rayne, P. M. Bath, B. S. Meldrum, and S. C. Williams, *Brain Res.* **905,** 91 (2001).

[8] N. F. Kalkers, F. Barkhof, E. Bergers, R. van Schijndel, and C. H. Polman, *Mult. Scler.* **8,** 532 (2002).

[9] T. Asanuma, H. Ishibashi, A. Konno, Y. Kon, O. Inanami, and M. Kuwabara, *Neurosci. Lett.* **329,** 281 (2002).

[10] T. Neumann-Haefelin, A. Kastrup, A. de Crespigny, M. A. Yenari, T. Ringer, G. H. Sun, and M. E. Moseley, *Stroke* **31,** 1965 (2000).

[11] R. M. Dijkhuizen, A. B. Singhal, J. B. Mandeville, O. Wu, E. F. Halpern, S. P. Finklestein, B. R. Rosen, and E. H. Lo, *J. Neurosci.* **23,** 510 (2003).

[12] K. Matsumoto, E. H. Lo, A. R. Pierce, H. Wei, L. Garrido, and N. W. Kowall, *Am. J. Neuroradiol.* **16,** 1107 (1995).

[13] F. A. van Dorsten, L. Olah, W. Schwindt, M. Grune, U. Uhlenkuken, F. Pillekamp, K. A. Hossmann, and M. Hoehn, *Magn. Reson. Med.* **47,** 97 (2002).

[14] Y. Assaf, E. Beit-Yannai, E. Shohami, E. Berman, and Y. Cohen, *Magn. Reson. Imaging* **15,** 77 (1997).

neuroprotective (both prophylactic and treatment) effects of aminoguanidine (AG), a selective inducible nitric oxide synthase inhibitor, on the brain pathogenesis and neurologic performance following brain injury were evaluated. The efficacy of AG was determined by combining the serial MRI and histology for monitoring brain lesion progression over a 72-h period. We also sought to determine whether the changes in behavioral measures could be correlated with the neurologic changes after TBI.

Materials and Methods

Animals and Chemicals

Male Sprague-Dawley rats (250–350 g) were used for the study. They were acclimatized for at least a week before the experiment. The rats were subjected to the following experimental procedures: Group 1: normal; Group 2: sham-operated; Group 3: TBI + normal saline; Group 4: AG (100 mg/kg) before TBI; Group 5: AG after TBI ($n = 12$ for each group). AG was obtained from Sigma Chemical (A-7009 Sigma, St. Louis, MO) and was dissolved in 0.9% sodium chloride solution (Sigma). Daily treatment of AG at the dosage of 100 mg/kg or normal saline was given intraperitoneally (IP) into rats starting 2 h before or 30 min after the brain injury. This dosage has been shown to be effective for traumatic brain injury and ischemia in previous studies.[7,15–18] In the handling and care of all animals, the international guiding principles for animal research as stipulated by the World Health Organization (WHO) *Chronicle*[19] and as adopted by the Laboratory Animal Centre, National University of Singapore, were followed.

Lateral Fluid-Percussive Brain Injury

The rats were anesthetized with sodium pentobarbital (70 mg/kg, IP) and were then ventilated and placed in a stereotaxic frame. A 1.5-cm sagittal incision from the midpoint between the ears toward the nose was made. The scalp and temporal muscle were then reflected to expose the cranium.

[15] K. Wada, K. Chatzipanteli, S. Kraydieh, R. Busto, and W. D. Dietrich, *Neurosurgery* **43,** 1427 (1998).
[16] M. Stoffel, M. Rinecker, N. Plesnila, J. Eriskat, and A. Baethmann, *Acta Neurochir. Suppl.* **76,** 357 (2000).
[17] C. Iadecola, F. Zhang, and X. Xu, *Am. J. Physiol.* **268,** R286 (1995).
[18] F. Y. Zhang, R. M. Casey, M. E. Ross, and C. Iadecola, *Stroke* **27,** 317 (1996).
[19] World Health Organization, "International Guiding Principles for Animal Research." *Chronicle*, Vol. 39, p. 51, 1985.

A ~5-mm-diameter burr hole positioned 2 mm lateral to the sagittal suture and 3 mm posterior to the coronal suture was determined and drilled with a 1-mm round head drill bit. The remaining bone was then removed, and any bleeding was arrested. This positioned the burr hole parasagittally over the cerebral sensorimotor cortex. The dura was left intact. A modified head cannula from a female Luer-Loc fitting (4.5 mm) was introduced into the burr hole until it abutted the dural surface. The head cannula was then affixed rigidly to the animal's skull with dental cement. Brain injury was performed, using a lateral fluid percussion model as previously described,[20] in which brief displacement and deformation of brain resulted from the rapid epidural injection of saline into the closed cranial cavity. Animals were subjected to a 3.7-atm pressure pulse, which produces severe tissue damage in ipsilateral cerebral cortex and hippocampus.[21] Sham-operated animals received anesthesia and surgery, but were not subjected to trauma treatment.

Magnetic Resonance Imaging Methods and Analysis

A total of 18 animals from Group 3 (TBI, $n = 6$), Group 4 (AG before TBI, $n = 6$), and Group 5 (AG after TBI, $n = 6$) were imaged at 24, 48, and 72 h after injury (Fig. 1). MRI was performed with a Bruker 2T MRI machine with custom-made rat coil. Coronal images were collected from eight contiguous, 1.2-mm-thick slices with echo time (TE) 52.6 ms, mixing time (TM) 29.2 ms, big delta 50 ms, and b values of 300 s/cm^2 and 30,000 s/cm^2.

Lesion measurements were performed following a direct Fourier transform (FT) of the raw time domain data. Images were displayed in the coronal plane, as collected (Fig. 2). Lesions were detected as hyperintense signals in the MRI. Selection of appropriate seed points by the operator permitted the application of a contour-tracing algorithm to delineate the boundaries of lesions. For each selected area, in each slice, pixel counts were generated. These values were then multiplied by the slice thickness and summed to yield total volume in mm^3. To limit subjective bias, the operator was blinded to the treatment strategy. The effect of AG on the changes in lesion volumes over time was analyzed by analysis of variance (ANOVA) with repeated measures. *P* values of <0.05 were considered significant.

[20] T. K. McIntosh, R. Vink, L. Noble, I. Yamakami, S. Fernyak, H. Soares, and A. L. Faden, *Neuroscience* **28**, 233 (1989).
[21] M. L. Prins, S. M. Lee, C. L. Cheng, D. P. Becker, and D. A. Hovda, *Brain Res. Dev. Brain Res.* **95**, 272 (1996).

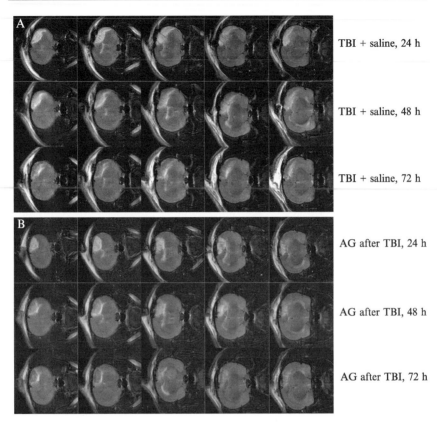

TBI + saline, 24 h

TBI + saline, 48 h

TBI + saline, 72 h

AG after TBI, 24 h

AG after TBI, 48 h

AG after TBI, 72 h

FIG. 1. An example of a series of coronal, T2-weighted MRI of rat brain of (A) saline-treated, (B) AG after TBI at 24, 48, and 72 h. Slices shown are approximately 2.2, 1, −0.8, −2, and −3.2 mm from bregma. The region of hyperintensity in the right (top) hemisphere of each image delineates the area of injury lesion. Notice that the total area of hyperintensity appears to increase in all time intervals in the saline-treated rat (A), whereas the size is significantly reduced in the AG-treated rat (B).

Neurologic and Behavioral Tests

Neurologic function was evaluated at 24, 48, and 72 h after TBI. When these time points coincided with MRI scanning time points, neurologic evaluation was performed just before MRI scanning. Briefly, the performance was based on three independent tests: locomotor activity, acoustic startle response, forelimb grip-strength test, and rotametric test.

Locomotor Activity. The locomotor activity of the animal was recorded using Columbus Instruments (Columbus, OH) Opto-Varimex-Mini, a

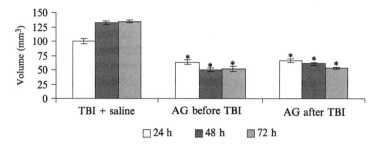

FIG. 2. Comparison of injury lesion volume, derived from the T_2-weighted MRI of saline-treated, AG before TBI and AG after TBI rats. Data are expressed as mean ± standard deviation (SD) ($n = 6$ in each group). Significant difference in lesion volume between the groups is revealed by one-way ANOVA. *denotes significant difference from TBI + saline group ($p < 0.05$).

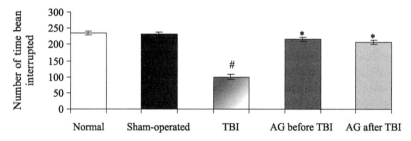

FIG. 3. Total locomotor response action in different groups of rats. Total locomotor response is reduced in rats subjected to TBI with saline injection. This adverse effect is reversed when the animals are administered with AG. # denotes significant difference from normal and sham-operated groups ($p < 0.05$); * denotes significant difference from TBI + saline ($p < 0.05$).

device that consists of a rectangular perspex container (17″ by 17″) with photocell-based sensors along its length. The rat was placed at one specific corner of the container and acclimatized for a 2-min period in a sound-attenuated room with a 60-dB background noise. Any locomotor activity made by the animal was then picked up by the sensor and was recorded for 5 min (Figs. 3 and 4). Total and ambulatory locomotor activities were recorded separately using two counters. Total locomotor activity refers to any movement made by the rat that is picked up by the sensors. This can include repeated activation of a single sensor. Ambulatory locomotor activity refers to the movement made by the rat as it moves between distinct sensors, excluding repeated activation of a single sensor. By subtraction of the measure for ambulatory locomotor activity from the measure of total locomotor activity, a score of the grooming and rearing behavioral patterns

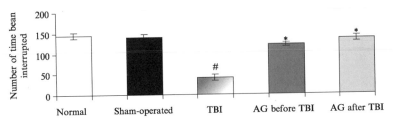

Fig. 4. Ambulatory locomotor response action in different groups of rats. Animals subjected to TBI with saline injection show significantly lower scores. Administration of AG significantly improves the locomotor performance. # denotes significant difference from normal and sham-operated groups ($p < 0.05$); * denotes significant difference from TBI + saline ($p < 0.05$).

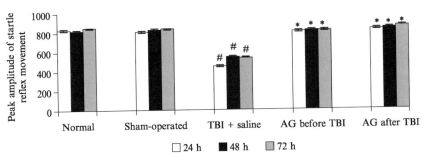

Fig. 5. Acoustic startle reflex in different groups of rats. A marked decrease in acoustic startle response was observed in rats subjected to TBI with saline injection. An improved response was observed in rats treated with AG before or after TBI. # denotes significant difference from normal and sham-operated groups ($p < 0.05$); * denotes significant difference from TBI + saline ($p < 0.05$).

of the animal can be calculated, given that no convulsive activity was observed. Locomotor activity was carried out only at 24 h after TBI.

Acoustic Startle Response. The animal was placed inside a Plexiglas cylinder in the acoustic startle apparatus (San Diego, CA) compartment. The animal was first acclimatized for 3 minutes in the presence of 70 dB of broadband white noise. After this period, the animal was immediately exposed to 50-ms bursts of white noise at 120 dB with a 5-s period of 70 dB of noise between each burst (Fig. 5). The peak amplitude of each startle reflex movement was recorded using the SR-lab software (San Diego, CA).

Forelimb Grip-Strength Test. Forelimb grip strength was determined using a grip strength meter (Columbus Instruments, Columbus, OH). The animals were placed on the electronic digital force gauge that measured the peak force exerted on it by the action of the animal. While drawing along a straight line leading away from the sensor, the animal was released at some point and the maximum force attained was stored on the display. The highest reading (in Newtons) of three successive trials was taken from each animal (Fig. 6).

Rotametric Test. A rotametric device (Rotamex 4/8 system, Columbus Instruments) was used to examine the ability of the animal to coordinate while being placed on a rotating rod. The rotating speed of the rod was set at 5 rpm (start speed) and 30 rpm (end speed) for a period of 240 s. An internal microcontroller was used to detect the time when a subject fell from the rod. The average reading (in seconds) of three successive trials was taken from each animal (Fig. 7).

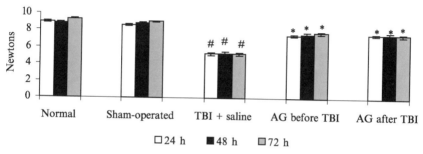

FIG. 6. Grip strength in different groups of rats. Rats subjected to TBI with saline injection show significantly lower average score, but this is significantly enhanced when the animals are administered with AG before or after TBI. # denotes significant difference from normal and sham-operated groups ($p < 0.05$); * denotes significant difference from TBI + saline, normal and sham-operated groups ($p < 0.05$).

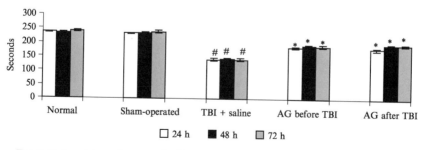

FIG. 7. Rotametric performance in different groups of rats. There is significant difference in rotameric performance among the groups. An improved rotameric performance is observed in the group with AG treatment when compared with rats subjected to TBI with saline injection. # denotes significant difference from normal and sham-operated groups ($p < 0.05$); * denotes significant difference from TBI + saline, normal and sham-operated groups ($p < 0.05$).

Perfusion

To characterize the tissue changes as a result of TBI, animals from time points corresponding to MRI data were sacrificed after scanning. Animals were deeply anesthetized with sodium pentobarbital (70 mg/kg, IP). For frozen sections, the rats were perfused with Ringer's solution transcardially for a few minutes until the liver and lungs were clear of blood, followed by an aldehyde fixative composed of a mixture of periodate–lysine–paraformaldehyde (0.01 M NalO$_4$, 0.075 M lysine, with a concentration of 2% paraformaldehyde, pH 7.4). After the perfusion, the brains were removed and kept in a similar fixative for 2 h. They were then kept in 0.1 M phosphate buffer containing 20% sucrose overnight at 4°. For paraffin sections, the rats were perfused with Ringer's solution, followed by 10% neutral formalin. The brains were then kept in a similar fixative overnight.

Caspase-3 Immunohistochemistry

We have tried two different methods for caspase-3 immunohistochemistry. For paraffin sections, the brains were dehydrated in an ascending series of alcohol, cleared with xylene, and embedded in paraffin wax. Ten-μm-thick coronal serial sections were cut. The sections were dewaxed in xylene and hydrated with descending series of alcohol, followed by distilled water. Slides were then microwaved in 10 mM of citrate buffer (pH 6), (NeoMarkers, AP-9003) for 10 min in the micromed oven, followed by cooling at room temperature for another 10 min. Slides were washed with 1× transcription buffered saline (TBS) for 10 min and then incubated with 3% H$_2$O$_2$ (BDH 101284N) in methanol for 10 min. After washing in 1× TBS for 10 min, the slides were incubated in normal goat serum (Vectastain ABC kit [Rabbit IgG], Vector Lab, PK 4001) for 20 min for blocking. Caspase-3 (CPP32) Ab-4 (NeoMarkers RB-1197) was diluted 1:100 with 0.3% bovine serum albumin (BSA) (Vector Lab, SP-5050) and applied on the sections for 1 hr at room temperature. Slides were washed with 1× TBS for 10 min, slightly blotted dry, and incubated with biotinylated goat anti-rabbit (Vectastain ABC kit [Rabbit IgG], Vector Lab, PK 4001) for 30 min. Slides were again washed with 1× TBS for 10 min, followed by

Fig. 8. Apoptotic cortical neurons after TBI by TUNEL (A–D) and caspase-3 immunohistochemistry (E–H). TUNEL positive (A) and caspase-3 immunopositive (E) neurons are absent in the contralateral cortex in rat following TBI. Many TUNEL positive (B) and caspase-3 immunopositive (F) neurons (arrows) are distributed in the ipsilateral parietotemporal cortex at 24 h after TBI. It was further increased at 48 and 72 h (C, D, G, H). Bar = 50 μm (for all).

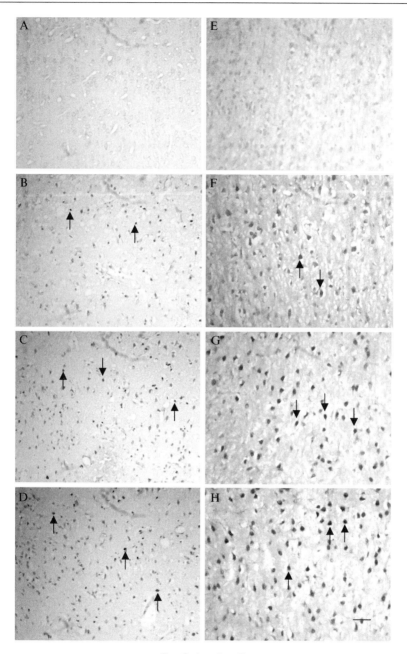

FIG. 8. (continued)

incubation with Vectastain ABC reagent (Vectastain ABC kit [Rabbit IgG], Vector Lab, PK 4001) for 30 min. Slides were washed with 1× TBS for 10 min, followed by detection with 3,3'-diaminobenzidine.

For frozen sections, coronal sections of the cerebrum of 10-μm thickness were cut and air-dried for 15 min at room temperature, followed by washing in 1× phosphate buffered saline (PBS) for 5 min. Slides were then incubated with 3% H_2O_2 (BDH 101284N) in methanol for 5 min and washed in 1× PBS for 5 min. The slides were then incubated in normal goat serum (Vectastain ABC kit [Rabbit IgG], Vector Lab, PK 4001) for 20 min for blocking. Caspase-3 (CPP32) Ab-4 (NeoMarkers RB-1197) was diluted 1:100 with 0.3% BSA (Vector Lab, SP-5050) and applied on the sections for 1 h at room temperature. Slides were washed with 1× PBS for 10 min, slightly blotted dry, and incubated with biotinylated goat anti-rabbit (Vectastain ABC kit [Rabbit IgG], Vector Lab, PK 4001) for 30 min. Slides were again washed with 1× PBS for 10 min, followed by incubation with Vectastain ABC reagent (Vectastain ABC kit [Rabbit IgG], Vector Lab, PK 4001) for 30 min. Slides were washed with 1× PBS for 10 min, followed by detection with 3,3'-diaminobenzidine (Fig. 8).

In Situ *Terminal Transferase d-UTP Nick-End Labeling (TUNEL)*

Degenerating neurons were identified in brain sections by *in situ* end labeling of nuclear DNA fragments using terminal deoxynucleotidyltransferase and biotin-16-dUTP as substrate (TdT-FragEL DNA Fragmentation Detection Kit, Oncogene, QIA 33). Paraffin sections were first dewaxed in xylene and hydrated with descending concentrations of alcohol, followed by washing in distilled water. Sections were slightly blotted dry and covered with 100 μl of Proteinase K (2 mg/ml Proteinase K diluted 1:100 in 10 mM Tris, pH 8). Slides were washed with 1× TBS for 10 min and then covered with 100 μl of 3% H_2O_2 (BDH 101284N) in methanol for 5 min. Slides were again washed with 1× TBS for 10 min, slightly blotted dry, and incubated with 100 μl of 1× TdT equilibration buffer for 20 min. At the same time, the working TdT labeling reaction mixture was prepared. After carefully blotting the equilibration buffer away from the specimen, taking care not to touch the sections, 60 μl of TdT labeling reaction mixture was immediately applied on the sections and incubated at 37° for 1.5 h, followed by washing in two changes of 1× TBS for 5 min each. Slides were dried around the sections and incubated with 100 μl of Stop Solution (TdT-FragEL DNA Fragmentation Detection Kit, Oncogene, QIA 33) at room temperature for 5 min. Slides were again rinsed with 1× TBS, followed by incubation with 100 μl of blocking buffer (TdT-FragEL DNA Fragmentation Detection Kit, Oncogene, QIA 33). After blotting the blocking buffer away from the specimen, 100 μl of diluted conjugate (50× conjugate diluted 1:300 in blocking buffer) was immediately applied to the specimen for 30 min. Slides were washed with 1× TBS for 10 min. Incorporated biotin was detected using streptavidin–peroxidase conjugate and 3,3'-diaminobenzidine (Sigma) as chromogen (Fig. 8).

Results and Discussion

In this study we used T_2W imaging to examine the temporal evolution of fluid percussion–induced TBI, and the results were then compared with histologic and neurologic evaluations. Our results showed that T_2W imaging is able to detect temporal- and region-specific changes associated with fluid percussion–induced TBI. MRI changes correlated with histologic evidence of tissue damage. Histologic examination of the brain after TBI showed apoptotic neurons in the cerebral cortex ipsilateral to the injury site, corresponding to the hyperintense area in T_2W image. The present T_2W results also showed that AG attenuated the progression of tissue damage and reduced the extent of cerebral lesion by 50% in rats following TBI when compared with those receiving saline injection. This correlates with our histologic findings, which also showed a marked reduction in the number of degenerating neurons (detected by caspase-3 immunohisto-chemistry and TUNEL) in the cerebrum in AG-treated rats. Moreover, neurobehavioral tests showed that administration of AG before or after TBI significantly improved rotametric performance, grip-strength score, total and ambulatory locomotor responses, and acoustic startle response. In conclusion, the utilization of T_2W MRI and histologic techniques provides important prognostic information on functional outcome. It is hoped that use of these approaches for evaluation of therapeutic strategies for head injuries will offer a chance for improved functional recovery following TBI.

Acknowledgments

Figures reprinted from J. Lu, S. Moochhala, M. Shirhan, K. C. Ng, A. L. Teo, M. H. Tan, X. L. Moore, M. C. Wong, and E. A. Ling, *Neuropharmacology* **44,** 253 (2003), with permission from Elsevier.

This study was supported by Defence Science & Technology Agency, Singapore.

[9] Using ^{99m}Tc-sestamibi to Evaluate the Effects of a Chemosensitizer on P-glycoprotein in Multidrug-Resistant Carcinoma Cells

By ZHENWEI ZHANG, XUEMEI ZHANG, HUA WU, and MING ZHAO

Introduction

Multidrug-resistance (MDR) is one of the main mechanisms of chemotherapy failure in cancer patients. Different cellular mechanisms of MDR have been identified. P-glycoprotein (P-gp) is a plasma membrane adenosine triphosphate (ATP)-binding cassette transporter, responsible for MDR in tumor cells. P-gp catalyzes the ATP hydrolysis-dependent efflux of numerous amphiphilic compounds of unrelated chemical structures. In the absence of any identified substrate, P-gp exhibits an apparently futile, basal ATPase activity.[1] A chemosensitizer, verapamil (VRP) may reverse the resistance induced by P-gp in the studies *in vitro*. However, such a chemosensitizer may lead to unexpected side effects in patients. At the present time, the effects of chemosensitizers are evaluated by observing the change of P-gp expression. However, the tissue sample needed in such studies and its clinical application are limited. Here we observe the changes of ^{99m}Tc-MIBI (sestamibi) uptake kinetics after using VRP in MDR human breast cells MCF-7/Adr, with the purpose of establishing a method to evaluate the effects of chemosensitizer on P-gp using ^{99m}Tc-MIBI scintigraphy.

Principle of Methods

Materials and Reagents

The human breast carcinoma cell line, MCF-7/Adr, characterized as MDR, was purchased from the Cancer Research Institute of Sun Yet-sen Medical University. Cells were cultured in DMEM with 2 mM L-glutamine (Gibco, Grand Island, NY) in 60-mm dishes.

^{99m}Tc-MIBI was prepared by using a Cardiolite kit (DuPont Pharmaceuticals, Beijing).

Verapamil (VRP) was purchased from North China Pharmaceuticals.

---

[1] A. Garrigues, A. E. Escargueil, S. Orlowski *et al.*, *Proc. Natl. Acad. Sci. USA* **99,** 10347 (2002).

Copyright 2004, Elsevier Inc.
All rights reserved.
0076-6879/04 $35.00

General Methods

Before treatment with VRP, cells were digested for the last time as in the following procedures: The incubation medium was replaced by 2 ml fresh medium. 1 μg/ml adriamycin was added to each dish to kill atavistic cells. A sterile slide (0.5 × 0.5 cm^2) was placed in each dish to attain a piece of cell slide for consequent immunohistochemical staining.

Effects of VRP on the Cellular Uptake of ^{99m}Tc-MIBI

Cells in the exponential growth phase cultured in 16 dishes were divided randomly into verapamil (VRP) and DMEM (control) groups. In the VRP group, 10 μM/L VRP was added into each dish; in the control group, the same volume of DMEM was added. Thirty-seven MBq ^{99m}Tc-MIBI were added to all dishes. After 2 h incubation at 37°, the dishes were rapidly washed three times with isotope-free cold (4°) saline to block all membrane exchanges. All slides in the dishes were removed for detecting P-gp expression by immunocytochemistry. Adherent cells were harvested with 0.05% trypsine (Gibco). ^{99m}Tc-MIBI uptake was measured using a gamma counter. The results were expressed as an uptake ratio (UR). UR = (counts per minute [CPM] of specific cell uptake/CPM of total radioactivity) × 100%.

Time Course of ^{99m}Tc-MIBI Uptake and P-gp Level During Administration of VRP

Cells in exponential growth phase were cultured with ^{99m}Tc-MIBI. After incubation with ^{99m}Tc-MIBI for 40, 60, 80, or 100 min (5 dishes per time point), 10 μM/L VRP was added into each group. The total incubation time for all dishes was 2 h (i.e., VRP incubated with cells for 80, 60, 40, and 20 min, respectively). Then the cells were harvested. The UR of ^{99m}Tc-MIBI and P-gp level was determined. To prolong the VRP treatment time, 10 μM/L VRP was added and all cells were incubated for 8, 24, 48, and 72 h (five dishes per time point). Two hours before the incubation ending, 37 MBq ^{99m}Tc-MIBI was added into all dishes. The uptake ratio of ^{99m}Tc-MIBI and P-gp level were determined accordingly.

Detection of P-gp Expression with Immunohistochemical Method

P-gp level was assessed by immunohistochemical stain. Positive controls were performed. Fifteen fields of view were selected in each slide, and the optical density was determined with MDS-120 Pathology Image Analysis System (Beijing University of Aviation and Aerospace, Beijing).

P-gp expression level (PL) was calculated as: PL = (specific optical density/ positive control value) $\times$ 100%.

Results

Effects of VRP on the Cellular Uptake of ^{99m}Tc-MIBI

After a 2-h incubation with VRP, the cellular uptake of ^{99m}Tc-MIBI in VRP groups increased significantly compared with the DMEM group ($t = 2.33, p < 0.05$). There was no difference in the P-gp level between both groups ($t = 1.82, p > 0.05$).

Difference Between UR and PL at Different Intervals

There was a significant difference between UR and PL at various durations (20 min–72 h) of incubation with VRP (F = 58.2, $p < 0.05$). When VRP treatment surpassed 8 h, the cellular uptake of ^{99m}Tc-MIBI remarkably increased (F = 77.9, $p < 0.05$) and PL also showed significant changes (F = 54.7, $p < 0.05$). However, in the early stage of VRP treatment, there was no difference of P-gp expression (F = 35.3, $p > 0.05$).

Correlation Between VRP Incubation Time and P-gp Expression Level (PL) and Cellular Uptake Ratio (UR) of ^{99m}Tc-MIBI

In the early stage of VRP treatment (20–80 min), UR correlated positively with the incubation time ($r = 0.62, p < 0.05$); that is, prolonging VRP treatment may induce higher intracellular accumulation of ^{99m}Tc-MIBI, whereas there was no alteration of P-gp expression ($r = 0.03, p > 0.05$). After VRP incubation time surpassed 8 h, UR correlated to VRP incubation time positively ($r = 0.75, p < 0.05$) and PL correlated to VRP incubation time negatively ($r = -0.81, p < 0.05$).

Correlation Between UR and PL

During the treatment of VRP (20 min–72 h), cellular UR of ^{99m}Tc-MIBI negatively correlated to P-gp expression level (PL), especially after the incubation time surpassed 8 h ($r = 0.71, p < 0.01$).

Remarks

MDR is one of the major obstacles to efficient chemotherapy of cancer. Many mechanisms contribute to MDR, such as P-gp, MDR-associated protein, and glutathione-S-transferase. P-gp is an ATP-consuming membrane glycoprotein that operates as an efflux pump to decrease the intracellular content of cytotoxic drugs. ^{99m}Tc-MIBI, a lipophilic phosphine cation that has been used for myocardial perfusing images, was found to be a substrate for P-gp and may be used to functionally image the P-gp status of tumors, although mechanisms are unidentified. The net cellular accumulation of ^{99m}Tc-MIBI is inversely proportional to the level of P-gp expression.

A variety of chemisensitizers, such as VRP and cyclosporin A, are capable of increasing the intracellular accumulation of cytotoxic drugs and thus may reverse MDR, which is related to P-gp. This property of VRP has been reported in several types of cancers, including breast, lung, and brain cancer. This effect of VRP may be due to the fact that VRP competed with chemotherapeutic drugs to combine with the two ATP-binding sites of P-gp. However, according to the findings of these workers,[2] prolonged exposure of MDR cells to VRP led to decreased P-gp expression. Moreover, as demonstrated by Biedler et al.,[3] VRP treatment (at high toxic concentration) decreased P-gp expression in a series of hamster lung cancer cell lines. They believed that it is possibly through deletion of amplified P-gp-encoding gene, in addition to down-regulation of gene expression.

This study showed that VRP significantly enhanced the cellular uptake of ^{99m}Tc-MIBI. In the early stage after VRP treatment (<2 h), uptake ratio of ^{99m}Tc-MIBI was higher than that of the control group, but P-gp expression did not decrease. These findings suggest that VRP may enhance ^{99m}Tc-MIBI uptake by competitive mechanism (i.e., VRP combines to the ATP-binding sites of P-gp, which leads to the decreased efflux of ^{99m}Tc-MIBI). The results of a dynamic study of kinetics of ^{99m}Tc-MIBI indicated that in the early stage (20–80 min), uptake of ^{99m}Tc-MIBI increased without alteration of P-gp. It is further suggested that during a relatively short period of VRP treatment, a competitive mechanism that induces the increased uptake of ^{99m}Tc-MIBI may be attributed to a competitive mechanism; that is to say, VRP changes the P-gp pump functionally rather than quantitatively.

Prolonged exposure to VRP not only increased ^{99m}Tc-MIBI uptake, but also decreased P-gp expression, and the UR correlated negatively to the P-gp level. Therefore sequential VRP treatment may induce decreased P-gp expression (i.e., so-called down-regulation), and the uptake of ^{99m}Tc-MIBI increased by changing the P-gp pump functionally as well as quantitatively.

Characterization of effects of chemosensitizers by routine techniques such as immunohistochemical analysis, polymerase chain reaction, and flow

[2] C. Muller, J. D. Bailly, K. Ozker et al., Int. J. Cancer **56,** 749 (1994).
[3] J. R. Biedler, J. H. Tsu, P. Firby et al., Cancer Res. **54,** 796 (1991).

cytometric analysis requires serial tissue biopsies. Furthermore, the utility of these approaches was limited because of sampling errors, invasiveness of the procedures, and technical complexity. The present study suggests the feasibility of using ^{99m}Tc-MIBI to evaluate both functional and quantitative effects of chemosensitizers on P-glycoprotein. Thus ^{99m}Tc-MIBI imaging potentially may be used to evaluate the effect of chemosensitizers noninvasively.

Section II

Preparation of Materials

[10] Vascular-Targeted Nanoparticles for Molecular Imaging and Therapy

By SAMIRA GUCCIONE, KING C. P. LI, and MARK D. BEDNARSKI

Introduction

Molecular imaging involves the noninvasive real-time observation of *in vivo* biologic events at the molecular level. The opportunity to visualize targets for the development of diagnostic and pharmaceutical agents over space and time has created wide interest in this marriage of physics, biology, and chemistry. The payoff for medicine can be enormous because proposed targets arising from high throughput screening can be validated *in vivo* and followed over both space and time. Also, pathways leading to the activity of new pharmaceutical agents can be delineated and followed before, during, and after therapy.

In nearly all cases, molecular imaging will require the delivery of a probe to the tissue site of interest. The design of probes for molecular imaging target two basic classes of biologic events: (1) alteration in metabolic processes and (2) changes in receptor expression. In the case of metabolic probes, small molecules are used that can perfuse most tissues and pathologic regions in the body (i.e., large volume of distribution). These small molecules can enter cells, get sequestered by forming compounds that are trapped in the cell, and be visible using imaging techniques. For example, ^{18}F labeled 2-deoxy-2-fluoroglucose (^{18}FDG) is a positron emitter that targets highly metabolic cells by entering the cell's glycolytic cycle and becomes trapped in the cell by enzymatic phosphorylation to form ^{18}F labeled 2-deoxy-2-fluoroglucose-6-phosphate. The initial probe is freely diffusable throughout all tissues in the body and has accessibility to pathologic regions in the body. The second class of biologic events involving changes in receptor expression has given rise to receptor target-based imaging. Here the protein of interest traditionally resides on the cell of pathologic origin. As in the first class of agents, the probe must first circulate in the bloodstream and pass through the endothelium into the target pathology, but it is retained in the region of pathology by binding to a target site in or on the surface of a cell. In some cases, binding to the receptor may induce endocytosis, but the probe usually does not enter metabolic pathways. The remaining probe materials must also clear from the body to reduce nonspecific background. This class of imaging agents has been dominated by monoclonal antibodies, antibody fragments, or large molecules labeled with technicium (^{99m}Tc) or indium (^{111}In) chelates.

Copyright 2004, Elsevier Inc.
All rights reserved.
0076-6879/04 $35.00

Recently, new discoveries in vascular biology have focused on new targeting sites that include cells and tissues adjacent to the site of pathology. This approach has been focused primarily on the changes that have been characterized in the blood vessels. Unlike targeting parenchymal cells, to reach the target sites on the surface of the vasculature, monoclonal antibodies and ligands do not need to be transported across the microvascular wall or through the interstitial space. This dramatically reduces the chances of losing these agents to metabolism and to nonspecific binding to proteins and other tissue components. Because of the significance and accessibility of endothelial receptors, they are superb molecular targets for developing new molecular imaging agents and therapeutics.

In this chapter we describe our approach to the design and construction of vascular receptor-targeted molecular imaging agents, with a focus on imaging the processes of inflammation and angiogenesis. The first section describes our rationale behind choosing a vascular target for molecular imaging derived from our studies on vascular permeability in tumors. The second section describes the design of our vascular-targeted probes and preclinical imaging studies. The last section demonstrates how imaging can be incorporated into vascular-targeted delivery systems to generate highly active therapeutic agents.

Rationale behind Choosing a Vascular Target for Molecular Imaging

Facing the Vascular Permeability Problem

Investigations into the changes during the development of pathology in human disease traditionally have focused on the cell of disease origin and not the supporting structures. For example, in cancer biology, extensive investigations on the precise genetic alterations, changes in the cancer cell, and its surface marker expression levels have been investigated for most major cancers such as colon, breast, prostate, and lung. Many molecules and a range of proteins have been developed for these targets, both as imaging and therapeutic agents. Despite intense investigations and a host of clinical trials, these materials have had only limited success, with major achievements limited to blood-borne diseases such as the lymphomas.

The poor outcome of many of these agents is primarily due to the unfavorable biodistribution of the targeting agent. There are many well-known physiologic barriers to the delivery of macromolecules, particles, and monoclonal antibodies to specific tissues *in vivo*. A blood-borne molecule that enters the circulation reaches the target cells via (1) distribution through vascular space, (2) transport across the microvascular wall, and (3) transport through the interstitial space. These materials may bind nonspecifically to proteins or other tissue components and get metabolized before reaching their target. If the probe is an antibody or antibody conjugate, it has been demonstrated that in rodent tumor models, less than 1% of injected antibody dose is distributed in the tumor. In human tumors this figure is even lower, where again less than 0.01% of the injected dose even gets to see its target.[1–3a] Therefore vascular permeability and transport of the agent across the endothelium is an important aspect for the successful outcome of using targets on the diseased tissues for molecular imaging. Also, it is not surprising that targeting the parenchymal cells of a disease has caused many clinical studies to fail primarily due to target accessibility and not target availability.

Therefore the development of molecular imaging agents toward the parenchymal cells of a disease will clearly be limited by the transport of these agents through the endothelial barriers to reach their respective target. On the flip side, the discovery of new blood markers for diseases (e.g., using genomic and proteomic and other microarray methods) will also be limited by the transport of these markers through the endothelium so they can enter the bloodstream. The consequences of vascular biology for research in these vastly large, diverse, and important areas can influence the way one thinks about molecular imaging probes and the design of pharmaceutical and diagnostic agents. If, for example, one discovers the most specific target for a disease, the use of this target for diagnosis and therapy can be limited by simple accessibility due to vascular permeability. This also holds true for those who discover selective blood markers of disease, since these markers can only be detected in clinical pathology when they gain access to the bloodstream by crossing endothelial barriers. An approach to circumvent this problem is to use a vascular cell surface marker as a target for developing receptor-based molecular imaging probes.

Endothelial-Targeted Imaging of Angiogenic Vessels

A critical early step in any inflammatory or immune response is the promotion of leukocyte adhesion to the vascular endothelium. The first realization that unique molecules called cell adhesion molecules (CAMs) are expressed by endothelial cells during a variety of physiologic and disease processes has led to extensive research in characterizing and manipulating

[1] S. Guccione, Y. S. Yang, G. Y. Shi, D. Y. Lee, K. C. P. Li, and M. D. Bednarski, *Radiology* **228,** 560 (2003).
[2] Y. Yang, S. Guccione, K. C. Li, and M. D. Bednarski, *Acad. Radiol.* **10,** 1165 (2003).
[3] A. M. Herneth, S. Guccione, and M. Bednarski, *Eur. J. Radiol.* **45,** 208 (2003).
[3a] H. Zhu, L. T. Baxter, and R. K. Jain, *J. Nucl. Med.* **38,** 731 (1997).

these molecules.[4–7] Multiple endothelial ligands and receptors are now known to be upregulated during various pathologies[6,8] and can potentially serve as molecular targets for diagnostic and/or therapeutic agents. For example, the delivery of nutrients to tissue is a fundamental necessity to maintain life in multicellular species. To facilitate the growth of tumors, neoplastic cells are capable of elaborating a host of factors that result in the development of channels to supply these nutrients in a process called "tumor angiogenesis." Neocapillaries, as opposed to normal capillary beds, overexpress specific cell markers such as Factor V2I-related antigen, acidic and basic (heparin binding) fibroblastic growth factor (bFGF), insulin-like growth factor, platelet-derived growth factor, the CD31, CD34, Ulex endothelial antigens, vascular endothelial receptor 1 (KDR) (Flk-1), and the vascular endothelial growth factor receptor (VEGF-R2).[4,5] The expression of these factors is regulated by as yet not fully understood mechanisms and pathways.

Integrins as Vascular Targets

These and other factors result in the attraction of cells and production of proteins, including the integrins, a family of heterodimeric endothelial cell membrane proteins. These proteins serve as adhesion receptors for arginine, glycine, aspartate (RGD)-containing proteins such as laminin, fibronectin, collagens, and vitronectin.[6] The RGD-containing proteins are used to form the extracellular matrix of blood vessels. The integrins are one of the best characterized members of the adhesion molecule family that is upregulated in angiogenic endothelial cells found in tumors and certain inflammatory injuries. The integrins are transmembrane molecules that favor the anchorage of endothelial cells to a wide variety of extracellular matrix proteins with an exposed arginine, glycine, aspartate (single letter coding RGD) amino acid sequence. In the adult human, the specific integrin $\alpha_v\beta_3$ has a limited tissue distribution. It is not expressed on quiescent epithelial cells and appears at minimal levels on smooth muscle cells. In contrast, activated endothelial cells in tumor-associated blood vessels express high levels of $\alpha_v\beta_3$.[7] In addition, other integrins such as $\alpha_1\beta_5$ have been characterized by antibody binding to tumor vessels. Staining for these antibodies have indicated the presence of this class of proteins in high concentration on the

[4] N. Weider, *J. Pathol.* **184,** 119 (1998).
[5] R. Folberg, V. Rummelt, R. Parys-Van Ginderdeuren, T. Hwang, R. F. Woolson, J. Pe'er, and L. M. Gruman, *Opthalmology* **100,** 1389 (1993).
[6] E. Ruoslahti and E. Engvall, *J. Clin. Invest.* **100,** S53 (1997).
[7] B. P. Eliceiri and D. A. Cheresh, *J. Clin. Invest.* **103,** 1227 (1999).
[8] J. Folkman, *N. Engl. J. Med.* **333,** 1757 (1995).

luminal side of the vessel.[9] Therefore we choose the integrins as our first target for a molecular imaging agent targeted to the vasculature.

Nanoparticles as Molecular Imaging Agents

The ideal vehicle for targeted delivery to vascular receptors should have the following properties: (1) biocompatibility, (2) sufficiently long intravascular half-life to allow for repeated passage through and interactions with the activated endothelium, (3) the ability to have ligands and proteins conjugated on the surface in multivalent configuration to increase the affinity and avidity of interactions with endothelial receptors, (4) the ability to have functional groups for high-affinity surface metal chelation or radiolabeling for imaging, (5) the ability to encapsulate drugs, and (6) the capability to have both imaging and therapeutic agents loaded on the same vehicle. Liposomes have been extensively studied as drug delivery vehicles, and stealth liposome-encapsulated doxorubicin (Doxil) is now commercially available.[10–13] A major drawback of conventional liposomal particles, however, is their lack of stability. It is difficult for these particles to withstand the chemical modifications needed to attach antibodies, ligands, and imaging probes to their surface. Therefore, despite their attractiveness, we feel that the use of conventional liposomes in vascular receptor-targeted imaging would be problematic.

We have chosen to develop polymerized nanoparticles rather than conventional liposomes as our targeting vehicle because they are rigid and therefore do not easily fuse with cell membranes, and they possess surface functional groups that can be used to attach to other targeting molecules and sufficient numbers of metal ions. Polymerized nanoparticles are composed of amphiphilic lipid molecules with polar head groups and hydrophobic tails that form aggregated bilayer-type structures in aqueous solution.[14,15] A functional group (diacetylene) can be incorporated into the tail portion of the lipids, which upon irradiation with ultraviolet (UV) light, will cross-link, polymerizing the lipids into structurally stable particles. These particles then serve as a scaffold for the attachment of antibodies, ligands, contrast agents, radiopharmaceuticals, and therapeutic

[9] D. McDonald, "New Targets and Novel Strategies." Proceedings of the Second Annual Meeting of the Society of Molecular Imaging, p. 220. San Francisco, 2003.

[10] M. Brandl, *Biotechnol. Annu. Rev.* **7**, 59 (2001).

[11] D. D. Lasic, J. J. Vallner, and P. K. Working, *Curr. Opin. Mol. Ther.* **1**, 177 (1999).

[12] K. J. Harrington, *Expert Opin. Investig. Drugs* **10**, 1045 (2001).

[13] J. W. Park, *Breast Cancer Res.* **4**, 95 (2002).

[14] R. W. Storrs, F. D. Tropper, H. Y. Li *et al.*, *J. Am. Chem. Soc.* **117**, 7301 (1995).

[15] R. W. Storrs, F. D. Tropper, H. Y. Li *et al.*, *J. Magn. Reson. Imaging* **5**, 719 (1995).

agents for imaging and therapy.[16,17] Moreover, the size distribution and rigidity of these particles can be chosen to avoid rapid clearance by the reticuloendothelial system, and the particle surface can be modified with ethylene glycol to further increase intravascular recirculation times.[14,15]

Design and Preclinical Studies of a Vascular-Targeted Molecular Imaging Agent

To reduce delivery to nonrelevant tissues and increase delivery to the tumor endothelium, a novel vasculature-confined versatile imaging agent that targets the surface of tumor-associated vessels needs to be developed. Delivery to healthy tissue may be reduced by confinement of the agent to the vasculature using macromolecules, and targeting of chemotherapeutic agents has been demonstrated using macromolecular immunoconjugates. Therefore a new approach to imaging vascular targets is to utilize the changes in vascular endothelium to target imaging agents. In this section we describe in detail a recently developed molecular-based imaging technology targeted toward the integrins that can be used to noninvasively detect and monitor vascular changes in disease such as angiogenesis and inflammation.

We have developed a lipid-based, biocompatible polymerized vesicle (PV) specifically engineered to be labeled with metals and targeted to specific vascular endothelial cell surface receptors. Endothelial cell receptors are ideal targets for this class of materials because of the material's pharmacokinetic properties and persistence in the intravascular compartment due to their size ($\sim$60–80 nm in diameter). The PVs are readily produced, inexpensive, and show no signs of toxicity *in vivo*. Contrast-enhanced magnetic resonance imaging (MRI) and scintigraphy have also been used to study the biodistribution of these nanoparticles for transition to the development of therapeutics.

Specifically, in this section we describe (1) the synthesis and characterization of these polymerized vesicles (PVs) and show that they have prolonged recirculation time, (2) demonstrate that PVs have good *in vivo* molecular imaging properties using MRI and gamma scintigraphy, and (3) give examples of how *in vivo* endothelial targeting can be achieved using an experimental autoimmune encephalitis (EAE) animal model and a rabbit V2 carcinoma model.[15–18]

[16] D. A. Sipkins, D. A. Cheresh, M. R. Kazemi, L. M. Nevin, M. D. Bednarski, and K. C. P. Li, *Nat. Med.* **4,** 623 (1998).

[17] D. A. Sipkins, K. Gijbels, F. D. Tropper, M. Bednarski, K. C. P. Li, and L. Steinman, *J. Neuroimmunol.* **104,** 1 (2000).

[18] J. Folkman, *Semin. Cancer Biol.* **3,** 65 (1992).

PV Synthesis and Characterization

To form antibody-conjugated paramagnetic polymerized vesicles (ACPVs), we initially constructed a particle containing biotinylated lipids. Via biotin molecules on the particle surface, an avidin bridge was used to attach biotinylated antibodies.[15,16]

Paramagnetic polymerized vesicles (PPVs) containing 0.5% biotinylated lipid, 29.5% Gd^{+3} chelator lipid, 10% amine-terminated lipid, and 60% filler lipid (PDA) were formed. The resulting PPVs are red (absorption maxima at 498 nm and 538 nm) and stable at room temperature even in the presence of serum. ACPVs are formed by addition of biotinylated antibody and avidin in a ratio of 2.7 to 1 to the PPVs (Scheme 1).

Figure 1 demonstrates the attachment of avidin and antibody to ACPV using immunodetection. Lane 1 in Fig. 1 (panel A) shows intense staining of 0.5 μg avidin, which, at its isoelectric point, moves slowly from the loading well. Lane 2 (panel A) is a 5-μl sample of PPVs, which move as a discrete band toward the positive pole. Lane 3 (panel A) is a 5-μl sample of PPVs, preincubated with avidin and unbiotinylated antivascular cell adhesion molecule (VCAM) antibody. Avidin now comigrates with the vesicle band (arrow), indicating that it has bound to the surface of the PPV; no free avidin is detected near the loading well. Lane 4 (panel A) represents a 5-μl sample of ACPVs. This preparation is similar to that described in Lane 3 (panel A); however, the anti-VCAM antibody is now biotinylated, allowing conjugation of antibody to the avidin–PPV complex to form the ACPV. As in Lane 3 (panel A), no free avidin is detected, indicating that avidin is now bound to the PPV. Interestingly, no avidin band appears

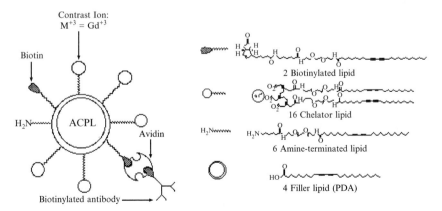

Scheme 1. Schematic representation of an antibody-conjugated polymerized vesicle (ACPV).

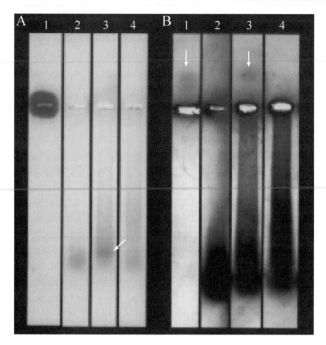

Fig. 1. Demonstration of anti-VCAM antibody–avidin conjugation to biotinylated vesicles by immunodetection. (See color insert.)

with the liposomes, suggesting that antibody conjugation to the particle surface sterically hinders binding of the antiavidin alkaline phosphatase immunodetection antibody to the complex.

Figure 1 (panel B) shows the results of immunodetection by anti-immunoglobulin G (IgG) alkaline phosphatase to assess antibody binding to PPVs. PPV preparations and antibody–avidin incubations were performed as described earlier. Lane 1 (panel B) shows a 2.5-μg aliquot of biotinylated anti-VCAM antibody that moves as a distinct band toward the negative pole (arrow). Lane 2 in Fig. 1 (panel B), as Lane 2 in Fig. 1 (panel A), is a 5-μl PPV sample that travels toward the positive pole; Lane 3 (panel B) is a 5-μl sample of PPVs, preincubated with avidin and unbiotinylated antibody, which contains 2.2 μg total antibody. As in Lane 1 (panel B), a free antibody band is detected (arrow), indicating that unbiotinylated antibody does not bind to the avidin–PPV complex. Finally, Lane 4 (panel B) is, 5-μl sample of PPVs, preincubated with avidin and biotinylated antibody, which again contains 2.2 μg total antibody. Note that the free antibody band is no longer detectable, demonstrating conjugation of biotinylated antibody to the avidin-coated liposome and formation of the ACPV. We have also shown that the ACPV is functional in a competitive inhibition assay.

For these studies, anti-intercellular cell adhesion molecule-1 (ICAM-1) ACPVs incubated on enzyme-linked immunosorbent assay (ELISA) plates coated with soluble ICAM-1 demonstrated inhibition of free monoclonal anti-ICAM-1 antibody binding. Further evidence that the anti-ICAM-1 antibody-conjugated paramagnetic vesicles could recognize antigens *in vitro* was provided by cell-binding assays using fluorescently-tagged ACPVs. PPVs containing the same ratios of lipids as described previously were coupled to Texas Red fluorophore (Pierce, Rockford, IL) according to a standard protocol. The fluorescent PPVs were then conjugated to anti-ICAM-1 antibodies as described earlier. Endothelial cells (bEnd 3) were plated onto 100-mm plastic SPECTri dishes and grown until confluent. Cells were stimulated with 1 μg/ml bacterial lipopolysaccharide approximately 24–48 h before use to elicit expression of ICAM-1 and other cell adhesion molecules. Unstimulated cells constitutively expressing only low levels of CAMS were designated as controls. ACPVs were incubated with the cells for 2 h at room temperature. Using fluorescence microscopy, anti-ICAM-1 ACPVs can be seen bound to the stimulated cells, outlining the morphology of individual cell membranes (Fig. 2). Minimal binding to unstimulated cells occurred.

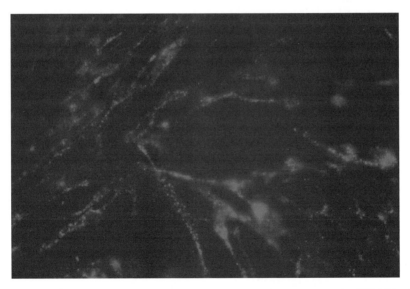

Fig. 2. Fluorescent ACPV labeling of stimulated endothelial cells expressing ICAM-1 cell adhesion molecules. (See color insert.)

ACPV Targeting in EAE Mice

To visualize CAM expression using a clinical imaging technique, a particle must carry a sufficient number of contrast materials for enhancement, recirculate in the blood pool, maintain its integrity *in vivo,* and be easily attached to a monoclonal antibody for specific receptor targeting. Fluorescently labeled anti-ICAM-1 ACPVs were next used in *in vivo* experiments so that the localization of the particle, as seen in contrast changes in imaging, could be confirmed with fluorescence microscopy. EAE was induced in SJL/J mice according to a proteolipid protein (PLP) immunization protocol (animals were cared for in accordance with institutional guidelines). When clinical signs of grade 2 disease were apparent (tail paralysis and limb weakness), the anti–ICAM-1 ACPVs (prepared as described earlier) were injected via a tail vein (10 μl/g representing 100 mg/kg Gd^{+3} and 15 μg total antibody) and allowed to recirculate for 24 h. Mice were then sacrificed and perfused with phosphate buffered saline (PBS). The brains were removed and cut in half sagittally, one half frozen for direct fluorescence-microscope analysis of thin sections and the other half fixed in paraformaldehyde and used for fluorescence and gamma imaging.

Fluorescence microscopy was used to examine areas of cerebellum in which fluorescent anti-ICAM-1 ACPVs were observed attached to the endothelium of small vessels (Fig. 3). Two controls, an anti-Vβ11 T-cell receptor ACPV (targeted to an antigen not expressed in the SJL/J mouse) injected in diseased animals and an anti-ICAM-1 ACPV injected in healthy animals, showed no ACPV binding. In Fig. 3, a fluorescent micrograph of cerebellum counterstained with hematoxylin shows three vessels in cross-section. Small vessels (sv) are bound by fluorescent anti-ICAM-1 ACPVs (arrows) while a central arteriole (LV) is negative for fluorescence. This is consistent with the pattern of expression of ICAM-1, which is upregulated on endothelium of venules and capillaries, but is not expressed on arterioles or larger vessels.

High-resolution T_1-weighted (T_1W) and T_2-weighted (T_2W) images of the intact brains were obtained on a 9.4-T MR scanner General Electric (Waukesha, WI) using 3DFT spin–echo pulse sequences. Parameters for T_1-weighted images were repetition time (TR) = 200 ms, echo time (TE) = 4 ms, number of excitations (NEX) = 1, and matrix = 256 × 256 × 256, resulting in a voxel size of approximately 50 μ^3. T_2-weighted parameters were TR = 1000 ms, TE = 20 ms, NEX = 8, and matrix = 256 × 256 × 256. Coronal and axial images are shown in Fig. 4. Figure 4A shows a T_2-weighted scan of the EAE mouse cerebrum (rostral) and cerebellum (caudal) to define the normal anatomy. Figure 4B shows a T_1-weighted scan

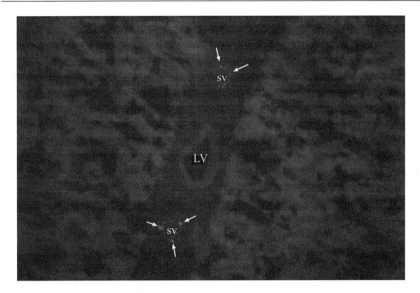

FIG. 3. Fluorescent anti-ICAM-1 ACPVs bind *in vivo* to cerebellar vasculature of a mouse with grade 2 EAE. Anti-ICAM-1 ACPVs (arrows) are shown bound to small vessels (sv), but not to the central large vessel (LV). (See color insert.)

	Anti-ICAM-1 PV	Anti-vβ-11 PV	Anti-ICAM-1 PV
A T₂W EAE	B T₁W EAE	C T₁W EAE	D T₁W Normal

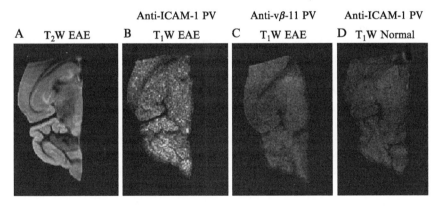

FIG. 4. High-resolution MR images of the EAE mouse brain with anti-ICAM-1 PV contrast enhancement versus controls.

of the same EAE mouse injected with anti-ICAM-1 ACPVs. Diffuse enhancement is present throughout the brain, lending particularly significant contrast between white and gray matter in the cerebellum. Small punctate lesions are observed throughout the image, but seem to be concentrated in the cerebellum (lower part of Fig. 4B). Figure 4C is a T_1-weighted scan of an EAE mouse injected with control anti-Vβ11 ACPVs. No significant enhancement is observed. Similarly, a scan of a healthy mouse injected with anti-ICAM-1 ACPVs showed no enhancement (Fig. 4D).

We have demonstrated that ACPVs, a new target-specific molecular imaging agent, can be successfully delivered to CAMs upregulated in disease. This result lays the groundwork for *in vivo* imaging studies of other endothelial antigens, particularly those related to angiogenesis.

Molecular Imaging of Angiogenesis

Murine antibodies against the $\alpha_v\beta_3$ integrin (LM609) were conjugated to PVs and evaluated in a rabbit tumor model (V2 carcinoma) that has previously shown upregulation of the integrin on the vasculature. Rabbit tumor models were used because antibodies against the $\alpha_v\beta_3$ integrin do not cross-react with murine models. V2 carcinoma cells were inoculated into the thigh muscle or placed subcutaneously in New Zealand white rabbits. The rabbits were closely monitored until a palpable tumor was established. For *in vivo* MR studies, rabbits with palpable tumors (approximately 1–3 cm in diameter) were injected intravenously with either 5 ml/kg anti-$\alpha_v\beta_3$ (LM609)-labeled ACPVs (1 mg antibody/kg, 0.005 mmol Gd^{+3}/kg) or control ACPVs with isotype-matched control antibodies. MRI was performed using a 1.5-T GE Signa MR imager using an extremity coil and the following imaging parameters: TR = 300 ms, TE = 18 ms, NEX = 2, Field of View (FOV) = 16 cm, 256 × 256 matrix, slice thickness = 3 mm. MR images were obtained immediately before contrast injection and at immediate, 30 min, 1 h, and 24 h post-contrast injection in the coronal plane. The rabbits were euthanized immediately after the last MRI experiment, and the tumor tissues were harvested for immunohistochemical studies. Figure 5 illustrates the MR findings of a V2 carcinoma-carrying rabbit injected with LM609-labeled ACPVs. At immediate, 30 min, and 1 h post-contrast injection, no noticeable enhancement of the tumor or tumor margin occurs as compared with the pre-contrast image (Fig. 5A,1), whereas at 24 h post-contrast injection (Fig. 5A,2), enhancement of the tumor margin is clearly visible. Isotype-matched controls showed low contrast enhancement in 24-h post-contrast injection in both tumor models (compare images 1 and 2 in Fig. 5B).

Figure 6 illustrates immunohistochemical slides taken at the tumor margin of the same rabbit shown in Fig. 5. Figure 6A was stained using anti-$\alpha_v\beta_3$ antibodies, and Fig. 6B was stained using polyclonal anti-mouse antibodies, which should bind to the LM609 antibodies on the ACPVs. Notice that the locations of the stained regions in Fig. 6A and 6B are

LM609-PV   Isotype-matched ACPV

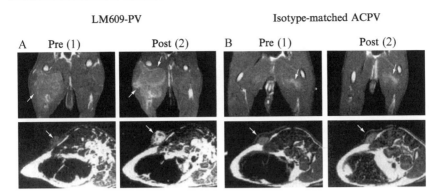

FIG. 5. (A) MR images of V2 carcinoma in the thigh muscle of a rabbit and subcutaneously before (1), and at 24 hours (2) after anti-$\alpha_v\beta_3$-labeled ACPV injection. (B) MR images of isotype-matched controls.

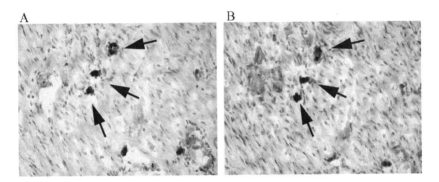

FIG. 6. Immunohistochemical slides taken at the tumor margin of the same rabbit shown in Fig. 5. Figure 6A was stained using anti-$\alpha_v\beta_3$ antibodies, and Fig. 6B was stained using polyclonal anti-mouse antibodies, which should bind to the LM609 antibodies on the ACPVs. Arrows indicate the most prominently stained areas.

identical, illustrating that the LM609-labeled ACPVs have indeed localized to the sites of $\alpha_v\beta_3$ upregulation. Our results also show that the zone of $\alpha_v\beta_3$ upregulation seen on immunohistochemistry corresponds to the zone of enhancement seen on LM609-labeled ACPVs enhanced MRI. Also note that the stain is localized to the vessels and is not distributed throughout the tumor cells.

We have also recently reported the imaging and therapy of a solid tumor with LM609-ACPVs labeled with indium 111 (^{111}In). Targeting of LM609-ACPVs was demonstrated by parenteral injection in the V2 carcinoma

A Unconjugated PV (^{111}In)

t = 0   t = 8 h   t = 24 h   t = 48 h   t = 72 h

B LM609-PV (^{111}In)

t = 0   t = 8 h   t = 24 h   t = 48 h   t = 72 h

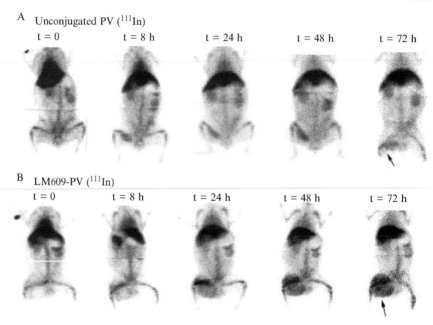

FIG. 7. Imaging of a solid tumor with LM609-ACPVs labeled with indium 111 (^{111}In).

rabbit model by scintographic imaging of the tumor with LM609-^{111}In ACPVs. Scintographic imaging of the V2 carcinoma implanted in the rabbit thigh was performed using 100 nm LM609-^{111}In PVs after a single IV injection. Before imaging, the tumor volume was approximately 3 cm^3. Gamma emission was monitored over a 72-h period and showed accumulation of 22% of the total injected radiation in the tumor at 72 h relative to 3% for the ^{111}In ACPVs, which lacks the anti-$\alpha_v\beta_3$ integrin antibody LM609 (Fig. 7). The accumulation of the targeted vesicle represents approximately 1% of the injected dose per gram of tumor tissue. However, this dose is a severe underestimation of the dose to the tumor vasculature because of the confinement of the vesicle to the tumor vasculature, which represents a small fraction of the total tumor mass. The blood pool half-life was 18 h for the LM609-^{111}In ACPVs and was similar to the complex without antibody.

Molecular Imaging and Vascular-Targeted Therapeutics

Since the seminal work by Folkman and co-workers, there has been tremendous interest in using changes in tumor vasculature as target for therapy.[19] We have recently developed a cationic form of the targeted

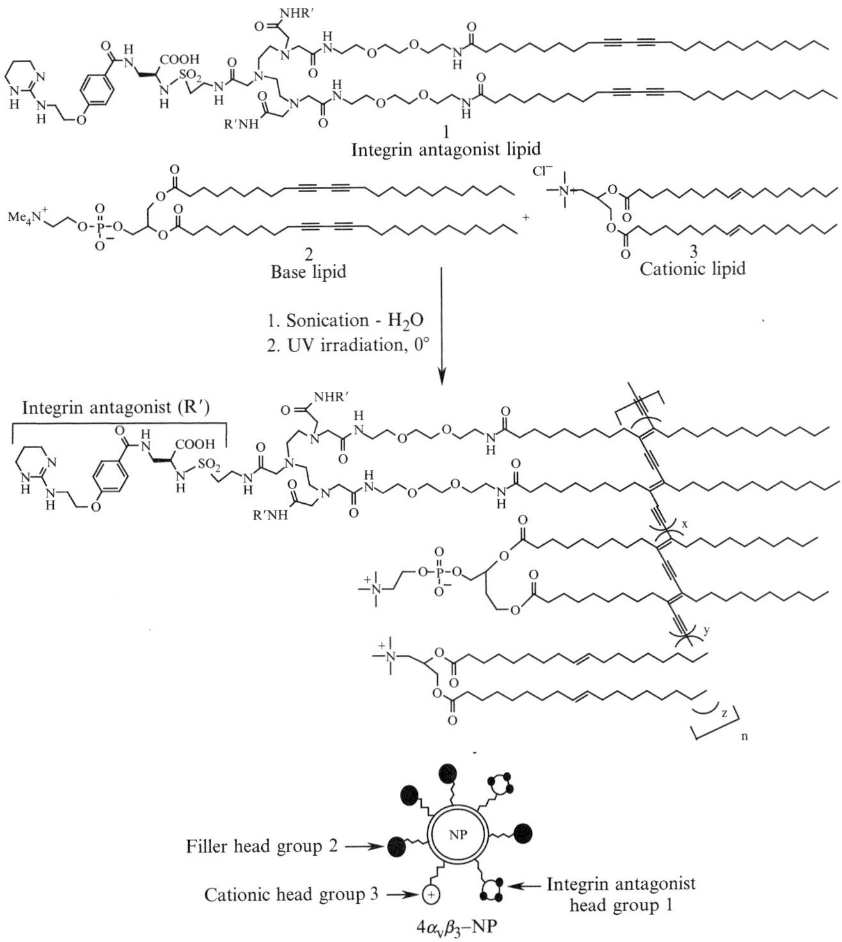

FIG. 8. Formation of targeted NPs for gene delivery to the tumor endothelium. Modified with permission from Fig. 1 of J. D. Hood.[20]

nanoparticle (NP) based on the previously described PV format that can bind tumor vasculature with high specificity, transport genetic material into the tumor endothelium, and be followed using molecular imaging techniques (Fig. 8). In this application, we used an integrin antagonist (IA) that

[19] T. J. Passe, D. A. Bluemke, and S. S. Siegelman, *Radiology* **203,** 593 (1997).

[20] J. D. Hood, M. D. Bednarski, R. Frausto, S. Guccione, R. A. Reisfeld, R. Xiang, and D. A. Cheresh, *Science* **296,** 404 (2002).

binds the integrin $\alpha_v\beta_3$[21] with high specificity as the targeting moiety. We first imaged these materials to screen for a highly selective targeting system and then turned our attention to delivering therapeutic genes to the vasculature. For therapy, a plasmid with the mutant form of *Raf-1* was used as the therapeutic agent. This was used because the disruption of the Ras-Raf-MEK-ERK pathway suppresses angiogenesis *in vivo,* and suppression of *Raf-1* activity has been reported to promote apoptosis.[22,23] The mutant form of *Raf-1* that we use fails to bind adenosine triphosphate (ATP) (ATP$^\mu$-Raf), and it has a dominant negative effect. To evaluate the therapeutic efficacy of this targeted NP–plasmid complex [IA-NP–Raf(−)], systemic injection of the complex in six mice bearing established 400-mm³ M21-L tumors was performed. The M21-L tumors do not express the integrin $\alpha_v\beta_3$, so any observed therapeutic effect should be from the effect on endothelial cells alone. Control groups received systemic injections of the nontargeted complex [nt-NP–Raf(−)], or PBS (six mice with established 400-mm³ M21-L tumors per group). A blocking experiment was also performed with the coinjected of IA-NP–Raf(−) and a 20-fold molar excess of the soluble IA. Our results demonstrated that control mice injected with PBS or nt-NP–Raf(−) formed large tumors (>1200 mm³) within 25 days and were euthanized. In contrast, mice injected with a single injection of IA-NP–Raf(−) showed rapid tumor regression (Figs. 9 and 10), with four of the six mice showing no evidence of tumor 6 days after NP–Raf treatment.

The two remaining mice showed a >95% reduction in tumor mass. It is important to note that the tumor regressions were sustained for >250 days, suggesting that the indirect effect of this approach may be widespread. It is also significant that injection of excess soluble $\alpha_v\beta_3$ ligand completely blocked the antitumor effect of IA-NP–Raf(−), demonstrating that this is a specific effect based on the ability of the IA-NP–plasmid complex to promote apoptosis of the angiogenic endothelium.[24]

Established syngeneic pulmonary and hepatic metastases of colon carcinoma were used for further evaluation of this therapeutic approach. Murine CT-26 carcinoma cells were injected either intravenously to induce pulmonary metastases or intrasplenically into Balb/C mice to induce hepatic metastases. NP–gene complexes were given after these metastases were allowed to establish for 10 days. Control groups included mice treated

[21] C. A. Hulka, W. B. Edmister, B. L. Smith *et al., Radiology* **205,** 837 (1997).
[22] N. A. Mayr, W. T. Yuh, V. A. Magnotta *et al., Int. J. Radiat. Oncol. Biol. Phys.* **36,** 623 (1996).
[23] H. Hawighorst, P. G. Knapstein, W. Weike *et al., Cancer Res.* **57,** 4777 (1997).
[24] P. C. Stomper, J. S. Winston, S. Herman *et al., Breast Cancer Res. Treat.* **45,** 39 (1997).

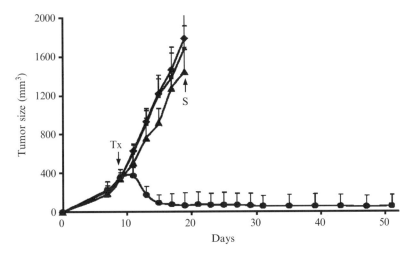

FIG. 9. Tumor size measurements as a function of time. ◆, PBS; ●, NP–plasmid; ▲, blocking experiment; ■, targeted NP–Raf(−) plasmid; Tx, treatment; S, sacrifice. Modified with permission from Fig. 4 of J. D. Hood.[20]

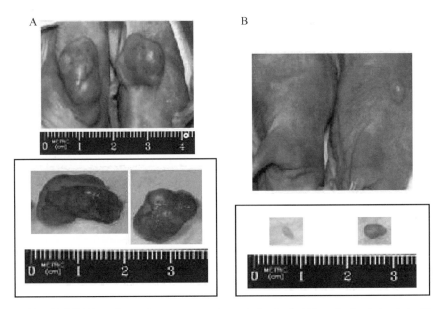

FIG. 10. M21-L murine melanoma tumors in control animals (A), and animals treated with a single injection of IA-NP–Raf(−) complex (B). Modified with permission from Supplemental Fig. 2 of J. D. Hood.[20]

with PBS, IA-NP complexed to a control vector, or a nt-NP–Raf(−). Our results showed that in the control mice, extensive tumor burden in their lungs or livers were observed, whereas in mice treated with IA-NP–Raf(−), little or no visible tumor metastasis was present, resulting in a drastic reduction of the wet lung or liver weight. Blocking experiments in which mice received a coinjection of IA-NP–Raf(−), along with a 20-fold molar excess of the soluble targeting ligand, showed tumor burden similar to that in control mice.[24]

Summary

In summary, we have shown *in vivo* imaging of angiogenic tumors using anti-avb3-targeted polymerized vesicles comprised of the murine antibody LM609 attached to PVs labeled with the MR contrast agent gadolinium (Gd) in the V2 carcinoma model in rabbits. MRI studies using this targeted contrast agent revealed large areas of avb3 integrin expression in tumor-associated vasculature that conventional MRI failed to show. Immunohistochemical staining of the tumors showed colocalization of the ACPVs with the avb3 integrin in the tumor blood vessels without penetration into the surrounding tumor tissue. Other investigators have used microemulsions conjugated to an antibody targeted against avb3 as imaging agents. These materials also show contrast enhancement of tumor vasculature undergoing angiogenesis.[25] Other markers such as the PECAM-1 (CD-31), VCAM-1 (CD54), and VEGF receptor (flk-1) have been shown to be upregulated on tumor endothelium and associated with angiogenesis, but have not been used in imaging studies.

[25] S. A. Wickline and G. M. Lanza, *Circulation* **107**, 1092 (2003).

[11]   Production and Applications of Copper-64 Radiopharmaceuticals

By XIANKAI SUN and CAROLYN J. ANDERSON

Introduction

Copper is the third most abundant natural trace metal in humans (after iron and zinc), and its roles in biochemistry and metabolism have been well defined.[1,2] There is a long history of application of the longer-lived copper radionuclides, [67]Cu and ^{64}Cu, as biomedical tracers to probe copper absorption, metabolism, and excretion,[3] and for assessment of patients with Wilson's disease.[4] More recently, ^{64}Cu has been utilized as a tracer to determine the role of copper chaperone proteins as transporters of copper across the cell membrane,[5] and to the secretory pathway in cells.[6]

Along with the use of copper radionuclides to probe the cellular biochemistry of copper, there are a relatively large number of copper radionuclides that are potentially suitable for use in diagnostic imaging and/or targeted radiotherapy, specifically: [67]Cu, ^{64}Cu, [62]Cu, [61]Cu, and [60]Cu (Table I). These copper radionuclides present diverse nuclear properties, including half-lives from 10 min to 62 h, and decay by positron (β^+) and/or beta minus (β^-) emission, and gamma (γ) emission. The regional tissue distribution of all of these copper radionuclides can be externally assessed with gamma or positron imaging techniques. Copper-67 and ^{64}Cu were the subject of early investigations in tumor imaging, either as porphyrin[7] or citrate[8] complexes. Over the past 2 decades, there has been an increased interest in the use of copper radionuclides in the development of radiopharmaceuticals for positron emission tomography (PET) imaging

[1] K. P. Karlin and J. Zubieta, "Copper Coordination Chemistry: Biochemical and Inorganic Perspectives." Adenine Press, New York, 1983.

[2] M. C. Linder, "Biochemistry of Copper," Vol. 10. Plenum Press, New York, 1991.

[3] C. A. Owen, "Biochemical Aspects of Copper: Copper Proteins, Ceruloplasmin, and Copper Protein Binding." Noyes Publications, Park Ridge, NJ, 1982.

[4] C. A. Owen, "Wilson's Disease: The Etiology, Clinical Aspects, and Treatment of Inherited Copper Toxicosis." Noyes Publications, Park Ridge, NJ, 1981.

[5] J. Lee, M. M. O. Pena, Y. Nose, and D. J. Thiele, *J. Biol. Chem.* **277**, 4380 (2002).

[6] I. Hamza, M. Schaefer, L. W. Klomp, and J. D. Gitlin, *Proc. Nat. Acad. Sci. USA* **96**, 13363 (1999).

[7] R. Bases, S. S. Brodie, and S. Rubenfeld, *Cancer* **11**, 259 (1963).

[8] C. Raynaud, D. Comar, M. Dutheil, P. Blanchon, O. Monod, R. Parrot, and M. Rymer, *J. Nucl. Med.* **14**, 947 (1973).

Copyright 2004, Elsevier Inc.
All rights reserved.
0076-6879/04 $35.00

TABLE I
DECAY CHARACTERISTICS OF COPPER RADIONUCLIDES

Isotope	$T_{1/2}$	β^- MeV (%)	β^+ MeV (%)	EC (%)	γ MeV (%)
^{60}Cu	23.4 min	–	3.92 (6)	7.4	0.85 (15)
			3.00 (18)		1.33 (80)
			2.00 (69)		1.76 (52)
					2.13 (6)
^{61}Cu	3.32 h	–	1.22 (60)	40	0.284 (12)
					0.38 (3)
					0.511 (120)
^{62}Cu	9.76 min	–	2.91 (97)	2	0.511 (194)
^{64}Cu	12.7 h	0.578 (39)	0.653 (17.4)	41	1.35 (0.6)
					0.511 (34.8)
^{67}Cu	62.0 h	0.577 (20)	–	–	0.184 (40)
		0.484 (35)			0.092 (23)
		0.395 (45)			

of various pathologies, including cardiovascular disease, inflammation, and cancer. There have been major advances in the areas of copper radio-nuclide production and copper radiopharmaceutical development for PET imaging and targeted radiotherapy. In this chapter, we address these aspects of copper radiopharmaceutical production and applications.

Production of Copper Radionuclides

One of the major challenges in the production of radionuclides for use as biologic tracers or diagnostic imaging and targeted radiotherapy applications is the production of radionuclides with high specific activity. This means that the amount of mass of nonradioactive nuclide must be kept to a minimum. In producing copper radionuclides, this is an even bigger challenge than for other radionuclides, since copper is ubiquitous in the environment. For all of the copper isotopes described as follows, a noncopper target is used to produce no-carrier-added copper radionuclides. In using a target that has a different atomic number, a chemical separation of the copper radionuclide from the target material is possible. In addition to using noncopper target material, the experimental conditions for preparing the target and separating the copper radionuclides from the target must be as metal-free as possible. In this section, we describe the production of copper radionuclides that are currently of interest as biologic tracers and for nuclear medicine applications.

Production of ^{67}Cu

Copper-67 is the longest-lived copper radionuclide, with practical applications in imaging and therapy. With a 2.6-d half-life, ^{67}Cu emits beta particles with energy maxima ranging from 0.4–0.6 MeV, which are ideal for cancer therapy. Along with 100% beta emission, ^{67}Cu emits gamma photons of 92 and 184 keV that are suitable for gamma scintigraphy. The combination of the half-life, beta, and gamma emissions makes ^{67}Cu a highly attractive radionuclide for cancer therapy. However, the production of amounts of ^{67}Cu suitable for nuclear medicine studies requires a higher-energy accelerator or cyclotron. Currently in the United States, ^{67}Cu is produced on a high-energy accelerator at Brookhaven National Laboratory at the Brookhaven Linac Isotope Producer (BLIP) by $^{68}Zn(n,2p)$ (neutron, 2 protons) reactions with 200 MeV protons.[9] Because of the high cost of producing ^{67}Cu, its availability is limited, and therefore this radionuclide is not currently routinely used in nuclear medicine.

There are reports in the literature describing the production of ^{67}Cu on lower energy accelerators and cyclotrons, which would reduce the cost and increase the availability of this radionuclide. Jamriska *et al.*[10] describe preliminary experiments on producing ^{67}Cu via a (p,α) reaction on ^{70}Zn targets using 20.4 MeV protons and currents up to 17 μA. The ^{70}Zn targets were fabricated from compressed oxide, electrodeposited zinc metal, or stacked zinc metal foils.[10] Short bombardments (up to 1 h) were performed, and small amounts of ^{67}Cu were produced [up to 385 kBq (10.4 μCi)]. A more recent publication investigated the feasibility of producing ^{67}Cu on a lower energy (20.5 MeV) cyclotron via the $^{70}Zn(p,\alpha)^{67}Cu$ Cu reaction.[11] In these experiments, ^{70}Zn was electroplated onto a gold backing. A disadvantage of this method is the high cost of the enriched ^{70}Zn target material (U.S. \$240/mg), although an efficient recovery process would minimize this expense. Only preliminary experiments were reported on the production of ^{67}Cu by this method,[11] and further experiments have not yet been reported.

Production of ^{64}Cu

Copper-64 can be effectively produced by both reactor-based and accelerator-based methods. One method of ^{64}Cu production is the $^{64}Zn(n,p)^{64}Cu$ reaction in a nuclear reactor.[12] Most reactor-produced radionuclides are

[9] S. Mirzadeh, L. F. Mausner, and S. C. Srivastave, *Appl. Rad. Isot.* **37,** 29 (1986).
[10] D. L. Jamriska, W. A. Taylor, M. A. Ott, R. C. Heaton, D. R. Phillips, and M. M. Fowler, *J. Radioanal. Nucl. Chem. Art.* **195,** 263 (1995).
[11] S. Kastleiner, H. H. Coenen, and S. M. Qaim, *Radiochimica Acta* **84,** 107 (1999).

produced using thermal neutron reactions, or (n,γ) reactions, where the thermal neutron is of relatively low energy, and the target material is of the same element as the product radionuclide. For producing high-specific activity ^{64}Cu, fast neutrons are used to bombard the target in a (n,p) reaction. Unlike a thermal neutron reaction, a fast or highly energetic neutron has sufficient energy to eject a particle from the target nucleus. Using fast neutrons, ^{64}Cu can be produced from ^{64}Zn by the reaction ^{64}Zn(n,p)^{64}Cu. By utilizing fast neutrons, high-specific activity ^{64}Cu was produced at the Missouri University Research Reactor (MURR) in amounts averaging 250 mCi.[12]

There are a few disadvantages to this method of ^{64}Cu production, however. Access to the fast flux trap at MURR is available only on a weekly basis, and limitations on sample size limit the amount of ^{64}Cu that can be made per production run. In addition, fast neutron reactions using reactor neutrons are always accompanied by thermal neutron reactions. Investigators at MURR used high-purity natural zinc metal, which is 48.6% ^{64}Zn. There are several thermal neutron reactions that occur on the different zinc isotopes, and this produces significant quantities of some undesirable impurities, such as ^{65}Zn ($T_{1/2} = 245$ d). These caveats have led to MURR discontinuing the production of ^{64}Cu.

Smith and co-workers[13] separated large amounts of ^{64}Cu by-product from cyclotron production of ^{67}Ga via the ^{68}Zn(p,2n)^{67}Ga reaction at the National Medical Cyclotron, Sydney, Australia. This mode of production has the advantage of being very economical and allows for production of very large amounts (>3 Ci) of reasonably high-specific activity material ($\sim$860 Ci/mmol). The disadvantage is that on-demand production would be problematic, since the major radionuclide produced is longer lived ^{67}Ga ($T_{1/2} = 72$ h).

The production of no-carrier-added ^{64}Cu via the ^{64}Ni(p,n)^{64}Cu reaction on a biomedical cyclotron was proposed by Szelecsenyi et al.[14] In this study, small irradiations were performed demonstrating the feasibility of ^{64}Cu production by this method. Subsequent studies by McCarthy et al.[15] were performed, and this method is now used to provide ^{64}Cu to researchers throughout the United States. The following sections outline the

[12] K. R. Zinn, T. R. Chaudhuri, T. P. Cheng, J. S. Morris, and W. A. Meyer, *Cancer* **73,** 774 (1994).

[13] S. V. Smith, D. J. Waters, and N. Di Bartolo, *Radiochimica Acta* **75,** 65 (1996).

[14] F. Szelecsenyi, G. Blessing, and S. M. Qaim, *Appl. Radiat. Isot.* **44,** 575 (1993).

[15] D. W. McCarthy, R. E. Shefer, R. E. Klinkowstein, L. A. Bass, W. H. Margenau, C. S. Cutler, C. J. Anderson, and M. J. Welch, *Nucl. Med. Biol.* **24,** 35 (1997).

FIG. 1. Gold disk *(top)* and gold disk electroplated with ^{64}Ni *(bottom)* for use as a target to prepare ^{64}Cu on a biomedical cyclotron via the ^{64}Ni(p,n)^{64}Cu reaction. Courtesy of Lucie Tang and Michael J. Welch. (See color insert.)

methods for production of large amounts [up to 37 GBq (1 Ci)] of ^{64}Cu for diagnostic imaging and cancer therapy applications.

The target for producing ^{64}Cu is enriched ^{64}Ni (99.6%) (ISOFLEX, San Francisco, CA). The ^{64}Ni is electroplated onto a gold disk (ESPI Metals, Ashland, OR) using a procedure modified from Piel *et al.* (Fig. 1).[16] Because of the low natural abundance of ^{64}Ni, the cost for ^{64}Ni is approximately \$35/mg. Appropriate quantities of nickel metal are dissolved in 6 *M* nitric acid and evaporated to dryness. Typical quantities of ^{64}Ni are

[16] H. Piel, S. M. Qaim, and G. Stocklin, *Radiochim. Acta* **57**, 1 (1992).

10–50 mg. The residue is treated with concentrated sulfuric acid, diluted with deionized water, and evaporated to almost dryness. The residue is cooled and diluted with deionized water. The pH is then adjusted to 9 with concentrated ammonium hydroxide, and ammonium sulfate electrolyte is added. The final volume of the solution is adjusted to approximately 10 ml with deionized water. This solution is transferred to the cells for electroplating. The cells are typically operated at 2.4–2.6 V and at currents between 10 and 50 mA. Electroplating is accomplished in 12–24 h. Gold disks are used as cathodes and graphite rods are used as anodes in the electroplating experiments. The anode is rotated at about 100 rpm during electrodeposition.

At the Washington University School of Medicine, ^{64}Cu is produced on a Cyclotron Corporation CS-15 cyclotron using 15.5 MeV protons by the ^{64}Ni(p,n)^{64}Cu reaction. This accelerator is capable of delivering external beams of up to 60 μA of 15.5 MeV protons. Generally, for ^{64}Cu production experiments, enriched ^{64}Ni targets are bombarded with 15–45 μA current. The target holder was specifically designed for the CS-15 cyclotron,[15] and different target holders will be required, depending on the make and model of the cyclotron used.

After bombardment, the ^{64}Cu is separated from the target nickel in a one-step procedure using an ion-exchange column (Fig. 2). The irradiated ^{64}Ni is dissolved off of the gold disk in 5 ml 6 N HCl at 90°. The solution is then eluted through a 4×1 cm Bio-Rad (Hercules, CA) AG1-X8 anion exchange column (treated with 6.0 N HCl prior to use). An additional 10 ml 6 N HCl is eluted through the column. The nickel fraction is eluted in the first 15 ml of 6 N HCl. Upon switching to 0.5 N HCl, the copper is eluted in the first 2–6 ml.

Utilizing the cyclotron conditions described earlier, the yields of ^{64}Cu are dependent on the amount of ^{64}Ni on the target and the length of the bombardment. Typically, 18.5 GBq (500 mCi) ^{64}Cu are produced with a 40-mg ^{64}Ni target and a bombardment time of 4 h. The specific activity of the ^{64}Cu ranges from 47.4 to 474 GBq/μmol (1280 to 12,800 mCi/μmol). The typical yields for ^{64}Cu productions are 0.2 mCi/μA $\times$ h per mg ^{64}Ni. As previously mentioned, it is challenging to produce high-specific activity copper radionuclides due to the extensive presence of copper in the environment. Extra care is taken to use highly pure reagents and metal-free glassware and plasticware.

Recently, Obata et al.[17] reported the production of ^{64}Cu on a 12-MeV cyclotron, which is more representative of the modern cyclotrons currently

---

[17] A. Obata, S. Kasamatsu, D. W. McCarthy, M. J. Welch, H. Saji, Y. Yonekura, and Y. Fujibayashi, *Nucl. Med. Biol.* **30,** 535 (2003).

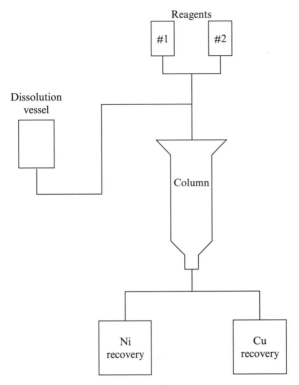

FIG. 2. Schematic for processing the ^{64}Ni target to separate the ^{64}Cu for radiopharmaceutical production and ^{64}Ni recycling. The steps in processing the ^{64}Ni are: (1) dissolve the ^{64}Ni from the gold disk target in hot 6 N HCl in the dissolution vessel; (2) add the contents of the dissolution vessel to the AG1 × 8 ion exchange column; (3) elute the column with reagent 1 (6 N HCl) to elute the ^{64}Ni target material into the Ni recovery vessel; (4) elute the column with reagent 2 (0.5 N HCl) to elute the ^{64}Cu from the ion exchange column and collect the contents in the ^{64}Cu recovery vessel. The 0.5 N HCl is then evaporated from the ^{64}CuCl$_2$ collected using heat under nitrogen, and is resolubilized in 0.1 N HCl for radiolabeling studies. Courtesy of Lucie Tang and Michael J. Welch.

in operation. They utilized very similar methods to those previously published.[15] A remote system was described for separation of the ^{64}Cu from the ^{64}Ni target.

At the Washington University School of Medicine, the ^{64}Ni is recovered by evaporating the nickel fraction to dryness and treating the residue first with 6 N HNO$_3$ and evaporating to dryness, and then with 1 ml concentrated sulfuric acid.[15] The solution is diluted with deionized water, and the pH is adjusted to 9 with concentrated ammonium hydroxide.

Ammonium sulfate is added to the solution, and the final volume is adjusted to 10 ml with deionized water. The solution is then transferred to the electroplating cells as described earlier. Obata et al.[17] reported a somewhat different method for recycling the ^{64}Ni. The ^{64}Ni in 6 N HCl was heated to 150° and evaporated to dryness in a silica glass flask. High-purity water was added, and complete evaporation was performed. The residue was heated to 900° in an oven. After heating for over 24 h, the ^{64}Ni was converted to ^{64}NiO and was ready for target preparation.

Production of ^{60}Cu and ^{61}Cu

A similar procedure is used to prepare ^{60}Cu and ^{61}Cu at the Washington University School of Medicine.[15,18] Isotopically enriched ^{60}Ni and ^{61}Ni are purchased from ISOFLEX. Nickel targets are prepared as described using ^{64}Ni. Targets are irradiated with approximately 14.7 MeV protons on the Cyclotron Corporation CS-15 cyclotron at the Washington University School of Medicine. The yields for ^{60}Cu and ^{61}Cu are typically 1.68 and 0.34 mCi/μA × h per mg nickel, respectively.

Production of ^{62}Cu

Copper-62 is obtained from the decay of cyclotron-produced ^{62}Zn ($T_{1/2}$ = 9.2 h). Although this generator only has a limited life span of 1–2 days, there has been significant interest in ^{62}Cu radiopharmaceuticals. Use of the ^{62}Zn–^{62}Cu parent–daughter generator as a source of ^{62}Cu for radiopharmaceuticals was first suggested by Robinson et al.[19] Several different modes of generator production of ^{62}Cu have been reported. The Robinson generator consists of ^{62}Zn loaded onto a Dowex 1 × 10 anion exchange column, and the generator was eluted with 0.1 N HCl with 100 mg/ml NaCl with 1 μg/ml CuCl$_2$. The carrier copper is not required for efficient separation of ^{62}Cu from ^{62}Zn, and Robinson-type no-carrier-added ^{62}Cu generators have been used extensively in assessment of ^{62}Cu-PTSM as a PET perfusion tracer.[20–22] Fujibayashi et al.[23] loaded ^{62}Zn in 2 ml water at pH 5.0 with

[18] D. W. McCarthy, L. A. Bass, P. D. Cutler, R. E. Shefer, R. E. Klinkowstein, P. Herrero, J. S. Lewis, C. S. Cutler, C. J. Anderson, and M. J. Welch, Nucl. Med. Biol. 26, 351 (1999).
[19] G. D. Robinson, Jr., F. W. Zielinski, and A. W. Lee, Int. J. App. Rad. Isot. 31, 111 (1980).
[20] M. A. Green, C. J. Mathias, M. J. Welch, A. H. McGuire, D. Perry, F. Fernandez-Rubio, J. S. Perlmutter, M. E. Raichle, and S. R. Bergmann, J. Nucl. Med. 31, 1989 (1990).
[21] C. J. Mathias, W. H. Margenau, J. W. Brodack, M. J. Welch, and M. A. Green, Appl. Radiat. Isot. 42, 317 (1991).
[22] C. J. Mathias, M. J. Welch, D. J. Perry, A. H. McGuire, X. Zhu, J. M. Connett, and M. A. Green, Nucl. Med. Biol. 18, 807 (1991).
[23] Y. Fujibayashi, K. Matsumoto, Y. Yonekura, J. Konishi, and A. Yokayama, J. Nucl. Med. 30, 1838 (1989).

≤ 2 μg/ml of Cu carrier on CG-120 Amberlite cation exchange resin. The generator eluent was 200 mM glycine. The Cu carrier was eluted from the column during the first elution. Zweit et al.[24] developed a ^{62}Zn–^{62}Cu generator where the ^{62}Zn was loaded onto a AG1-X8 column that was eluted with 0.3 M HCl/40% ethanol to give >90% ^{62}Cu yield in 3 ml with extremely low ^{62}Zn breakthrough of $<3 \times 10^{-7}$%. Bormans et al.[25] developed a ^{62}Zn–^{62}Cu generator based on a small (25 mm × 5 mm) ion exchange column (Dowex 1 × 16) that is eluted with 1.7 M NaCl and 0.1 M HCl. Currently, a commercially available ^{62}Zn–^{62}Cu generator is available[26] based on the generator described by Herrero et al.[27] This generator has been used for Phase III clinical trials.[28]

Properties of Copper Complexes for Radiopharmaceutical Applications

Unlike technetium and rhenium, which can exist in a wide range of oxidation states (I–VII), copper radioisotopes exhibit relatively simple chemistry in both coordination and redox behavior, which is dominated by two principal oxidation states (I and II). The longer-lived copper radionuclides, such as ^{64}Cu and ^{67}Cu, have been utilized for development of copper-labeled biologic molecules for tumor targeting using monoclonal antibodies and peptides, whereas the shorter-lived copper radionuclides are typically used to form lipophilic copper complexes for measuring blood flow and hypoxia.[29]

Within the last decade, there has been considerable research interest in the field of radiometal-labeled receptor-targeted agents. Receptor ligands that target specific receptors or antigens on tumors can be biomolecules such as monoclonal antibodies and peptides, or small organic molecules such as folic acid or peptide mimetics. The radiometal is connected to these targeting molecules via a bifunctional chelator (BFC), which consists of a chelator to complex the radiometal and a functional group for attachment to the targeting molecule. A large variety of chelating agents have been evaluated for potential applications in copper radiopharmaceuticals, such

[24] J. Zweit, R. Gordall, M. Cox, J. W. Babich, G. A. Petter, H. L. Sharma, and R. J. Ott, *Eur. J. Nucl. Med.* **14**, 418 (1992).

[25] G. Bormans, A. Janssen, P. Adriaens, D. Crombez, A. Witsenboer, J. De Goeij, L. Mortelmans, and A. Verbruggen, *Appl. Radiat. Isot.* **43**, 1437 (1992).

[26] J. L. Lacy, S. C. Chien, J. K. Lim, C. J. Mathias, and M. A. Green, *J. Nucl. Med.* **36**, 49P (1995).

[27] P. Herrero, J. Markham, C. J. Weinheimer, C. J. Anderson, M. J. Welch, M. A. Green, and S. R. Bergmann, *Circulation* **87**, 173 (1993).

[28] N. G. Haynes, J. L. Lacy, N. Nayak, C. S. Martin, D. Dai, C. J. Mathias, and M. A. Green, *J. Nucl. Med.* **41**, 309 (2000).

[29] C. J. Anderson and M. J. Welch, *Chem. Rev.* **99**, 2219 (1999).

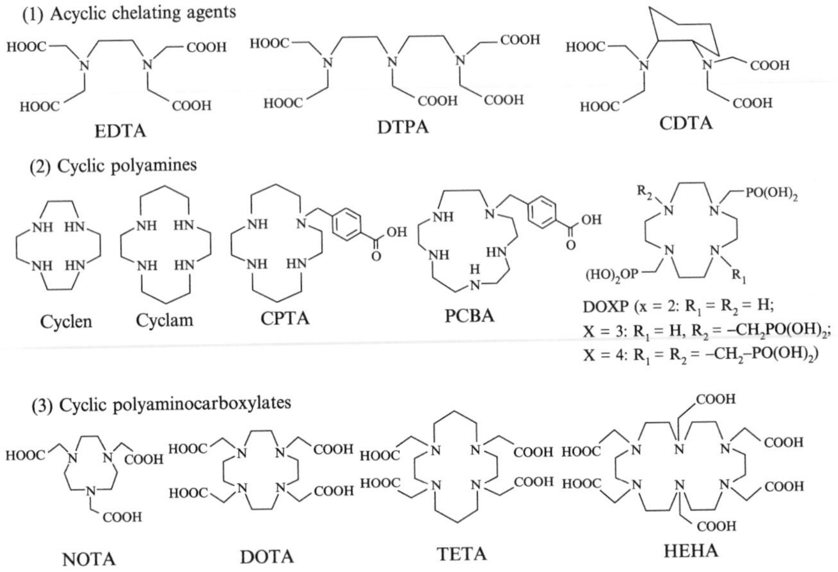

Fig. 3. Representatives of chelating agents evaluated as BFCs for copper radiopharmaceuticals.

as acyclic polyaminocarboxylates,[30–33] cyclic polyamines,[34–36] and cyclic polyaminocarboxylates (Fig. 3).[35,37] Because stability toward metal loss is a highly desirable property for copper chelates utilized in the *in vivo* delivery of copper radionuclides, the kinetic stability of a metal complex *in vivo* is of special importance in the consideration for BFC candidacy. Among the structural factors that affect the affinity between a ligand and a metal ion, topologic and rigidity constraints in the ligand structures before and after the complex formation

[30] D. J. Hnatowich, W. W. Layne, R. L. Childs, D. Lanteigne, and M. A. Davis, *Science* **220**, 613 (1983).
[31] C. F. Meares, M. J. McCall, D. T. Reardan, D. A. Goodwin, C. I. Diamanti, and M. McTigue, *Anal. Chem.* **142**, 68 (1984).
[32] M. K. Moi, C. F. Meares, M. J. McCall, W. C. Cole, and S. J. DeNardo, *Anal. Biochem.* **148**, 249 (1985).
[33] D. Parker, *Chem. Soc. Rev.* **19**, 271 (1990).
[34] P. V. Bernhardt and P. C. Sharpe, *Inorg. Chem.* **39**, 4123 (2000).
[35] I. Lukes, J. Kotek, P. Vojtisek, and P. Hermann, *Coord. Chem. Rev.* **216–217**, 287 (2001).
[36] X. Sun, M. Wuest, Z. Kovacs, A. D. Sherry, R. Motekaitis, Z. Wang, A. E. Martell, M. J. Welch, and C. J. Anderson, *J. Biol. Inorg. Chem.* **8**, 217 (2003).
[37] P. J. Blower, J. S. Lewis, and J. Zweit, *Nucl. Med. Biol.* **23**, 957 (1996).

play critical roles in kinetic stability. For ligands of the polyamine system, the macrocyclic chelators are more topologically constrained than the acyclic ones because of the absence of "head" and "end" groups in a ring,[38] which provides the copper macrocyclic complexes with extra kinetic stability and thermodynamic stability as well. This is the reason why macrocyclic chelators have historically been used as BFCs to bind ^{67}Cu(II) to antibodies rather than acyclic chelators such as ethylenediaminetetraacetic acid (EDTA) and diethylenetetraaminepentaacetic acid (DTPA).[32,39–41] Currently, the application of 1,4,8,11-tetraazacyclotetradecane-1,4,8,11-tetraacetic acid (TETA) as a BFC for copper radiopharmaceuticals has advanced to clinic trials.[42,43] However, it has been reported that the transchelation of ^{67}Cu(II) from a TETA antibody conjugate to ceruloplasmin presented in the plasma of lymphoma patients[44] and dissociation of ^{64}Cu(II) from TETA-octreotide (TETA-OC) to the protein superoxide dismutase occurred in rat liver.[45] Therefore there is a strong need to develop new chelating agents that are more resistant to *in vivo* transchelation.

Various classes of copper chelators have been under evaluation for radiopharmaceutical development (Fig. 4). Brechbiel and co-workers designed and synthesized a series of tachpyr [N,N',N''-tris(2-pyridylmethyl)-1,3,5-cis,cis-triaminocyclohexane] and IM {1,3,5-cis,cis-triaminocyclohexane-N,N,N''-tris-[2-methyl-(N-methylimidazole)]} and their derivatives for the kinetic stability enhancement of copper chelates, and they have reported that both tachpyr and IM copper complexes have copper chelation capabilities that are superior to TETA.[46] Smith and co-workers reported a new cage ligand [SarAr; 1-N-(4-aminobenzyl)-3,6,10,13,16,19-hexaazabicyclo(6.6.6)eicosane-1,8-diamine] that can form

[38] D. H. Busch, *Chem. Rev.* **93**, 847 (1993).

[39] C. J. Anderson, P. A. Rocque, C. J. Weinheimer, and M. J. Welch, *Nucl. Med. Biol.* **20**, 461 (1993).

[40] W. C. Cole, S. J. DeNardo, C. F. Meares, M. J. McCall, G. L. DeNardo, A. L. Epstein, H. A. O'Brien, and M. K. Moi, *Int. J. Rad. Appl. Inst. Part B, Nucl. Med. Biol.* **13**, 363 (1986).

[41] W. C. Cole, S. J. DeNardo, C. F. Meares, M. J. McCall, G. L. DeNardo, A. L. Epstein, H. A. O'Brien, and M. K. Moi, *J. Nucl. Med.* **28**, 83 (1987).

[42] C. J. Anderson, F. Dehdashti, P. D. Cutler, S. W. Schwarz, R. Laforest, L. A. Bass, J. S. Lewis, and D. W. McCarthy, *J. Nucl. Med.* **42**, 213 (2001).

[43] R. T. O'Donnell, G. L. DeNardo, D. L. Kukis, K. R. Lamborn, S. Shen, A. Yuan, D. S. Goldstein, C. E. Carr, G. R. Mirick, and S. J. DeNardo, *J. Nucl. Med.* **40**, 2014 (1999).

[44] G. R. Mirick, R. T. O'Donnell, S. J. DeNardo, S. Shen, C. F. Meares, and G. L. DeNardo, *Nucl. Med. Biol.* **26**, 841 (1999).

[45] L. A. Bass, M. Wang, M. J. Welch, and C. J. Anderson, *Bioconj. Chem.* **11**, 527 (2000).

[46] D. Ma, F. Lu, T. Overstreet, D. E. Milenic, and M. W. Brechbiel, *Nucl. Med. Biol.* **29**, 91 (2002).

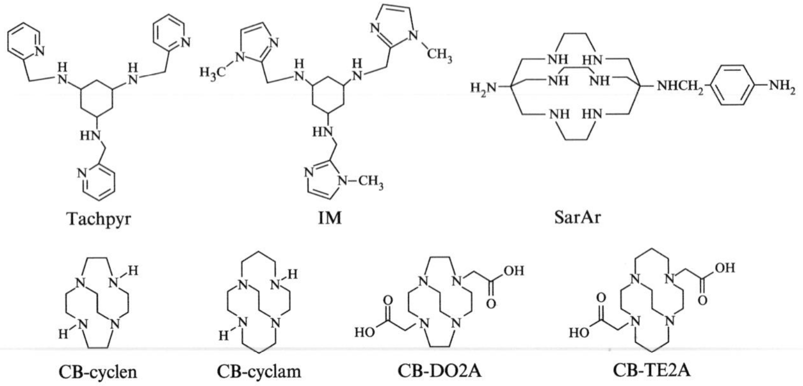

FIG. 4. Three types of new chelating agents currently under evaluation: (1) Tachpyr and IM and their derivatives; (2) Hexaazanacrobicyclic ligands (e.g., SarAr); (3) Cross-bridged tetraazamacrocyclic ligands.

kinetically inert complex with copper(II),[47] whose extraordinary stability may come from the more constrained cage structure and the so-called cryptate effect.[38]

The group at Washington University, in collaboration with investigators at the University of New Hampshire, is investigating a new type of tetra-azamacrocyclic ligands with nonadjacent nitrogens bridged by CH_2CH_2 (Fig. 4). Weisman and Wong have communicated the syntheses of CB-cyclam [1,4,8,11-tetraazabicyclo(6.6.2)hexadecane] and CB-TE2A [4,11-bis-(carboxymethyl)-1,4,8,11-tetraazabicyclo (6.6.2) hexadecane][48,49] and their coordination chemistry with transition metal ions.[49–52] These cross-bridged ligands [bicyclo(6.6.2), (6.5.2), and (5.5.2) tetraamines] were designed to be capable of adopting conformations having all four nitrogen lone pairs convergent upon a cleft (*in,in* at the bridgehead nitrogens), for

[47] N. M. D. Bartolo, A. M. Sargeson, T. M. Donlevy, and S. V. Smith, *J. Chem. Soc. Dalton Trans.* 2303 (2001).

[48] G. R. Weisman, M. E. Rogers, E. H. Wong, J. P. Jasinski, and E. S. Paight, *J. Am. Chem. Soc.* **112,** 8604 (1990).

[49] G. R. Weisman, E. H. Wong, D. C. Hill, M. E. Rogers, D. P. Reed, and J. C. Calabrese, *J. Chem. Soc. Chem. Commun.* 947 (1996).

[50] W. Niu, E. H. Wong, G. R. Weisman, K.-C. Lam, and A. L. Rheingold, *Inorg. Chem. Commun.* 361 (1999).

[51] W. Niu, E. H. Wong, G. R. Weisman, R. D. Sommer, and A. L. Rheingold, *Inorg. Chem. Commun.* **5,** 1 (2002).

[52] E. H. Wong, G. R. Weisman, D. C. Hill, D. P. Reed, M. E. Rogers, J. P. Condon, M. A. Fagan, J. C. Calabrese, K.-C. Lam, I. A. Guzei, and A. L. Rheingold, *J. Am. Chem. Soc.* **122,** 10561 (2000).

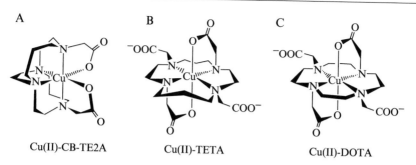

FIG. 5. Structural illustration of (A) Cu(II)-CB-TE2A, (B) Cu(II)-TETA, and (C) Cu(II)-DOTA.

complexation of small hard metal ions. As illustrated in Fig. 5, the crystal structure of Cu(II)-CB-TE2A reveals that Cu(II) is enveloped in the "clam shell" of the cross-bridged ligand, where the Jahn-Teller distortions only amount to 2–3% of the Cu-N and Cu-O bond lengths,[52] whereas the axial Cu-O bonds of Cu(II)-TETA are elongated by 6–13% of the equatorial Cu-N bond lengths.[53] Therefore Cu(II) is well protected from external attacks. The *in vivo* evaluation of ^{64}Cu-labeled cross-bridged ligand complexes has demonstrated that they are significantly more resistant to trans-chelation in rat liver and more rapidly cleared through blood, liver, and kidney than the currently used TETA–DOTA (1,4,7,10-tetraazacyclodode-cane-1,4,7,10-tetraacetic acid) analogues.[54,55]

Preparation of Select Tetraazamacrocyclic Bifunctional Chelators

TETA and DOTA are commercially available from Sigma-Aldrich (St. Louis, MO), Fluka Chemie AG (Buchs, Switzerland), Strem (New-buryport, MA), and Macrocyclics, Inc. (Dallas, TX) at reasonable prices. Macrocyclics, Inc. also provides other BFCs, such as DOTA-NHS (N-hydroxysuccinimide)-ester and DOTA-tris(t-butyl ester), which can be easily used for the conjugation to biomolecules. The first two cross-bridged tetraazamacrocyclic ligands, 4,11-dimethyl-1,4,8,11-tetraazabicyclo[6.6.2]-hexadecane and 4,10-dimethyl-1,4,7,10-tetraazabicyclo[5.5.2]tetradecane, were synthesized by Weisman and co-workers in 1990[48] and Bencini and

[53] A. Riesen, M. Zehnder, and T. A. Kaden, *J. Chem. Soc. Chem. Comm.* 1336 (1985).
[54] C. A. Boswell, X. Sun, M. Wang, B. Ramos, G. R. Weisman, E. H. Wong, and C. J. Anderson, *J. Lab. Compd. Radiopharm.* **44**, S770 (2001).
[55] X. Sun, M. Wuest, G. R. Weisman, E. H. Wong, D. P. Reed, C. A. Boswell, R. Motekaitis, A. E. Martell, M. J. Welch, and C. J. Anderson, *J. Med. Chem.* **45**, 469 (2002).

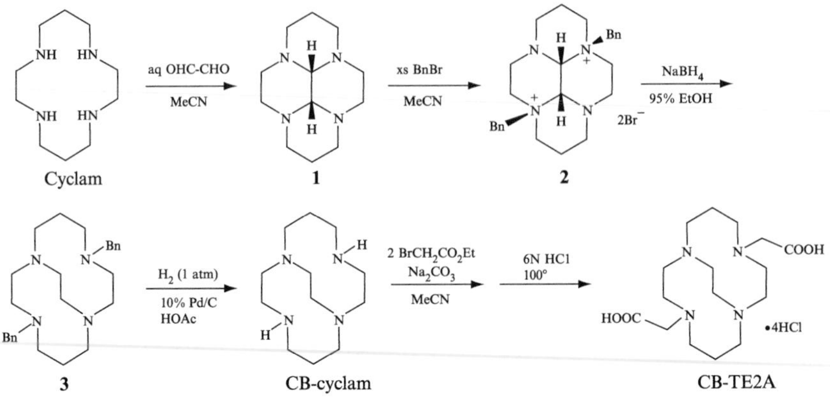

SCHEME 1.

co-workers in 1994, respectively.[56] In 1996 Weisman and co-workers reported the synthesis of the two parent ligands, CB-cyclam and CB-cyclen [1,4,7,10-tetraazabicyclo(5.5.2)tetradecane]. Now modified syntheses have been described to prepare the parent ligands and their derivatives in gram quantities.[49–52] Here we take the synthesis of CB-cyclam and its derivatives as an example to illustrate the preparation.

The efficient route to the parent cross-bridged cyclam (Scheme 1) involves four steps: (1) condensation of glyoxal and cyclam to give tetracyclic bisaminal 1; (2) regioselective dialkylation of 1 to give a bis-quaternary ammonium halide 2; (3) reduction to give double-ring-expanded cross-bridged cyclam 3; and (4) debenzylation by atmospheric hydrogenolysis to give the parent ligand CB-cyclam.[52] CB-cyclam can then easily be converted to a number of N,N'-difunctionalized pendant-arm derivatives (e.g., CB-TE2A) by standard secondary amine functionalization methods (Scheme 1). This synthesis can be done in any organic chemistry laboratory.

Radiochemistry of Copper-Labeled Tetraazamacyclic Bifunctional Chelators

The following procedures are outlined for use with ^{64}Cu; however, they can be applied to the longer-lived copper radionuclides ^{61}Cu and ^{67}Cu. Most of the procedures take longer than 30 minutes, which would be challenging for applications with ^{60}Cu or ^{62}Cu.

[56] A. Bencini, A. Bianchi, C. Bazzicalupi, M. Ciampolini, V. Fusi, M. Micheloni, N. Nardi, P. Paoli, and B. Valtancoli, *Supramol. Chem.* **3,** 141 (1994).

Radiolabeling Chelators with ^{64}Cu

Most tetraazamacrocyclic chelators can be directly radiolabeled with ^{64}Cu by following one general protocol with minor modifications.[36,55,57,58] To radiolabel the chelator efficiently, the concentration and pH of the buffer is very important. In general, ammonium acetate or ammonium citrate buffer is prepared at a concentration of 0.1 M, with the optimal pH range being 5.5–8.0. The freshly prepared buffer should be further treated with Chelex 100 (100–200 mesh, sodium form, ~5 g/L) (Bio-Rad) overnight and filtered through a disposable Corning 1 L filter system (Corning, NY) to eliminate trace metal ions prior to the ^{64}Cu labeling. The buffers can be stored at 4° for up to 2 months. The chelator solution is prepared freshly at a concentration of 5 mM. The radiolabeling yield of a ^{64}Cu complex is not affected as much by the ^{64}Cu specific activity as by its structural nature. Therefore a large amount of radioactivity can generally be used without the risk of radiolysis; for example, 100 μl of 5 mM ligand can be labeled with 15 mCi of ^{64}Cu. The labeling conditions vary with chelators, and for a specific ligand, the reaction conditions should be optimized to use the shortest possible time, under mild temperature and pH conditions. The radiochemical purity can be easily determined by radio thin-layer chromatography (radio-TLC) using reversed-phase C18 or normal-phase silica gel plates (Waters Corp, Milford, MA). Normally no purification or separation of unlabeled ligand is needed for the evaluation of ^{64}Cu-labeled chelators. Typical reaction times for chelators such as DOTA or TETA are from 15–30 min at room temperature.

The radiolabeling of cross-bridged tetraazamacrocyclic ligands with ^{64}Cu following the previous protocol in ammonium citrate buffer failed to give satisfactory results. This is probably because of the increased basicity of the nitrogen atoms of the cross-bridged ligands. To overcome this problem, a new approach has been developed to label ligands with highly basic nitrogen atoms that cannot be deprotonated in aqueous solutions to complex ^{64}Cu.[55] The protocol for this new approach is as follows: (1) Deprotonate the cross-bridged ligand in pure ethanol solution (10 mM ligand) at 95° using cesium carbonate (20 min); (2) add ^{64}CuCl$_2$ to the ligand/Cs$_2$CO$_3$ mixture, and incubate the resulting solution at 95° for 15 min. The radio-TLC conditions are C18 plates as the stationary phase and 9:1 (v/v) MeOH:0.1 M ammonium citrate as the eluant.

[57] C. S. Cutler, M. Wuest, C. J. Anderson, D. E. Reichert, Y. Sun, A. E. Martell, and M. J. Welch, *Nucl. Med. Biol.* **27,** 375 (2000).
[58] T. M. Jones-Wilson, K. A. Deal, C. J. Anderson, D. W. McCarthy, Z. Kovacs, R. J. Motekaitis, A. D. Sherry, A. E. Martell, and M. J. Welch, *Nucl. Med. Biol.* **25,** 523 (1998).

When the radiolabeling is very slow (over 4 h) or the reaction does not go to completion using the previous approaches, using a microwave cavity may be a feasible way to speed up and complete the labeling. In our experience of labeling DO2P [1,4,7,10-tetraazocyclododecane, 1,7-*di*(methanephosphonic acid)], DO3P [1,4,7,10-tetraazacyclododecane-1,4,7-*tri*(methanephosphonic acid)], and DOTP [1,4,7,10-tetraazacyclododecane-1,4,7,10-*tetra*(methanephosphonic acid)] (Fig. 3), DO2P could be only labeled with ^{64}Cu at high radiochemical purity (>97%) after 16 h of incubation at 70° or 4 h at 90°, whereas both DO3P and DOTP could be labeled at comparative yields after 2 h at room temperature. Thus a microwave cavity (*ca.* 180 W) was employed to hasten the radiolabeling of DO2P with [64]Cu. The labeling was complete within 90 sec at 99.7% radiochemical purity.[36]

Characterization of [64]Cu Complexes

A novel ^{64}Cu-labeled complex is generally characterized by chromatography, either radio-TLC or high-performance liquid chromatography (HPLC) as described earlier. Lipophilicity is often determined by Log P measurements, and charge can be determined by electrophoresis. Biologic stability is first determined in serum *in vitro,* and then *in vivo* experiments, such as microPET imaging or biodistribution, are performed. The next sections describe these experiments.

Determination of Partition Coefficients. The Log P measurement can be done by the following protocol: (1) Add 5–10 μl of ^{64}Cu-labeled complex to a biphasic solution containing 500 μl of octanol and 500 μl of water (both octanol and water are obtained from presaturated octanol–water solution); (2) mix the resulting solutions on a shaker for 1–4 h at room temperature; (3) remove an aliquot of 100 μl from each phase and count them separately; and (4) calculate the partition coefficient as a ratio of counts in the octanol fraction to counts in the water fraction.

Complex Charge Measurement. The ^{64}Cu-labeled complex charge can be determined by electrophoresis using a Helena Laboratories electrophoresis chamber (Beaumont, TX) with Sepraphore III cellulose polyacetate strips (Gelman Sciences, Inc., Ann Arbor, MI) presoaked in either 0.1 M HEPES [4-(2-hydroxyethyl)-1-piperazinethanesulfonic acid] buffer, pH 7.4; 0.1 M ammonium acetate buffer, pH 5.5; or 0.1 M citrate buffer, pH 5.5. The procedures can be briefly summarized as: (1) spot the ^{64}Cu-complex solution to the center line of the strips and wait until the spots are dry; (2) apply a power supply (Bio-Rad model 1000/500; Richmond, CA) to develop the strips at a constant current of 5–50 mA for 1–2 h; and (3) analyze the strips using a radio-TLC imaging scanner (Bioscan,

Washington, DC) to determine the migration of radioactivity and overall charge of the complex. Freshly prepared radiolabeled compounds with known overall charges, such as ^{64}Cu-DO2A (neutral), ^{64}Cu-DOTA (2−), ^{64}Cu-Cyclam (2+), and ^{111}In-DTPA (2−) are good standards for the charge determination.[36,58]

It has been shown that the biologic activity of radiometal labeled BFCs are indicative of the *in vivo* behavior of their corresponding peptide[58,59] and protein[60] conjugates. For a newly developed BFC candidate, the first step is to evaluate its *in vitro* and *in vivo* properties, from which it will be justified whether or not it is necessary to pursue further studies. Some of the chelating agents may not be suitable for BFC applications due to their structural features (lack of functionalizable groups for biomolecule attachment).

Serum Stability. The *in vitro* stability can be evaluated by the following procedure: (1) add 10 to 50 μl of ^{64}Cu-labeled complex to 500 μl rat serum (Sigma); (2) incubate the solutions at 37°; and (3) analyze samples by radio-TLC at various time points (10 min, 30 min, 1 h, 2 h, 4 h, and 24 h).

Chemistry and Applications of ^{64}Cu-Labeled BFC–Biomolecule Conjugates

Copper-labeled bioconjugates have been successful in PET imaging and targeted radiotherapy of cancer. A few of the reports in the literature are ^{67}Cu- and/or ^{64}Cu-labeling of monoclonal antibodies, including anti-L1-CAN antibody chCE7,[61] anti-colorectal carcinoma antibody 1A3 and its fragments,[62–64] anti-human B-cell lymphoma antibody Lym-1,[65–69] anti-carcinoembryonic antigen (CEA) antibody T84.66 and its fragments,[57]

[59] C. J. Anderson, T. S. Pajeau, W. B. Edwards, E. L. C. Sherman, B. E. Rogers, and M. J. Welch, *J. Nucl. Med.* **36**, 2315 (1995).

[60] B. E. Rogers, C. J. Anderson, J. M. Connett, L. W. Guo, W. B. Edwards, E. L. C. Sherman, K. R. Zinn, and M. J. Welch, *Bioconj. Chem.* **7**, 511 (1996).

[61] K. Zimmermann, J. Grunberg, M. Honer, S. Ametamey, P. A. Schubiger, and I. Novak-Hofer, *Nucl. Med. Biol.* **30**, 417 (2003).

[62] C. J. Anderson, J. M. Connett, S. W. Schwarz, P. A. Rocque, L. W. Guo, G. W. Philpott, K. R. Zinn, C. F. Meares, and M. J. Welch, *J. Nucl. Med.* **33**, 1685 (1992).

[63] J. M. Connett, C. J. Anderson, L. W. Guo, S. W. Schwarz, K. R. Zinn, B. E. Rogers, B. A. Siegel, G. W. Philpott, and M. J. Welch, *Proc. Nat. Acad. Sci. USA* **93**, 6814 (1996).

[64] G. W. Philpott, S. W. Schwarz, C. J. Anderson, F. Dehdashti, J. M. Connett, K. R. Zinn, C. F. Meares, P. D. Cutler, M. J. Welch, and B. A. Siegel, *J. Nucl. Med.* **36**, 1818 (1995).

[65] G. L. DeNardo, S. J. DeNardo, D. L. Kukis, R. T. O'Donnell, S. Shen, D. S. Goldstein, L. A. Kroger, Q. Salako, D. A. DeNardo, G. R. Mirick, L. F. Mausner, S. C. Srivastava, and C. F. Meares, *Anticancer Res.* **18**, 2779 (1998).

and anti-glucose-6-phosphate isomerase (GPI) immunoglobulin G (IgG).[70] The majority of publications on copper-labeled peptides involve somatostatin analogues[42,59,71–73]; however, a recent paper describes a ^{64}Cu-labeled bombesin analog for imaging prostate cancer.[74]

To conjugate with biomolecules, the tetraazamacrocylic BFCs need to be derivatized at either the carbon or nitrogen backbone or the pendant arm to introduce a reactive functional group that can form a covalent linkage with the thiol or amine groups on proteins or peptides in the form of thioether, thiourea, or amide. The reactive functional groups generally include p-(bromoacetamido)benzyl,[60,62,75] p-(isothiocyanato)benzyl,[60,76] N-hydroxysulfosuccinimidyl,[70,77–79] and activated maleimidyl (Fig. 6).[80–82] Coupling BFCs to peptides is currently performed through standard automated solid phase peptide synthesizer.

Conjugations between BFCs and proteins are mainly conducted in solution by using four typical conjugation methods (Fig. 7). The activated

[66] G. L. DeNardo, S. J. DeNardo, C. F. Meares, D. Kukis, H. Diril, M. J. McCall, G. P. Adams, L. F. Mausner, D. C. Moody, and S. Deshpande, *Antibod. Immunoconj. Radiopharm.* **4**, 777 (1991).

[67] G. L. DeNardo, S. J. DeNardo, C. F. Meares, Q. Salako, D. L. Kukis, L. F. Mausner, J. P. Lewis, L. F. O'Grady, and S. C. Srivastava, *J. Nucl. Med.* **34**, 93P (1993).

[68] G. L. DeNardo, D. L. Kukis, S. Shen, D. A. DeNardo, C. F. Meares, and S. J. DeNardo, *Clin. Cancer Res.* **5**, 533 (1999).

[69] G. L. DeNardo, D. L. Kukis, S. Shen, L. F. Mausner, C. F. Meares, S. C. Srivastava, L. A. Miers, and S. J. DeNardo, *Clin. Cancer Res.* **3**, 71 (1997).

[70] B. T. Wipke, Z. Wang, J. Kim, T. J. McCarthy, and P. M. Allen, *Nat. Immunol.* **3**, 366 (2002).

[71] C. J. Anderson, L. A. Jones, L. A. Bass, E. L. C. Sherman, D. W. McCarthy, P. D. Cutler, M. V. Lanahan, M. E. Cristel, J. S. Lewis, and S. W. Schwarz, *J. Nucl. Med.* **39**, 1944 (1998).

[72] J. S. Lewis, M. R. Lewis, P. D. Cutler, A. Srinivasan, M. A. Schmidt, S. W. Schwarz, M. M. Morris, J. P. Miller, and C. J. Anderson, *Clin. Cancer Res.* **5**, 3608 (1999).

[73] J. S. Lewis, M. R. Lewis, A. Srinivasan, M. A. Schmidt, J. Wang, and C. J. Anderson, *J. Med. Chem.* **42**, 1341 (1999).

[74] B. E. Rogers, H. M. Bigott, D. W. McCarthy, D. Della Manna, J. Kim, T. L. Sharp, and M. J. Welch, *Bioconj. Chem.* **14**, 756 (2003).

[75] M. J. McCall, H. Diril, and C. F. Meares, *Bioconj. Chem.* **1**, 222 (1990).

[76] T. M. Rana and C. F. Meares, *Bioconj. Chem.* **1**, 357 (1990).

[77] M. R. Lewis, C. A. Boswell, R. Laforest, T. L. Buettner, D. Ye, J. M. Connett, and C. J. Anderson, *Cancer Biother. Radiopharm.* **16**, 483 (2001).

[78] M. R. Lewis, J. Y. Kao, A.-L. J. Anderson, J. E. Shively, and A. Raubitschek, *Bioconj. Chem.* **12**, 320 (2001).

[79] M. R. Lewis, A. Raubitschek, and J. E. Shively, *Bioconj. Chem.* **5**, 565 (1994).

[80] J. R. Morphy, D. Parker, R. Kataky, A. Harrison, M. A. W. Eaton, A. Millican, A. Phipps, and C. Walker, *J. Chem. Soc. Chem. Commun.* 792 (1989).

[81] D. Parker, J. R. Morphy, K. Jankowski, and J. Cox, *Pure Appl. Chem.* **61**, 1637 (1989).

[82] Y. Yasushi, H. Matsushima, M. Tagawa, M. Koizumi, K. Endo, J. Konishi, and A. Yokoyama, *Bioconj. Chem.* **2**, 71 (1991).

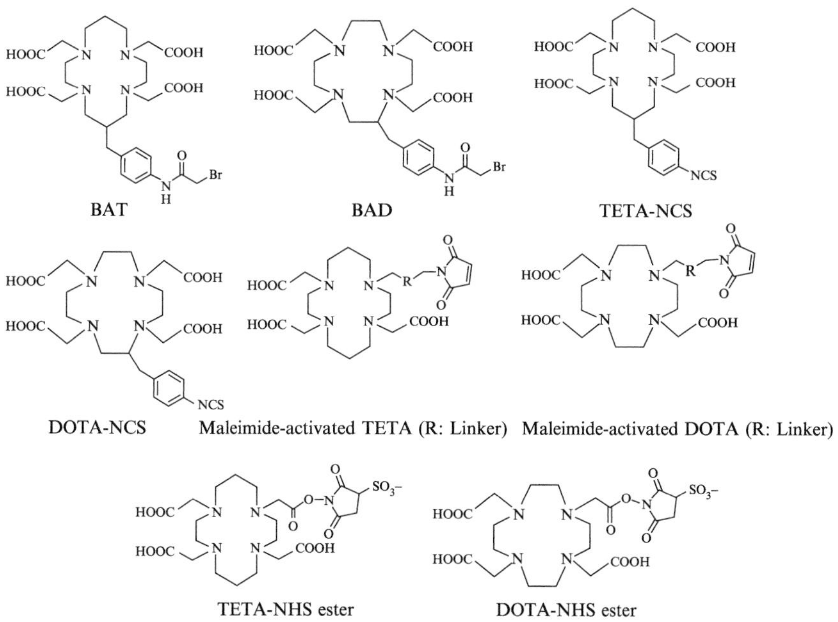

FIG. 6. Activated BFCs with a mono-reactive functional group for conjugation to a protein or peptide.

sulfo-NHS-ester approach (route 4) is preferred because it has the following advantages: (1) the reactions are carried out in aqueous solution at physiologic pH and 4°, which conserves antibody immunoreactivity; (2) the BFC activation and antibody conjugation can be done in a one-pot reaction; (3) the excess of BFC reduces cross-linking; and (4) the resulting peptide bond is relatively stable under physiologic conditions.

The mono-activated esters of BFCs can be prepared via reported synthetic routes, and most of them are also commercially available or can be custom-synthesized at Macrocyclics, Inc. In the following sections, we focus on the procedures for construction of BFC–biomolecule conjugates by two examples: (1) TETA-octreotide and (2) DOTA-mAb coupling via the *sulfo*-NHS-ester route.

Chemistry of BFC–Peptide Conjugation

During the past 10 years, our group at Washington University has been interested in ^{64}Cu-labeled somatostatin analogues (Fig. 8) for PET imaging and targeted radiotherapy of somatostatin receptor-positive

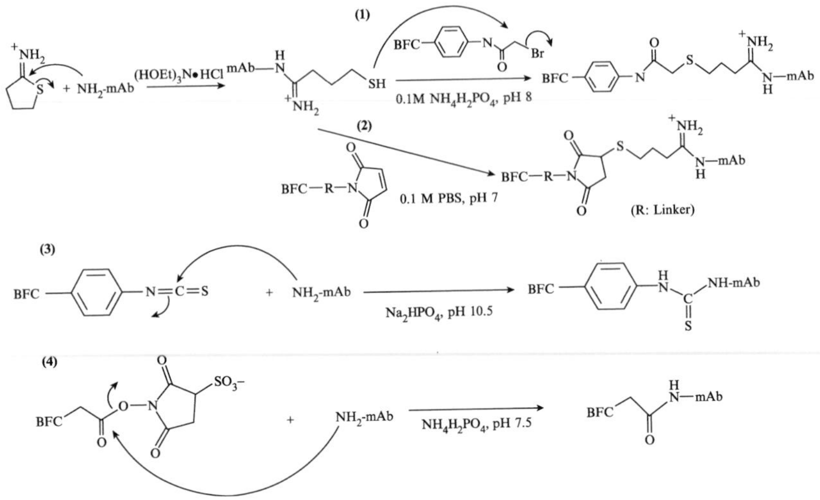

FIG. 7. Four typical routes for BFC–protein conjugation.

tumors. Somatostatin, a 14-amino acid peptide, is involved in the regulation and release of a number of hormones (e.g., growth hormone, thyroid-stimulating hormone, and prolactin).[83,84] Somatostatin receptors (SSRs) have been found in many different normal organ systems such as the central nervous system, the gastrointestinal tract, and the exocrine and endocrine pancreas.[83–85] A large number of human tumors also overexpress SSRs,[86] which makes SSR a good target for diagnostic imaging and therapeutic purposes. Because somatostatin has a very short biologic half-life, the analogue octreotide (OC) was developed with a longer biologic half-life than the native somatostatin.[87] In more recent years, other analogues have been developed, such as Tyr³-octreotide (Y3-OC), octreotate (TATE), and Tyr³-octreotate (Y3-TATE) (Fig. 8).[72,88,89] ^{111}In-DTPA-OC

[83] S. Reichlin, N. Engl. J. Med. 309, 1495 (1983).
[84] S. Reichlin, N. Engl. J. Med. 309, 1556 (1983).
[85] R. Guillemin, Science 202, 390 (1978).
[86] J. C. Reubi, L. K. Kvols, E. P. Krenning, and S. W. J. Lamberts, Metabolism 39(Suppl. 2), 78 (1990).
[87] W. Bauer, U. Briner, W. Doepfner, R. Haller, R. Huguenin, P. Marbach, T. J. Petcher, and J. Pless, Life Sci. 31, 1133 (1982).
[88] M. de Jong, B. F. Bernard, E. de Bruin, A. van Gameren, W. H. Bakker, T. J. Visser, H. R. Maecke, and E. P. Krenning, Nucl. Med. Comm. 19, 283 (1998).
[89] M. de Jong, W. A. P. Breeman, W. H. Bakker, P. P. M. Kooij, B. F. Bernard, L. J. Hofland, T. J. Visser, A. Srinivasan, M. Schmidt, J. L. Erion, J. E. Bugaj, H. R. Maecke, and E. P. Krenning, Cancer Res. 58, 437 (1998).

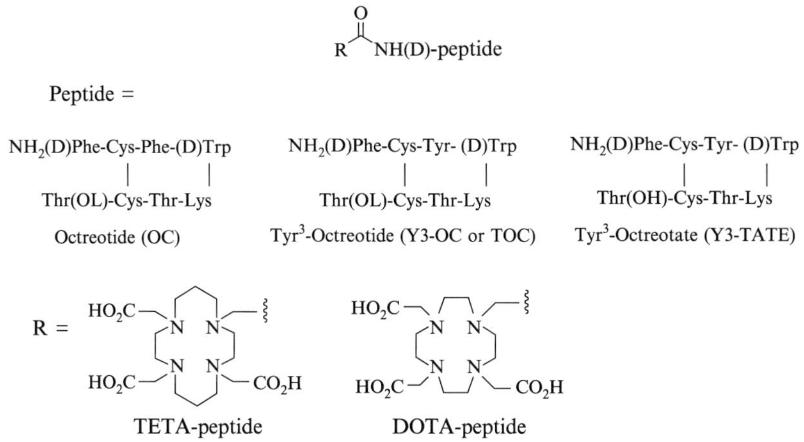

$$R \overset{O}{\underset{}{\|}} NH(D)\text{-peptide}$$

Peptide =

NH$_2$(D)Phe-Cys-Phe-(D)Trp
　　|　　　　|
Thr(OL)-Cys-Thr-Lys

Octreotide (OC)

NH$_2$(D)Phe-Cys-Tyr- (D)Trp
　　|　　　　|
Thr(OL)-Cys-Thr-Lys

Tyr^3-Octreotide (Y3-OC or TOC)

NH$_2$(D)Phe-Cys-Tyr- (D)Trp
　　|　　　　|
Thr(OH)-Cys-Thr-Lys

Tyr^3-Octreotate (Y3-TATE)

R =

TETA-peptide   DOTA-peptide

FIG. 8. Somatostatin analogues for labeling with copper radionuclides.

was one of the first clinically approved peptide-based tumor receptor imaging agents for neuroendocrine tumors.[90] Our studies have shown the potential of these somatostatin analogues to be utilized for both PET diagnostic imaging and targeted radiotherapy.[42,59,71,73]

Solid-Phase Synthesis of Somatostatin Analogues

In the following method, somatostatin analogues are used as examples of how BFC–peptide conjugates are constructed in solid phase as previously described.[73,91–93] Generally, it involves two steps: (1) preparation of the linear protected peptide, Fmoc-D-Phe-Cys(Acm)-Tyr(OtBu)-D-Trp(Boc)-Thr(OtBu)-Cys(Acm)-Thr(OtBu), using the 9-fluorenylmethoxycarbonyl (Fmoc) strategy; and (2) conjugation of the chelator to the peptide. The peptide can be synthesized by an automated peptide synthesizer (Applied Biosystems Model 432A "Synergy" Peptide synthesizer, San Francisco, CA). The instrument protocol requires 25 mmol of subsequent

[90] E. P. Krenning, W. H. Bakker, P. P. M. Kooij, W. A. P. Breeman, H. Y. Oei, M. de Jong, J. C. Reubi, T. J. Visser, C. Bruns, D. J. Kwekkeboom, A. E. M. Reijs, P. M. van Hagen, J. W. Koper, and S. W. J. Lamberts, *J. Nucl. Med.* **33,** 652 (1992).

[91] S. Achilefu, H. N. Jimenez, R. B. Dorshow, J. E. Bugaj, E. G. Webb, R. R. Wilhelm, R. Rajagopalan, J. Johler, and J. L. Erion, *J. Med. Chem.* **45,** 2003 (2002).

[92] W. B. Edwards, C. G. Fields, C. J. Anderson, T. S. Pajeau, M. J. Welch, and G. B. Fields, *J. Med. Chem.* **37,** 3749 (1994).

[93] W. P. Li, J. S. Lewis, J. Kim, J. E. Bugaj, M. A. Johnson, J. L. Erion, and C. J. Anderson, *Bioconj. Chem.* **13,** 721 (2002).

Fmoc-protected amino acids activated by a combination of N-hydroxyben-zotriazole (HOBt) and 2-(1-H Benzotriazol-1-yl)-1,1,3,3-tetramethyluro-nium hexafluorophosphate (HBTU). The prepacked amino acids are available from Perkin-Elmer (Norwalk, CT), BACHEM Bioscience (King of Prussia, PA), or Novabiochem (San Diego, CA). At the end of the syn-thesis, thallium(III) trifluoroacetate is used to remove the Acm-protecting groups on the cysteines, concomitant with disulfide formation to generate the cyclic peptide. The resin with the cyclic peptide is put back into the syn-thesizer for the conjugation of chelator. In the case of TETA, a derivative with a mono-functional group was used, *tri-tert*-butyl TETA. This TETA derivative can be obtained by either a published synthetic procedure[94] or from Macrocyclics, Inc. *To prevent chelation of the thallium, the thallium-mediated peptide cyclization must be done before the TETA conjugation.* The resulting peptide is cleaved from the resin and deprotected with tri-fluoroacetic acid:thioanisole:phenol:water (85:5:5:5) for 8–10 h. Final puri-fication is accomplished by C-18 reversed-phase HPLC [Solvent A: 0.1% trifluoroacetic acid (TFA)[95]/water, Solvent B: 0.1% TFA/10% H_2O/aceto-nitrile; gradient: 90% A/10% B to 30% A/70% B in 40 min; detection mode: ultraviolet (UV) at 230 nm]. Fractions are analyzed by analytical HPLC (detection mode: UV at 214 nm) and mass spectrometry prior to final lyophilization.

Radiolabeling of BFC–Peptide Conjugates

It is of great importance that receptor-targeting peptides be radiola-beled with copper radionuclides with high-specific activity [>27,750 MBq/μmol (750 mCi/μmol)]. Tumor receptors are generally easily saturated, and the presence of significant amounts of unlabeled peptide will affect the radiolabeled peptide uptake in the tumor.

The BFC–peptide conjugates can be directly radiolabeled with ^{64}Cu in ammonium acetate (or ammonium citrate) buffer and purified by previ-ously published protocols.[59,71–73,93,96] In a typical labeling of TETA-Y3-TATE as an example, 1.0 μg of TETA-Y3-TATE can be radiolabeled with more than 1000 μCi (37 MBq) of ^{64}Cu-acetate in 0.1 M ammonium ace-tate, pH 5.5, in 30 min incubation at room temperature. The ^{64}Cu-TETA-Y3-TATE purification is accomplished using a C18 SepPak cartridge

[94] A. K. Mishra, K. Draillard, A. Faivrechauvet, J. F. Gestin, C. Curtet, and J. F. Chatal, *Tet. Lett.* **37,** 7515 (1996).

[95] E. D. Yorke, L. E. Williams, A. J. Demidecki, D. B. Heidorn, P. L. Roberson, and B. W. Wessels, *Med. Phys.* **20,** 543 (1993).

[96] J. S. Lewis, R. Laforest, M. R. Lewis, and C. J. Anderson, *Cancer Biother Radiopharm.* **15,** 593 (2000).

(Waters Corp.). For a 1-ml ethanol elution, >90% of the product is in the fractions between 150–500 μl. The radiochemical purity is confirmed by radio-TLC. The TLC conditions are C18-coated plates eluted in 7:3 (v/v) MeOH:10% NH_4OAc. The radiochemical purity is typically >95%. The ^{64}Cu-TETA-Y3-TATE in ethanol may need to be evaporated under nitrogen or argon prior to dilution with saline (ethanol content must be less than 5%) for PET imaging and/or radiotherapy in tumor-bearing rodent models or humans.

BFC–Antibody Conjugation

The protocol for BFC–antibody conjugation is via the *sulfo*-NHS route (see Fig. 7).[77–79] This conjugation is done in aqueous solution at 4° and pH 7.5, which maintains the immunoreactivity of the antibody. The protocol consists of two steps: (1) preparation of *sulfo*-NHS–DOTA ester; and (2) DOTA or TETA conjugation. The *sulfo*-NHS–DOTA ester can be purchased from Macrocyclics, Inc. or prepared by a published method where N-hydroxysulfosuccinimide (*sulfo*-NHS) is reacted with DOTA in 0.1 M phosphate buffered saline solution (PBS, pH 5.5) at 4° for 30 min in the presence of 1-ethy-3-[3-(dimethylamino)propyl] carbodiimide (EDC).[78,79] The molar ratio of *sulfo*-NHS:DOTA:EDC is 10:10:1. It is not necessary to purify the *sulfo*-NHS–DOTA ester prior to the conjugation to the protein. In the same reaction vessel, 0.2 M Na_2HPO_4 (pH 9.2) is used to raise the pH to 7.5. Then, the antibody in 0.1 M PBS (pH 7.4) is added to the *sulfo*-NHS–DOTA ester solution with a 1:1000 molar ratio of antibody: *sulfo*-NHS–DOTA ester. The reaction is again performed at 4° with continuous mixing for 12–16 h. The DOTA–antibody conjugate is then separated from unconjugated *sulfo*-NHS–DOTA ester by either repeated centrifugation (5×) through a Centricon-30 (Millipore Corp, Bedford, MA), or dialysis against the buffer with five buffer changes to eliminate unreacted ester. The concentration of DOTA–antibody conjugate is determined by a UV spectrophotometer, and its purity is monitored by size-exclusion HPLC or fast protein liquid chromatography (FPLC) using a Superose 12 column (Amersham Biosciences, Uppsala, Sweden) eluting with 20 mM HEPES, 150 mM NaCl, pH 7.3, at an isocratic flow rate of 0.5 ml/min.

Radiolabeling of BFC–Antibody Conjugates

The BFC–antibody conjugates can be directly radiolabeled with ^{64}Cu in ammonium acetate (or ammonium citrate) buffer and purified by the methods reported.[70,78] In a typical radiolabeling, 100 μg of DOTA (or TETA) conjugated antibody (or protein) can be radiolabeled with more than 1 mCi of ^{64}Cu after 30–60 min incubation at 43°. DTPA or EDTA is normally used to minimize the nonspecific ^{64}Cu binding by incubating with the reaction mixture at room temperature for 15 min prior to the separation of ^{64}Cu-labeled conjugate. Although the separation can be done by centrifugation through a Centricon-30 as in the conjugation protocol, it is more conveniently carried out using Bio-Spin 6 size-exclusion columns (Bio-Rad, Richmond, CA). The latter method only takes about 5 min, as compared to at least 2 h if using the former, and both methods result in comparable separation efficiency. The radiochemical purity is determined by either size-exclusion HPLC or FPLC using a Superose 12 column (Amersham Biosciences) eluting with 20 mM HEPES, 150 mM NaCl, pH 7.3, at an isocratic flow rate of 0.5 ml/ min, and integrated radiodetector.

Biologic Evaluation of BFC–Biomolecule Conjugates

After the ^{64}Cu-labeled bioconjugate is prepared, its biologic activity and effectiveness as a radiopharmaceutical can be determined by several methods. Determining that the copper-labeled BFC–biomolecule conjugate retains its biologic activity, and evaluation of its biodistribution and/ or therapeutic efficacy are essential prior to human trials.

In Vitro *Receptor Binding Assay for Peptide Conjugates.* The binding affinity is measured using an appropriate cell line that expresses the desired receptor to which the peptide conjugate is designed to bind. The method for this assay for somatostatin receptors has been well established.[71,93] Generally, a competitive binding assay is performed by displacing the ^{64}Cu-BFC-peptide with the "cold" competing ligand complexes using the Millipore Multiscreen system. The best-fit IC$_{50}$ values, the dissociation constant (K_d), and the number of binding sites (B_{max}) for the cell line can be calculated using PRISM (Graphpad, San Diego, CA).

Immunoreactivity Determination of Antibody Conjugates. It is critical that the BFC-conjugated antibody maintain immunoreactivity (IR) for targeting antigens present on tumors or other tissues of interest. IR determination is generally accomplished by monitoring the interaction of the antibody conjugate and the antigen to which the antibody conjugate is designed to attack.[78] In the case of determining the IR of a radiolabeled anti-CEA monoclonal antibody, the radiolabeled antibody and the CEA antigen are incubated at 37° for 15 min with continuous end-over-end mixing. Then an FPLC integrated with a radio-detector is employed for the assay using a Superose 12 column (Amersham Biosciences) eluting with an appropriate buffer at an isocratic flow rate of 0.5 ml/min. Immunoreactivity is calculated as the percentage of the total radioactivity shifted to the antibody–antigen complex, which has an apparently higher molecular weight than the antibody conjugate.

Biodistribution Studies

The *in vivo* evaluation of ^{64}Cu-labeled bioconjugates is typically carried out using an appropriate tumor-bearing animal model at several time points after injection (usually between 1 and 48 h). The targeting capability of the ^{64}Cu-labeled compound can be evaluated as the ratio of tumor to surrounding normal tissue (e.g., tumor:muscle), or tumor:blood. In addition to determining the target tissue uptake, clearance of the agent through the blood, liver, and kidneys is also measured, with more rapid clearance through the nontarget tissues being desirable.

MicroPET Imaging

MicroPET imaging has recently become a popular method to evaluate PET radiopharmaceuticals using animal models of disease.[70,74,93] For studies involving expensive transgenic rodent models, this noninvasive technique is greatly preferred, since repeat studies can be performed without sacrificing the animals. Copper-64 labeled bioconjugates can be quickly evaluated by both static and dynamic imaging. The activity concentrations in the regions of interest (ROI) are quantified by measuring the radioactivity in the selected volumetric regions and averaging the activity concentration over these regions. For the microPET-R4 rodent scanner (Concorde Microsystems, Knoxville, TN), the ROI analysis software consists of two programs: Analyze AVW 3.0 (Biomedical Imaging Resource, Mayo Foundation, Rochester, MN) and a viewing application program developed by Laforest at the Washington University School of Medicine in St. Louis using International Data Language (Research Scientific, Boulder, CO).

Summary

The goal of this chapter is to discuss the production and applications of copper radionuclides in nuclear medicine. The interest in copper radionuclides for PET imaging and targeted radiotherapy of cancer, particularly the use of ^{64}Cu, is increasing. Copper-64 has attractive decay properties for PET imaging and therapy, and can be used in a wide variety of applications. This chapter provides information related to the production of the radionuclides, the radiochemistry for labeling them to chelators and chelator–biomolecule conjugates, and also describes the biologic methods for evaluating compounds as radiopharmaceuticals.

Acknowledgments

The authors would like to acknowledge assistance from Todd Perkins, Lucie Tang, and Michael J. Welch, Ph.D., for assistance with the copper radionuclide production section.

[12] Generation of DOTA-Conjugated Antibody Fragments for Radioimmunoimaging

By Peter M. Smith-Jones and David B. Solit

Introduction

Since the identification of antibodies as mediators of the immune response to foreign antigens, efforts have been made to harness the specificity and cytotoxicity of these proteins as targeted cancer agents. Technological advances, in particular hybridoma technology, which allows for the efficient generation of large quantities of epitope-specific monoclonal antibodies,[1] have led to the realization of this promise, and several monoclonal antibodies are now approved for use as molecular imaging probes or as therapeutic agents (Table I). Full-length immunoglobulin G (IgG) antibodies, which consist of two identical heavy and light chains, are used for most clinical applications. Unconjugated ("naked") antibodies such as Herceptin (traztuzumab)[2] and Rituxan (rituximab)[3] have modest activity in patients with breast cancer and lymphoma. These antibodies kill cancer cells by enhancing complement fixation or initiating antibody-dependent cell-mediated cytotoxicity.[4] They also may exert antitumor activity by modulating intracellular signaling pathways activated by the cell surface target to which they bind.

To improve on the promising results achieved with naked antibodies, several conjugated monoclonal antibodies have been generated. In this strategy, the antibodies are used as a delivery vehicle to target drugs, immunotoxins, or radiation directly to the tumor cells.[5–7] Antibodies labeled with gamma- and positron-emitting radionuclides can also be imaged. This technology allows for the identification of distant sites of cancer spread and has found clinical application in prostate, ovarian, and colorectal cancers (Table I). Radioimmunoimaging may also have a role in cancer therapy selection by allowing for real-time noninvasive quantitation of target expression at metastatic tumor sites. A critical challenge in the development of

[1] G. Kohler and C. Milstein, *Nature* **256,** 495 (1975).
[2] J. Baselga, D. Tripathy *et al., J. Clin. Oncol.* **14,** 737 (1996).
[3] B. Coiffier, C. Haioun *et al., Blood* **92,** 1927 (1998).
[4] M. X. Sliwkowski, J. A. Lofgren *et al., Semin. Oncol.* **26,** 60 (1999).
[5] T. E. Witzig, C. A. White *et al., J. Clin. Oncol.* **17,** 3793 (1999).
[6] M. S. Kaminski, A. D. Zelenetz *et al., J. Clin. Oncol.* **19,** 3918 (2001).
[7] E. L. Sievers, R. A. Larson *et al., J. Clin. Oncol.* **19,** 3244 (2001).

Copyright 2004, Elsevier Inc.
All rights reserved.
0076-6879/04 $35.00

TABLE I
SELECTED FDA-APPROVED MONOCLONAL ANTIBODIES IN CLINICAL USE

	Disease	Target	Conjugate
Cold antibodies			
Rituximab (Rituxan)	Lymphoma	CD20	–
Traztuzumab (Herceptin)	Breast cancer	HER2	–
Alemtuzumab (Campath)	CLL	CD52	–
Conjugated antibodies			
Ibritumomab tiuxetan (Zevalin)	Lymphoma	CD20	^{111}In, ^{90}Y
Tositumomab (Bexxar)	Lymphoma	CD20	^{131}I
Radioimmunoimaging reagents			
Capromab pendetide (ProstaScint)	Prostate cancer	PSMA	^{111}In
Satumomab pendetide (OncoScint)	Colorectal/ovarian	TAG-72	^{111}In
Arcitumomab (CEA-Scan)	Colorectal/ovarian	CEA	^{99m}Tc

CLL, Chronic lymphocytic leukemia.

novel biologic therapies that inhibit single transduction pathways in patients with solid tumors has been the difficulty in identifying the subset of patients whose tumors express a particular molecular target. The identification of such patients is particularly challenging when the expression pattern of the target changes during the course of a patient's disease, as has been observed with transmembrane tyrosine kinases such as epidermal growth factor receptor (EGFR) and HER2/neu.[8,9] Radioimmunoimaging technology therefore may provide an advantage in this setting over the use of traditional immunohistochemical techniques, which require an invasive procedure to obtain tissue for analysis.

This chapter outlines the methods used for generation of radiolabeled monoclonal antibodies for use as imaging probes. The radionuclides most commonly used are reviewed, and the methods for the generation and conjugation of full length and F(ab')$_2$ antibody fragments are outlined.

Selection of a Radionuclide and Chelating Agent

Tables II and III list some of the more commonly used positron- and gamma-emitting metallic radionuclides. Positron-emitting isotopes are typically produced in a cyclotron and have relatively short half-lives (Table II). Therefore incorporation of the radioisotope into the imaging probe must be performed shortly before its injection into the subject.

[8] H. I. Scher, A. Sarkis et al., Clin. Cancer Res. **1**, 545 (1995).
[9] S. Signoretti, R. Montironi et al., J. Natl. Cancer Inst. **92**, 1918 (2000).

TABLE II
PROPERTIES OF COMMONLY USED POSITRON-EMITTING ISOTOPES

Isotope	Half-life	Decay energy	Positron abundance (%)	Positron energy (%)	Gamma-ray energy (%)
^{68}Ga	67.6 min	2.91 MeV	89	0.82 MeV (1.2)	1.08 MeV (3)
				1.89 MeV (88)	
^{66}Ga	9.5 h	5.18 MeV	55	0.92 MeV (3.8)	
				4.15 MeV (50)	
^{64}Cu	12.7 h	1.68 MeV	19	0.65 MeV (17.4)	1.35 MeV (0.5)
				1.67 MeV (43.1)	
^{86}Y	14.7 h	5.27 MeV	26	0.9 MeV (1.1)	0.44 MeV (16.9)
				1.03 MeV (1.9)	0.52 MeV (4.9)
				1.13 MeV (1.3)	0.58 MeV (94.8)
				1.22 MeV (11.9)	0.63 MeV (32.6)
				1.55 MeV (5.6)	0.65 MeV (9.2)
				1.74 MeV (1.7)	0.70 MeV (15.4)
				1.99 MeV (3.6)	0.78 MeV (22.4)
				3.14 MeV (2.0)	1.08 MeV (83)
					1.15 MeV (30.5)
					1.85 MeV (17.2)
					1.92 MeV (20.8)

Reference: http://ie.lbl.gov/education/isotopes.htm

TABLE III
PROPERTIES OF SELECTED COMMERCIALLY AVAILABLE METALLIC RADIONUCLIDES
SUITABLE FOR GAMMA CAMERA IMAGING

Isotope	Half-life	Decay mode (%)	Gamma-ray energy (%)
^{99m}Tc	6.01 h	Isomeric transition (100)	140.5 keV (89)
^{111}In	67.3 h	Electron capture (100)	171.3 keV (90)
			245.4 keV (94)
^{67}Ga	78.3 h	Electron capture (100)	93.3 keV (39.2)
			184.6 keV (21.2)
			300.2 keV (16.8)
			393.5 keV (4.7)

Reference: http://ie.lbl.gov/education/isotopes.htm

For dynamic imaging in which the goal is to noninvasively monitor changes in a cell surface protein over time in response to a novel drug or other intervention, we prefer the use of the positron emitter ^{68}Ga. It has a short 68-minute half-life and a high abundance of positrons (89%) (Table II). When designing a dynamic molecular probe for detecting treatment-induced changes in a cell surface protein, several criteria must be fulfilled. The molecular probe must have a high affinity for the imaging target and a short biologic half-life. It must cause no change in the imaging target. Furthermore, the pharmacokinetics of the imaging probe must not be altered by the targeted agent being studied. The antibody portion of the probe may be based upon a murine, chimeric, humanized, or human antibody. Because the goal of such preclinical studies is often the generation of molecular probes for repetitive human use, the use of murine monoclonal antibodies is potentially problematic. The formation of human antimouse antibodies could significantly alter the biodistribution of an antibody-derived molecular probe and thus significantly limit its clinical use. Therefore although murine monoclonal antibodies may suffice for preclinical studies, the use of human, humanized, or chimeric antibodies is theoretically preferable.

In our experience, the short half-life of ^{68}Ga allows repetitive daily imaging when conjugated to F(ab')$_2$ antibody fragments, and such probes have been used by our group to dynamically and noninvasively monitor drug-induced changes in cell surface receptor proteins. For example, we have used ^{68}Ga-labeled fragments of Herceptin, a monoclonal antibody that binds to the transmembrane tyrosine kinase HER2/neu,[2] to noninvasively image changes in HER2/neu expression following treatment with the Hsp90 inhibitor, 17-allylamino-geldanamycin (17-AAG).[10] Hsp90 is a cellular chaperone protein required for the maturation and stability of a subset of signaling proteins, including tyrosine kinases and steroid receptors.[11] Inhibition of Hsp90 by 17-AAG causes the proteasomal degradation of HER2 and other Hsp90-dependent client proteins.[12] We found that ^{68}Ga-DOTA–F(ab')$_2$ Herceptin was able to image HER2-expressing tumors and that adequate contrast between tumor and normal tissues was apparent within 3 hours of tracer administration (Fig. 1).[10] A linear correlation was observed between the data obtained by micro-positron emission tomography (microPET) and direct assessment of tissues by gamma counter. The short half-life of ^{68}Ga and the rapid clearance of the F(ab')$_2$ fragments allowed for daily imaging. Furthermore, the tracer amount of the ^{68}Ga-DOTA–F(ab')$_2$ Herceptin probe had no effect on HER2 expression in the tumor as confirmed by immunoblot analysis. This is a critical finding because antibodies to other cell surface tyrosine kinases have been reported to induce receptor downregulation.[13]

[10] P. M. Smith-Jones, D. Solit *et al.*, *J. Nucl. Med.* **44,** 361 (2003).
[11] W. B. Pratt and M. J. Welch, *Semin. Cell Biol.* **5,** 83 (1994).
[12] E. G. Mimnaugh, C. Chavany *et al.*, *J. Biol. Chem.* **271,** 22796 (1996).
[13] P. G. Kasprzyk, S. U. Song *et al.*, *Cancer Res.* **52,** 2771 (1992).

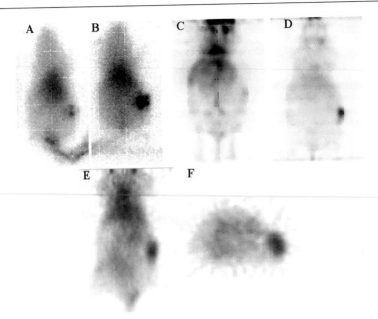

FIG. 1. (A, B) Planar gamma camera images of a BT-474 xenograft-bearing mouse 6 (A) and 30 (B) hours after injection with ^{111}In-Herceptin. (C, D) MicroPET images of a BT-474 xenograft-bearing mouse 2 (C) and 28 (D) hours after injection with ^{64}Cu-Herceptin. (E) Coronal and (F) transverse microPET images of a BT-474 xenograft-bearing mouse 3 hours after injection with ^{68}Ga-F(ab')$_2$ Herceptin. The greatest contrast between tumor to normal tissue is observed with intact Herceptin labeled with [64]Cu ($T_{1/2}$ = 12.7 h). The use of intact antibody results in limited tumor-specific uptake at early time points (C). ^{68}Ga-F(ab')$_2$ Herceptin provides adequate contrast between normal and tumor tissues at early time points (E, F), and the short half-lives of both ^{68}Ga and the F(ab')$_2$ antibody fragments allow for daily imaging of dynamic changes in HER2/neu expression. Tumors are implanted within the right flank.

[68]Ga can be generated from the decay of [68]Ge ($T_{1/2}$ = 271 days), and therefore an onsite cyclotron is not required. Several generator systems have been proposed to produce [68]Ga.[14,15] We are currently using a SnO_2-based system[14] (4-ml column) with an inline pyrogallol–formaldehyde resin column[16] (0.5 ml) to remove any eluted [68]Ge (Fig. 2). The ^{68}Ga is eluted in 5 ml of 1 M HCl and has a contamination of 10^{-6}% of [68]Ge.

For intact antibody biodistribution studies, we prefer a longer half-life isotope such as [64]Cu. For this application, repetitive imaging is not required

[14] G. J. Ehrhardt and M. J. Welch, *J. Nucl. Med.* **19**, 925 (1978).
[15] C. Loc'h, B. Maziere *et al.*, *J. Nucl. Med.* **21**, 171 (1980).
[16] J. Schumacher and W. Maier-Borst, *Int. J. Appl. Radiat. Isot.* **31**, 31 (1980).

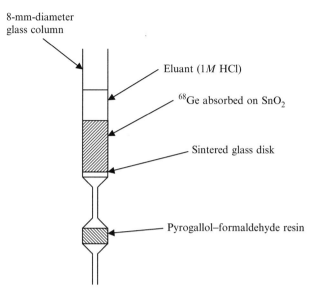

8-mm-diameter
glass column

Eluant (1 *M* HCl)

^{68}Ge absorbed on SnO$_2$

Sintered glass disk

Pyrogallol–formaldehyde resin

FIG. 2. ^{68}Ga can be produced without the use of a cyclotron using a ^{68}Ge generator. ^{68}Ge is absorbed on a SnO$_2$ solid support, and ^{68}Ga is eluted using 1 *M* HCl. A pyrogallol–formaldehyde resin is used to purify the product by removing traces of ^{68}Ge.

and the longer half-life of ^{64}Cu (12.7 h) allows for imaging at later time points. After 24 h, most full-length antibodies have been cleared from the blood pool and antigen-negative normal tissues but are still retained within antigen-expressing tumor tissues, providing the greatest contrast between tumor and normal tissue (see Fig. 1). ^{66}Ga is another positron-emitting isotope with a long half-life (9.5 h), but its use is less desirable in imaging applications than ^{64}Cu due to emission of high-energy beta particles that lower image resolution. Similarly, ^{86}Y (T$_{1/2}$ of 14.7 h) can be used for antibody biodistribution studies, but its high-energy gamma emissions interfere with the characteristic 511-keV gamma rays, which are characteristic of the positron annihilation that forms the basis for PET scanning technology, thus degrading image quality.

Gamma-emitting isotopes (^{99m}Tc, ^{111}In, ^{131}I) can also be used for *in vivo* imaging, but are not detected by PET and require the use of a planar gamma camera or a SPECT (single photon emission computed tomography) camera (Table III). Both PET and SPECT are able to generate 3-dimensional images. Although gamma camera technology may be preferable for some applications due to its lower cost and greater availability in the clinic, PET imaging currently delivers the highest resolution. For imaging purposes, ^{99m}Tc is preferable to other gamma-emitting isotopes due

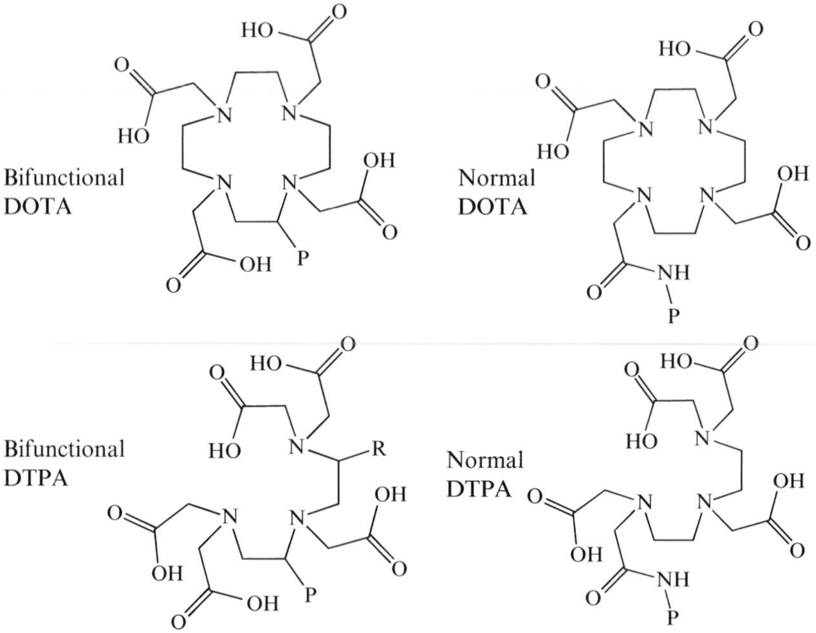

Fig. 3. Chemical structures of DOTA and DTPA conjugates. P is the site of conjugation to the protein structure. The DOTA and DTPA chelating agents normally attach to lysine residues.

to its generation of an optimal monoenergetic gamma ray (Table III). [111]In is limited by its high abundance of moderate-energy gamma rays, and ^{67}Ga is the least desirable choice for imaging applications due to its low abundance of higher-energy gamma rays.

Numerous chelators have been proposed to coordinate radioactive metal ions. Systems based on 1,4,7,10-tetraazacyclododecane-N,N′, N″,N‴-tetraacetic acid (DOTA)[17–19] and diethylenetriaminepentaacetic acid (DTPA)[20–22] have been shown to be the most versatile (Fig. 3). The primary advantage of DOTA over DTPA is that it produces a more stable chelate, but it has the disadvantage of slightly slower reaction kinetics. Bifunctional conjugates have more donor groups to chelate the metal ions and are therefore more stable than standard DOTA and DTPA chelators.

[17] S. V. Deshpande, S. J. DeNardo et al., J. Nucl. Med. 31, 473 (1990).
[18] P. M. Smith-Jones, S. Vallabahajosula et al., Cancer Res. 60, 5237 (2000).
[19] L. L. Chappell, D. Ma et al., Nucl. Med. Biol. 30, 581 (2003).
[20] S. J. Wagner and M. J. Welch, J. Nucl. Med. 20, 428 (1979).
[21] D. A. Westerberg, P. L. Carney et al., J. Med. Chem. 32, 236 (1989).
[22] T. K. Nikula, M. J. Curcio et al., Nucl. Med. Biol. 22, 387 (1995).

DOTA generally produces a sufficiently stable metal chelate and is commercially available, whereas other bifunctional forms of these two agents have to be synthesized. All of the radionuclides listed in Table II form a stable chelate with DOTA.

Generation of Antibody Fragments

As discussed earlier, antibody fragments may be preferable to intact antibodies as dynamic imaging probes due to their shorter biologic half-lives. As a result of their smaller size, they are cleared more quickly from nonantigen-expressing tissues and therefore provide greater tumor to normal tissue contrast at early time points (Fig. 1).[23] Furthermore, they may migrate more efficiently through the extravascular space into the interior of poorly vascularized tumors than intact antibodies and may be less immunogenic due to the removal of the Fc region. In this section, we outline a method for generating antibody fragments that retain the ability to specifically bind to their antigen target. The method involves selective digestion of the antibody with pepsin, followed by purification of the product using a protein A column to remove residual undigested intact antibody. The product is then purified by high-performance liquid chromatography (HPLC) to resolve the F(ab′)₂ fragments from the digested Fc portions. During fragmentation, the reaction must be closely monitored by HPLC to prevent overdigestion, which may reduce immunoreactivity.

Reagents

20 mM Sodium acetate (pH 4.5)
10 mM Tris–HCl (pH 8.0)
0.1 M Glycine (pH 2.8)
50 mM Sodium phosphate (pH 7.1)
Immobilized pepsin (Pierce Biotechnology, Inc., Rockford, IL)
Immobilized protein A (Pierce Biotechnology, Inc., Rockford, IL)
30 kDa Centrifugal filters (Amicon, Millipore, Bedford, MA)

Method

1. Suspend 0.25 ml immobilized pepsin with 5 ml of 20 mM sodium acetate buffer. Centrifuge and remove supernatant. Repeat washing step five times.
2. Add 10 mg of IgG to 3 ml of 20 mM sodium acetate buffer and concentrate to 1 ml in a 30-kDa centrifugal filter. Remove filtrate and repeat washing step five times.

[23] K. A. Harrison and M. A. Tempero, *Oncology (Huntingt.)* **9,** 625 (1995).

3. Add IgG in 1 ml of sodium acetate to immobilized pepsin and mix slurry on a shaker at 37° for 6–48 h.
4. At various times, remove a 5-μl fraction of the solution and analyze by size-exclusion HPLC (TSK 3000 column) to see the amount of intact IgG.
5. When the amount of intact IgG is <5%, add 3 ml of 10 mM Tris–HCl and separate the gel by centrifugation.
6. Pour a 2-ml protein A column and equilibrate it with 5 × 2 ml of 10 mM Tris–HCl.
7. Load digested IgG onto column and wash column with 10 mM Tris–HCl.
8. Collect the first eluted protein peak (measure absorbance at 280 nM) and concentrate this eluate with a 30-kDa centrifugal filter.
9. Undigested IgG Fc, and Fc fragments may be recovered by washing the column further with 0.1 M glycine.
10. Purify crude F(ab')$_2$ fraction by size-exclusion HPLC using a TSK 3000 column and an eluant of 50 mM sodium phosphate.
11. Concentrate HPLC purified F(ab')$_2$ with a 30-kDa centrifugal filter and then sterilize by filtration through a 0.22-micron filter. This product, when stored at 4°, may remain stable for up to several years.

Conjugation of DOTA to Intact Antibodies or Fragments

Conjugation of DOTA to intact antibodies or fragments involves first generation of the active ester of DOTA, which is then reacted with lysine residues on the protein. The conjugated antibody is then purified by ultrafiltration prior to radiolabeling.

Reagents

> 1,4,7,10-tetraazacyclododecane-N,N',N'',N'''-tetraacetic acid (DOTA; Strem Chemical, Newburyport, MA)
> N-hydroxysuccinimide
> 1-ethyl-3-(3-dimethylaminopropyl)carbodiimide
> 0.25 M Ammonium acetate
> Chelex 100 resin, 100–200 mesh (Bio-Rad, Hercules, CA)

Method

1. Pour a 10-ml column of Chelex 100 resin and wash with 5 × 10 ml of 0.25 M ammonium acetate. Pass remaining buffer through column and collect metal-free buffer in a plastic bottle.

2. Dissolve 146 mg DOTA (0.361 mM) and 36 mg N-hydroxysuccini-
   mide (0.313 mM) in 1 ml of water and adjust the pH to 7.3 with
   NaOH. Dilute solution to 2 ml and cool in an ice bath for 30 min
   before adding 10 mg of 1-ethyl-3-(3-dimethylaminopropyl)carbo-
   diimide. Let reaction proceed for 1 h on ice to form the active ester.
3. For each mg of F(ab')$_2$, add 80 μl of active ester solution and allow
   to react overnight at 4°.
4. Concentrate reaction mixture in a 30-kDa centrifugal filter to about
   10 mg/ml. Add an equal volume of 0.25 M ammonium acetate and
   concentrate to 10 mg/ml. Repeat washing step 20 times.

Radiolabeling of DOTA Conjugates

To radiolabel DOTA-conjugated antibodies with ^{68}Ga, ^{68}Ga eluted
from the ^{68}Ge generator is concentrated by extracting with ether and
then back extracting to a small volume of water. The ^{68}Ga or an alternative
isotope is then reacted with antibody in an ammonium acetate buffer.
Finally, the reaction is quenched with DTPA, with the product run over a
size-exclusion column to remove unchelated isotope.

Reagents

   1–5 mCi of ^{68}Ga in 5 ml of 1 M HCl
   12 M HCl
   Diethyl ether
   1 M Metal-free ammonium acetate
   General: All apparatus used should be plastic and free of any metal
      contamination. In particular, colored plastic tubes and pipette tips
      should be avoided, since metals are sometimes used to achieve the
      colored effect.

Method

1. In a glass tube place ^{68}Ga, an equal volume of 12 M HCl, and 1 ml
   of diethyl ether. Shake tube and allow layers to separate.
2. Add 50 μl of deionized water to a 1.5-ml conical polypropylene
   centrifuge tube and add ether layer containing ^{68}Ga. Place tube in a
   heating block and remove ether under a gentle stream of air.
3. Repeat ether extraction and repeat evaporation step.
4. Add 10 μl of 1 M ammonium acetate to back-extracted ^{68}Ga and
   spot-test 0.5 μl onto a pH indicator strip. The pH should be >5.
5. Add 10 μl of a 10 mg/ml solution of the DOTA–F(ab')$_2$ conjugate
   and incubate at 37° for 10 min.

6. Add 50 μl of 5 mM DTPA to quench reaction and load solution onto a 10-ml P6 column equilibrated with 0.5% bovine serum albumin in phosphate buffered saline (BSA–PBS).

7. Elute column with BSA–PBS and collect the ^{68}Ga-DOTA–F(ab')$_2$ fraction, which elutes after about 3 ml.

All the other metals listed in Table II may be used in an analogous manner to label the DOTA–F(ab')$_2$ conjugate as long as the labeling is performed within the pH range of 5–7 in the presence of 0.25–1.0 M ammonium acetate.

Characterization of DOTA–F(ab')$_2$ Conjugates

Upon initial generation of the DOTA–antibody conjugate, the quality of the product must be verified. This involves determining the number of DOTAs per antibody by labeling the antibody with a known quantity of nonradioactive indium spiked with ^{111}In and then examining the two species formed. The immunoreactivity of the antibody conjugate is also determined to confirm that the conjugation process has not disrupted binding of the antibody to its antigen target. This may result from overdigestion at the time of fragment generation. Alternately, it may occur with DOTA conjugation if a large number of lysine residues are present in or around the antigen-binding site.

For most antibodies the immunoreactivity should be >90% and the number of DOTAs bound/F(ab')$_2$ should be 3–6. If the immunoreactivity is too low and the number of DOTAs attached is less than four, then the conjugation procedure should be repeated using fresh F(ab')$_2$ and a lower volume of the active ester solution. Conversely, if the immunoreactivity is preserved and the number of DOTAs bound is less than one, then the conjugation step should be repeated using the same F(ab')$_2$, but with the volume of the active ester solution of DOTA increased. Once prepared and characterized, the conjugate may be stored in a plastic vial at 4° for up to several years. The immunoreactivity test should be repeated prior to radiolabeling each batch of DOTA–F(ab')$_2$ to confirm that the conjugate is stable. We have several conjugates that are still >90% immunoreactive after storage for >18 months at 4°.

Reagents

^{111}In (0.05 M HCl)

1 M Metal-free ammonium acetate

1 mM InCl$_3$ (0.05 M HCl)

5 mM DTPA (pH 7.0)

5 mM DTPA (pH 5.0)

Silica gel-impregnated glass fiber (ITLC-SG; 1 × 10 cm, Gelman, Ann Arbor, MI)

DOTA–F(ab′)$_2$

0.5% Bovine serum albumin in phosphate buffered saline (BSA–PBS)

Method for Number of Sites

1. Add 1 μl of [111]In (*ca* 200–500 μCi) to 100 μl of 1 mM InCl$_3$ and add 10, 20, and 30 μl portions of this solution to three 20-μl samples of DOTA–F(ab′)$_2$. Incubate solutions at 37° for 16 h.
2. Add 50 μl of 5 mM DTPA to the three reaction mixtures and reincubate at 37° for 6 h.
3. Spot-test 5 μl of the solutions onto a 10-cm ITLC-SG strip. Develop strip in 5 mM DTPA (pH 5.0).
4. Cut strip at an R$_f$ of 0.5 and count both parts in an gamma counter. F(ab′)$_2$ activity remains at the origin and [In-DTPA] moves with the solvent front.
5. Calculate the number of DOTAs attached to each F(ab′)$_2$ according to the following formula:

Number attached =

$$\frac{(\text{Activity in lower strip}) \times (\text{In concentration (in moles)})}{(\text{Total activity}) \times (\text{Antibody concentration (in moles)})}$$

Method for Immunoreactivity[24]

1. Add 1 μCi of ^{111}In-DOTA–F(ab′)$_2$ to 3 ml of 0.5% BSA–PBS and add 50 μl of this solution to 18 × 1.5 ml centrifuge tubes. Place three of these tubes to one side.
2. Prepare *ca* 20 × 10^6 cells expressing the correct antigen expression in 3.5 ml of PBS.
3. To three tubes, add 0.5 ml of suspended cells.
4. To three tubes, add 0.25 ml cells and 0.25 ml PBS.
5. To three tubes, add 0.15 ml cells and 0.35 ml PBS.
6. To three tubes, add 0.10 ml cells and 0.40 ml PBS.
7. To three tubes, add 0.50 ml PBS.
8. Briefly vortex all 15 samples and place on a rocker for 1 h so that the cells remain in suspension.
9. Centrifuge all 15 tubes and aspirate supernatant.

---

[24] T. Lindmo, E. Boven *et al.*, *J. Immunol. Methods* **72,** 77 (1984).

10. Add 1 ml of ice-cold PBS, centrifuge tubes, and aspirate supernatant.
11. Count all 18 tubes in a gamma counter.
12. Subtract the mean of the nonspecific binding from the 12 test samples and plot a reciprocal of the number of cells added (x-axis) to the total/bound ratio. The intercept gives the immunoreactive fraction, which binds at an infinite excess of antigen.

PET Imaging

As an example, outlined as follows is our standard procedure for imaging mice with the ^{68}Ga-DOTA–F(ab')$_2$ Herceptin probe using the Concorde Microsystems microPET scanner. This is a commercially available, dedicated small animal PET scanner. Flood phantom measurements are performed on a daily basis with a cylindrical ^{68}Ge phantom to ensure detector linearity and uniformity.

1. Animals are injected with 100–500 μCi of the ^{68}Ga-DOTA–F(ab')$_2$ Herceptin and returned to their cages, where they are fed and watered *ad libertum*.
2. 2–3 hours after injection, the animals are anesthetized using an isoflurane–oxygen (Baxter Healthcare, Deerfield, IL) gas mixture and placed in the microPET camera.
3. Coincident data are collected for the 511 KeV gamma rays with a 250–750 KeV window for 5–10 min.
4. After imaging, the data are reconstructed using back projection filtering.
5. Regions of interest (ROI) are drawn around the organs expressing the antigen (i.e., tumor), as well as major organs such as the heart, liver, and kidneys.
6. An average is taken of the maximum activity per voxel in the three consecutive slices that have the highest uptake of activity. This number is then corrected for the size of the organ–tumor, the efficiency of the microPET camera, and the time difference between injection and imaging before the uptake is expressed as a percentage of the injected dose per gram.

Upon generation of a novel antibody-based molecular probe, we identify the pattern of nonspecific tracer uptake by preinjecting animals with a large quantity of cold antibody (or fragment) to saturate all antigen sites. The radiolabeled antibody probe is then injected to determine the pattern of nonspecific renal and hepatobiliary excretion of the probe.

Conclusion

The previous procedures outline a method for the generation of radiolabeled $F(ab')_2$ fragments for *in vivo* imaging applications. The recent dramatic advance in our understanding of the pathophysiology of cancer has led to the identification of targets expressed preferentially on transformed cells. Using molecular imaging probes based upon conjugated antibody fragments, the expression of these targets can be noninvasively quantitated. This should allow for the real-time identification of patients whose tumors express a particular molecular target, thus facilitating their inclusion in clinical trials of novel targeted agents. Furthermore, this technology may aid in the development of such agents by allowing for the noninvasive monitoring of treatment-induced changes in target protein expression.[10] Such detailed pharmacodynamic monitoring is not currently feasible in patients with solid tumors using current technologies and will likely be critical for the successful development of novel targeted therapies.

[13] Preparation of Magnetically Labeled Cells for Cell Tracking by Magnetic Resonance Imaging

By Jeff W. M. Bulte, Ali S. Arbab, T. Douglas, and J. A. Frank

Introduction

Magnetic resonance (MR) tracking of magnetically labeled cells following transplantation or transfusion is a rapidly evolving new field. At the one hand, MR cell tracking, with its excellent spatial resolution, can be used as a noninvasive tool to provide unique information on the dynamics of cell movement within and from tissues in animal disease models. Alternatively, MR cell tracking may be applied in the future to monitor (stem) cell therapy in patients. Both approaches require magnetic labeling of the cells of choice for the particular study. It is the aim of this review to provide methods and protocols for the preparation of magnetically labeled cells as well as methods for analysis and evaluation of cell labeling. Due to its biocompatibility and strong effects on T2(*) relaxation, iron oxide nanoparticles are now the MR contrast agent of choice for cell labeling, and several methods are provided that shuttle sufficient amounts into cells. Now that magnetic labeling methods have been developed and optimized it is expected that MRI cell tracking will find broad applications in monitoring transplantation protocols for cell-based therapy.

Copyright 2004, Elsevier Inc.
All rights reserved.
0076-6879/04 $35.00

Cell Transplantation

The administration or transplantation of exogenous or autologous therapeutic cells has been pursued as a very active research area over the last decade, and, for progenitor and stem cell therapy, remarkable progress has been obtained in animal disease models. Because of its limited regenerative capacity, the central nervous sytem (CNS) so far has received the most interest for cell (replacement) therapy. For instance, transplantation of mouse embryonic stem (ES), which was directed *in vitro* toward differentiation into cells of the neural lineage in a rat model of spinal cord injury, resulted in a significant increase of the behavioral Basso–Beattie–Bresnahan (BBB) open field score.[1] The result of the ES transplantation translated into improvement of hindlimb support, including the ability to support weight, that was absent in sham-operated controls. When neural stem cells were seeded on a polymer scaffold and implanted in the hemisected rat spinal cord, the animals demonstrated coordinated, weight-bearing hindlimb stepping that was absent in lesion-control groups.[2] Stem cells also may provide a nearly unlimited supply of dopaminergic neurons for treatment of Parkinson's disease. Mouse ES cells have been successfully differentiated into midbrain neural stem cells and subsequently into functional dopamine neurons.[3] When undifferentiated mouse ES cells were transplanted into protein-deprived (PD) rats, behavioral recovery coincided with dopamine-associated imaging findings.[4] Human neural stem cells[5] and embryonic germ cells[6] have been transplanted in the parkinsonian 1-methyl-4-phenyl-1,2,3,6-tetrahydropyridine (MPTP) and 6-hydroxydopamine (6-OHDA)-lesioned mouse model, respectively.

Therapeutic cell transplantation also has been studied in myelin disorders. One approach is the use of exogeneous myelinating cells (e.g., stem cells and oligodendrocyte progenitors). New myelination in the CNS by transplanted cells will, however, only be therapeutically relevant if it results

[1] J. W. McDonald, X. Z. Liu, Y. Qu, S. Liu, S. K. Mickey, D. Turetsky, D. I. Gottlieb, and D. W. Choi, *Nat. Med.* **5,** 1410 (1999).
[2] Y. D. Teng, E. B. Lavik, X. Qu, K. I. Park, J. Ourednik, D. Zurakowski, R. Langer, and E. Y. Snyder, *Proc. Natl. Acad. Sci. USA* **99,** 3024 (2002).
[3] J. H. Kim, J. M. Auerbach, J. A. Rodriguez-Gomez, I. Velasco, D. Gavin, N. Lumelsky, S. H. Lee, J. Nguyen, R. Sanchez-Pernaute, K. Bankiewicz, and R. D. McKay, *Nature* **418,** 50 (2002).
[4] L. M. Bjorklund, R. Sanchez-Pernaute, S. Chung, T. Andersson, I. Y. Chen, K. S. McNaught, A. L. Brownell, B. G. Jenkins, C. Wahlestedt, K. S. Kim, and O. Isacson, *Proc. Natl. Acad. Sci. USA* **99,** 2344 (2002).
[5] M. A. Liker, G. M. Petzinger, K. Nixon, T. McNeill, and M. W. Jakowec, *Brain Res.* **971,** 168 (2003).
[6] J. M. Savitt, J. W. M. Bulte, V. L. Dawson, J. D. Gearhart, M. J. Shamblott, J. M. Hakumaki, J. Axelman, and T. M. Dawson, *Movement Disord.* **17,** P694 (2002).

in restoration of nerve function. Transplantation of adult human Schwann cells,[7] adult human neural precursor cells,[8] or pig olfactory ensheathing cells[9] in rat models of chemical demyelination or transection has resulted in restoration of nerve conduction. For the dysmyelinated rat, transplanted glial cells are able to restore the nerve conduction velocity to near-normal values.[10] An animal model where the myelin defect is similar in scale to the human disease multiple sclerosis (MS) is the shaking pup; transplantation of oligodendrocyte progenitors resulted in repair of large areas similar in size to human MS plaques.[11] Brüstle et al.[12] derived oligodendroglial precursor cells from mouse ES cells and showed that large areas in the spinal cord could be myelinated following intraventricular transplantation. Yandava et al.[13] transplanted clonal neural stem cells (NSCs) in the shiverer mouse model, and demonstrated that these cells, following differentiation into oligodendroglial cells, can myelinate large areas in the CNS. Moreover, a significant number of recipient animals showed a decrease in their symptomatic tremor.

In an animal stroke model that caused deficits in learning and motor function, transplantation of neuronal cells induced a significant improvement in learning behavior and motor tasks.[14] The transplanted cells were found to integrate within the host brain along with the formation of axonal processes and release of neurotransmitters.[15,16] A recent clinical study of neuronal transplantation in patients with basal ganglia stroke and fixed motor deficits did not show any adverse cell-related effects. Moreover, an improvement in the total stroke scale score was observed in half of the patients.[17]

[7] I. Kohama, K. L. Lankford, J. Preiningerova, F. A. White, T. L. Vollmer, and J. D. Kocsis, J. Neurosci. 21, 944 (2001).

[8] Y. Akiyama, O. Honmou, T. Kato, T. Uede, K. Hashi, and J. D. Kocsis, Exp. Neurol. 167, 27 (2001).

[9] T. Imaizumi, K. L. Lankford, W. V. Burton, W. L. Fodor, and J. D. Kocsis, Nat. Biotechnol. 18, 949 (2000).

[10] D. A. Utzschneider, D. R. Archer, J. D. Kocsis, S. G. Waxman, and I. D. Duncan, Proc. Natl. Acad. Sci. USA 91, 53 (1994).

[11] D. R. Archer, P. A. Cuddon, D. Lipsitz, and I. D. Duncan, Nat. Med. 3, 54 (1997).

[12] O. Brüstle, K. N. Jones, R. D. Learish, K. Karram, K. Choudhary, O. D. Wiestler, I. D. Duncan, and R. D. McKay, Science 285, 754 (1999).

[13] B. D. Yandava, L. L. Billinghurst, and E. Y. Snyder, Proc. Natl. Acad. Sci. USA 96, 7029 (1999).

[14] C. V. Borlongan, Y. Tajima, J. Q. Trojanowski, V. M. Lee, and P. R. Sanberg, Exp. Neurol. 149, 310 (1998).

[15] S. A. Kleppner, K. A. Robinson, J. Q. Trojanowski, and V. M. Lee, J. Comp. Neurol. 357, 618 (1995).

[16] J. Q. Trojanowski, J. R. Mantione, J. H. Lee, D. P. Seid, T. You, L. J. Inge, and V. M. Lee, Exp. Neurol. 122, 283 (1993).

From the previous examples, it is clear that progenitor cells and stem cells have significant potential for (partial) restoration of lost CNS function. It should be noted that many therapeutic cell applications have been described for diseased tissues other than the CNS; however, they are beyond the scope of this chapter. Human ES[18] and embryonic germ (EG)[19] cells have been successfully propagated and now derivatized into a wide variety of cell lines. This development has brought the prospect of human stem cell therapy closer to the bedside. To develop successful clinical (stem) cell-based therapies, it will be important to develop methods that can assess the fate and distribution of cells noninvasively. It is obvious that traditional histopathologic methods for cell detection used in animal studies, which require repeated biopsies or euthanasia and the removal of tissue, cannot be completely applied to patients. Among the various noninvasive imaging techniques that are currently available, magnetic resonance imaging (MRI) stands out in terms of resolution and whole body imaging capability. For cells to be visualized on the MR images, they need to be magnetically labeled in order to be discriminated from the surrounding native tissue. This chapter provides several methods and examples for the preparation of magnetically labeled cells. Other than its potential clinical applicability, MRI cell tracking as a tool in itself will be very useful for dynamic cell studies in animal models as well. While each animal can serve as its own baseline control and can be serially imaged over time, the total number of animals will not only be reduced, but unique information on the spatial temporal cell dynamics can be obtained that is not available by other methods.

Superparamagnetic Iron Oxides

As for MR contrast agents, gadolinium is the most effective paramagnetic contrast agent, owing to its seven unpaired electrons, but its relaxivity (effectiveness in increasing the MR relaxation rate per mM of metal) is far lower than the so-called superparamagnetic iron oxides (SPIOs). Unlike gadolinium, which is potentially toxic in unchelated form and is designed not to stay within the body for prolonged times, iron oxides are composed

[17] D. Kondziolka, L. Wechsler, S. Goldstein, C. Meltzer, K. R. Thulborn, J. Gebel, P. Jannetta, S. DeCesare, E. M. Elder, M. McGrogan, M. A. Reitman, and L. Bynum, *Neurology* **55,** 565 (2000).

[18] J. A. Thomson, J. Itskovitz-Eldor, S. S. Shapiro, M. A. Waknitz, J. J. Swiergiel, V. S. Marshall, and J. M. Jones, *Science* **282,** 1145 (1998).

[19] M. J. Shamblott, J. Axelman, S. Wang, E. M. Bugg, J. W. Littlefield, P. J. Donovan, P. D. Blumenthal, G. R. Huggins, and J. D. Gearhart, *Proc. Natl. Acad. Sci. USA* **95,** 13726 (1998).

of biocompatible iron.[20] For magnetic labeling of cells, they are therefore the contrast agent of choice. SPIOs provide the targeted cell with a large magnetic moment that creates substantial disturbances in the local magnetic field, leading to a rapid dephasing of protons, including those not directly in the vicinity of the targeted cell. MRI techniques, such as gradient–echo techniques that do not compensate for dephasing, are particularly sensitized to detect the presence of iron oxide magnetic nanoparticles. Dextran-coated SPIOs where first introduced for hepatic imaging.[21,22] Following intravenous injection, the particles are rapidly taken up by liver Kupffer cells, which appear hypointense or black on the MR images. In areas where the normal liver architecture is disturbed (i.e., lack of Kupffer cells), such as exists in the presence of a primary liver tumor or liver metastasis, the signal intensity remains unaltered and thus stands out from the normal surrounding tissue. Subsequently, ultrasmall SPIOs (USPIOs) were developed that have a longer blood half-life and are normally taken up by macrophages, including those in lymph nodes[23–25] and bone marrow.[26] Upon USPIO administration, normal lymph nodes will appear dark on the MR images, whereas tumor metastases remain isointense with the precontrast image (i.e., no USPIO uptake). Thus (U)SPIOs are rapidly taken up by macrophages following intravenous (IV) administration, which one can refer to as magnetic labeling of cells *in vivo*.

Preparation of Magnetically Labeled Cells

For efficient uptake of magnetic nanoparticles in nonphagocytic cells, the contrast agent needs to be optimized or modified to have an appropriate outer surface layer that not only binds to cellular membranes, but also induces internalization of the particles into the cytoplasm. Particles that do not internalize and stay attached to the outer cell membrane are likely to interfere with cell surface interactions (including cell homing into tissues),

[20] R. Weissleder, D. D. Stark, B. L. Engelstad, B. R. Bacon, C. C. Compton, D. L. White, P. Jacobs, and J. Lewis, *Am. J. Roentgenol.* **152**, 167 (1989).
[21] D. D. Stark, R. Weissleder, G. Elizondo, P. F. Hahn, S. Saini, L. E. Todd, J. Wittenberg, and J. T. Ferrucci, *Radiology* **168**, 297 (1988).
[22] R. Weissleder, *Radiology* **193**, 593 (1994).
[23] M. G. Harisinghani, J. Barentsz, P. F. Hahn, W. M. Deserno, S. Tabatabaei, C. H. van de Kaa, J. de la Rosette, and R. Weissleder, *N. Engl. J. Med.* **348**, 2491 (2003).
[24] R. Weissleder, G. Elizondo, J. Wittenberg, A. S. Lee, L. Josephson, and T. J. Brady, *Radiology* **175**, 494 (1990).
[25] R. Weissleder, J. F. Heautot, B. K. Schaffer, N. Nossiff, M. I. Papisov, A. Bogdanov, Jr., and T. J. Brady, *Radiology* **191**, 225 (1994).
[26] E. Seneterre, R. Weissleder, D. Jaramillo, P. Reimer, A. S. Lee, T. J. Brady, and J. Wittenberg, *Radiology* **179**, 529 (1991).

may detach easily from the membrane, or can be transferred to other cells. Unmodified (U)SPIOs have been used, at high concentrations, to label mononuclear (T) cells,[27,28] glioma cells and macrophages,[29,30] and oligodendrocyte progenitors[31] *in vitro*. Several modifications of (U)SPIOs, including monocrystalline iron oxide nanoparticles (MIONs) and cross-linked iron oxides (CLIOs), that induce an efficient internalization of the iron oxide nanoparticles have been described.

Use of Peptide- or Antibody-Coated SPIOs

A significant improvement of labeling of nonphagocytic cells has been achieved by linking the particles to the human immunodeficiency virus (HIV) tat peptide.[32,33] This peptide contains a membrane translocating signal that efficiently transports the iron oxides into cells, with a further transport and accumulation in the nucleus. Functional tests have shown no differences between labeled and unlabeled T cells.[34] By conjugating a higher number of tat peptides per iron oxide particle, the cellular uptake has recently been improved, showing an increased uptake with an increased tat peptide/CLIO ratio.[35] Labeled T cells could be seen homing to liver and spleen following systemic injection.[34,36] For antigen-specific T-lymphocyte subsets, specific homing following adoptive transfer has been seen in the pancreas in animal models of autoimmune diabetes.[37,38] CLIO–tat labeled neural stem cells have been transplanted into the embryonic

[27] J. C. Sipe, M. Filippi, G. Martino, R. Furlan, M. A. Rocca, M. Rovaris, A. Bergami, J. Zyroff, G. Scotti, and G. Comi, *Magn. Reson. Imaging* **17,** 1521 (1999).

[28] T. C. Yeh, W. Zhang, S. T. Ildstad, and C. Ho, *Magn. Reson. Med.* **30,** 617 (1993).

[29] A. Moore, R. Weissleder, and A. Bogdanov, Jr., *J. Magn. Reson. Imaging* **7,** 1140 (1997).

[30] R. Weissleder, H. C. Cheng, A. Bogdanova, and A. Bogdanov, Jr., *J. Magn. Reson. Imaging* **7,** 258 (1997).

[31] R. J. Franklin, K. L. Blaschuk, M. C. Bearchell, L. L. Prestoz, A. Setzu, K. M. Brindle, and C. French-Constant, *Neuroreport* **10,** 3961 (1999).

[32] L. Josephson, C. H. Tung, A. Moore, and R. Weissleder, *Bioconjug. Chem.* **10,** 186 (1999).

[33] M. Lewin, N. Carlesso, C. H. Tung, X. W. Tang, D. Cory, D. T. Scadden, and R. Weissleder, *Nat. Biotechnol.* **18,** 410 (2000).

[34] C. H. Dodd, H. C. Hsu, W. J. Chu, P. Yang, H. G. Zhang, J. D. Mountz, Jr., K. Zinn, J. Forder, L. Josephson, R. Weissleder, J. M. Mountz, and J. D. Mountz, *J. Immunol. Methods* **256,** 89 (2001).

[35] M. Zhao, M. F. Kircher, L. Josephson, and R. Weissleder, *Bioconjug. Chem.* **13,** 840 (2002).

[36] U. Schoepf, E. M. Marecos, R. J. Melder, R. K. Jain, and R. Weissleder, *Biotechniques* **24,** 642 (1998).

[37] O. Beuf, M. Janier, C. Asport, C. Thivolet, and C. Billotey, *Proc. Intl. Soc. Mag. Reson. Med.* **11,** 828 (2003).

[38] A. Moore, P. Z. Sun, D. Cory, D. Hogemann, R. Weissleder, and M. A. Lipes, *Magn. Reson. Med.* **47,** 751 (2002).

mouse brain using ultrasound biomicroscopy.[39] Migration of cells could be detected along the rostral migratory stream, a well-documented neuronal migration pathway from the subventricular zone of the lateral ventricles to the olfactory bulb.

Another approach for intracellular magnetic labeling is the use of internalizing monoclonal antibodies (moabs). The mouse antitransferrin receptor (Tfr) moab OX-26[40] induces internalization of the Tfr upon binding. Compounds have been conjugated to OX-26 to deliver the drug to cells by receptor-mediated endocytosis (RME). MION-46L–OX-26 complexes have been used to magnetically label oligodendrocyte progenitors[41] and neural precursor cells[42] by RME. Covalent constructs of (U)SPIOs or MION-46L and moabs can be made by linking the amine (lysine) groups of the antibody to the alcohol groups of the dextran coating using the periodate oxidation–borohydride reduction method (Fig. 1). To this end, MION-46-L in 0.01 M citrate buffer, pH = 8.4, is first oxidized at 4° for 24 h with 1 mg $NaIO_4$ per milligrams Fe. The oxidized nanoparticle preparation, containing reactive aldehyde groups, is then purified using a Sephadex G-100 column, removing any unbound or released free dextran. The moab, dialyzed against 0.2 M sodium bicarbonate buffer, pH = 6.5, is allowed to form Schiff's bases with MION-46L at 4° for 16 h at a 1:1 protein to Fe weight ratio. The unstable Schiff's bases are subsequently reduced to secondary amide bonds by adding 1 mg of $NaCNBH_3$ per mg Fe and incubation at room temperature for 4 h. This reduction step also blocks any reactive aldehydes on the dextran that did not form Schiff's bases with the protein. Finally, the resulting MION-46L–moab construct is purified over a Sephadex G-100 column.

Oligodendrocyte progenitors have been magnetically labeled using the anti-Tfr moab OX-26 and were transplanted in the spinal cord of 7-day-old *md* rats.[41] Ten to 14 days later, the spinal cord was removed and imaged at 4.7 Tesla at 78-μm resolution. Migration of labeled cells was easily identified on the MR images as streaks of hypointensity, primarily along the dorsal column, over a distance of up to 1 cm away from the injection site. It was shown that labeled cells retained their capacity to myelinate axons *in vivo* following magnetic labeling. Immunohistochemical analysis of

[39] D. H. Turnbull, Y. Z. Wadghiri, L. Josephson, and H. Wichterle, *Proc. Intl. Soc. Magn. Reson. Med.* **9,** 359 (2001).

[40] W. A. Jefferies, M. R. Brandon, S. V. Hunt, A. F. Williams, K. C. Gatter, and D. Y. Mason, *Nature* **312,** 162 (1984).

[41] J. W. M. Bulte, S. Zhang, P. van Gelderen, V. Herynek, E. K. Jordan, I. D. Duncan, and J. A. Frank, *Proc. Natl. Acad. Sci. USA* **96,** 15256 (1999).

[42] J. W. M. Bulte, T. Ben-Hur, B. R. Miller, R. Mizrachi-Kol, O. Einstein, E. Reinhartz, H. A. Zywicke, T. Douglas, and J. A. Frank, *Magn. Reson. Med.* **50,** 201 (2003).

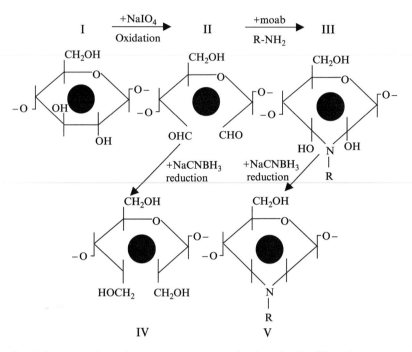

FIG. 1. Sequence of reactions to conjugate monoclonal antibodies (R) to dextran-coated SPIO using the periodate oxidation–borohydride reduction technique. III represents intermediate Schiff's bases, IV represents blocked unreacted dextran–SPIO, and V represents the resulting moab–SPIO construct.

newly formed myelin showed an excellent match between the MR contrast and staining for proteolipid protein, an essential component of myelin (Fig. 2). The density of magnetically labeled cells, as evaluated by Prussian Blue staining, was not as high as the density of myelinated axons, but this can be explained by the fact that one oligodendrocyte is believed to have the capacity to myelinate at least 30–40 axons.

Another example of cell labeling using internalizing moabs is the use of biotinylated anti-CD-11 moab in conjuction with streptavidinylated dextran-coated iron oxides, which are commercially available; a high degree of magnetic labeling was achieved for dendritic cells.[43] Dendritic cells are antigen-presenting cells that have recently been exploited for (tumor) antigen-specific vaccination studies. Their *in vivo* homing properties following local or systemic injections may be further investigated by MRI. A disadvantage

[43] E. T. Ahrens, M. Feili-Hariri, H. Xu, G. Genove, and P. A. Morel, *Magn. Reson. Med.* **49,** 1006 (2003).

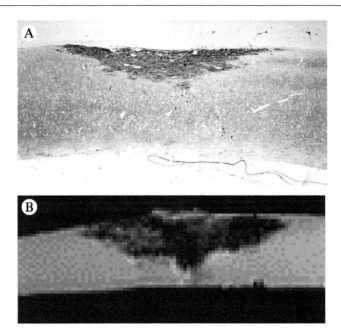

Fig. 2. Magnetic labeling of cells using a SPIO–moab construct. (A) Antiproteolipid protein (PLP) immunostaining of the spinal cord of a myelin deficient (*md*) rat 14 days after transplantation of OX-26–MION-46L-labeled oligodendrocyte progenitors. (B) Corresponding MR image (obtained at 78 µm resolution using a 4.7-T magnet) of specimen shown in (A) shows a good agreement between the area of cell distribution in (B) and new myelin formation in (A). For details, see Bulte *et al.*[41]

of the use of (internalizing) moabs, however, is that they are species specific, and a new synthesis is required when performing studies in different animals (e.g., mice or rats). Also, when these constructs are used clinically, there will be regulatory issues regarding the use of a xenogeneic (i.e., mouse) protein or development of a humanized moab.

Use of Magnetodendrimers

A convenient magnetic label that may have broad applications in MRI cell tracking ideally shows an intracellular uptake that is nonspecific regardless of cell origin or animal species. An example is the use of magneto-dendrimers (MD-100),[44,45] representing SPIOs coated with a generation

[44] J. W. M. Bulte, T. Douglas, B. Witwer, S. C. Zhang, E. Strable, B. K. Lewis, H. A. Zywicke, B. Miller, P. van Gelderen, B. M. Moskowitz, I. D. Duncan, and J. A. Frank, *Nat. Biotechnol.* **19,** 1141 (2001).

4.5 carboxylated dendrimer in lieu of dextran. For synthesis of MD-100, the $G = 4.5$ dendrimer (10 mg, 4×10^{-4} mM) is added to a reaction vessel and the initial dendrimer solvent (methanol) is removed by a stream of N_2 for 20 min. After this, 30 ml of deaerated H_2O (0.1 M NaCl, pH 8.5) is added and further deaerated by bubbling N_2 for 20 min and maintained under N_2 during the reaction. For an Fe loading of 100 (i.e., synthesis of MD-100), deaerated solutions of $(NH_4)_2Fe(SO_4)$ (25 mM, 1.58 ml, 4×10^{-2} mM) and the 2-electron oxidant Me_3NO (25 mM, 1.58 ml) are added at a fixed rate via a syringe pump. Fe(II) and Me_3NO solutions are added at a rate of approximately 0.5 ml/min. For synthesis of magnetodendrimers with a loading factor of 100 (MD-100), 1.58 ml of Fe(II) (25 mM) and 1.58 ml of Me_3NO (25 mM) are used. The solution pH is maintained by titration of the H^+ in a pH-Stat experiment using an autotitrator. This eliminates the need to use contaminating buffers and allows monitoring of the progress of the reaction. The change in pH that occurs is due in part to the acidic nature of the Fe(II) added and also due to H^+ generated by the oxidative hydrolysis reaction to form the iron oxide mineral. After the completion of the reaction, sodium citrate is added (to 35 mM) to remove any unmineralized Fe. After reaction, the MD-100 product is dialyzed exhaustively against double-distilled H_2O and concentrated 50-fold by ultrafiltration using a 100-kDa M_w cutoff filter.

The MD-100 preparations are fully soluble for several years following synthesis. Transmission electron microscopy revealed an oligocrystalline structure of 7–8-nm crystals separated by a somewhat smaller distance. Because the diameter of a $G = 4.5$ dendrimer is approximately 5 nm, the actual structure is believed to be several SPIO particles within a cluster coated on the outside with the dendrimer. This result is different from a few reports[46–48] that describe the presence of intradendrimeric metal particles, but is consistent with a later report[49] in which the mineral particles are not entrapped within the dendrimer but grow from the dendrimer surface. For intradendrimer synthesis, the initial binding of a critical number of metal ions to terminal amine groups or interior amide groups of the dendrimer[50] is of paramount importance. In the synthesis of the magnetodendrimer, the added iron atoms do not form strong interactions with the interior amide or terminal amine groups. Rather, it is thought that the

[45] E. Strable, J. W. M. Bulte, B. M. Moskowitz, K. Vivekanandan, M. Allen, and T. Douglas, *Chem. Mater.* **13**, 2201 (2001).
[46] L. Balogh and D. A. Tomalia, *J. Am. Chem. Soc.* **120**, 7355 (1998).
[47] M. Q. Zhao and R. M. Crooks, *Angew. Chem. Int. Ed. Engl.* **38**, 364 (1999).
[48] M. Q. Zhao, L. Sun, and R. M. Crooks, *J. Am. Chem. Soc.* **120**, 4877 (1998).
[49] M. E. Garcia, L. A. Baker, and R. M. Crooks, *Anal. Chem.* **71**, 256 (1999).
[50] M. Zhao and R. M. Crooks, *Chem. Mater.* **11**, 3380 (1999).

terminal carboxyl groups initially bind and sequester the ferrous ion, analogous to the use of dendrimer as uranyl ion sponges,[51] but with the dendrimer acting here as a template for iron oxide synthesis. This is supported by experimental findings that either an initial Fe loading factor per dendrimer of >200 (i.e., higher than the relative number of terminal carboxyl groups per dendrimer) or the use of amine group-terminated dendrimers resulted in near-immediate precipitation of insoluble material. It may be possible, however, that the use of higher generation dendrimers (G = 5 to G = 10) allows synthesis of iron oxides inside the dendrimer cavity: for G < 6, dendrimers aggregate when stabilizing gold metal particles; for G = 6 to G = 9, one gold colloid per dendrimer is templated; and for G = 10, multiple smaller gold particles per dendrimer were observed.[52]

Our interest in dendrimers stems from the fact that these polymers are currently used as agents to transfect oligonucleotides into cells.[53–55] The highly charged polymers bind on multiple sites on the cell membrane, induce membrane bending, and then induce endocytosis.[56] A variety of cells from mouse, rat, and human origin can be easily labeled by simple addition of MD-100 to the cell culture for 24–48 h at low concentrations of iron (10–25 μg Fe/ml).[42,44,57–59] This includes mouse ES cells, mouse 3T3 fibroblasts, mouse C2C12 muscle progenitors, rat neural stem cells, rat oligodendroglial progenitors, human small cell lung carcinoma cells, human cervix carcinoma cells, human neural stem cells, human mesenchymal stem cells, and human embryoid body-derived (EBD) pluripotent stem cells (Fig. 3). Regardless of origin or animal species, incubated cells show a comparable degree of particle uptake into endosomes, demonstrating the nonspecific nature of the magnetodendrimer uptake. There is no difference

[51] M. F. Ottaviani, P. Favuzza, M. Bigazzi, N. J. Turro, S. Jockusch, and D. A. Tomalia, *Langmuir* **16,** 7368 (2000).

[52] F. Grohn, B. J. Bauer, Y. A. Akpalu, C. L. Jackson, and E. J. Amis, *Macromolecules* **33,** 6042 (2000).

[53] L. H. Bryant, Jr. and J. W. M. Bulte, *in* "Focus on Biotechnology, Vol. 7: Physics and Chemistry Basis of Biotechnology" (M. de Cuyper and J. W. M. Bulte, eds.). Kluwer Academic, Dordrecht, The Netherlands.

[54] J. F. Kukowska-Latallo, A. U. Bielinska, J. Johnson, R. Spindler, D. A. Tomalia, and J. R. Baker, Jr., *Proc. Natl. Acad. Sci. USA* **93,** 4897 (1996).

[55] M. X. Tang, C. T. Redemann, and F. C. Szoka, Jr., *Bioconjug. Chem.* **7,** 703 (1996).

[56] Z.-Y. Zhang and B. D. Smith, *Bioconjug. Chem.* **11,** 805 (2000).

[57] J. W. M. Bulte, T. Douglas, P. van Gelderen, B. K. Lewis, and J. A. Frank, *Proc. Intl. Soc. Magn. Reson. Med.* **9,** 52 (2001).

[58] J. W. M. Bulte, J. Lu, H. A. Zywicke, P. van Gelderen, T. Douglas, J. W. McDonald, and J. A. Frank, *Proc. Intl. Soc. Magn. Reson. Med.* **9,** 130 (2001).

[59] G. A. Walter, K. S. Cahill, J. Huard, H. Feng, T. Douglas, H. L. Sweeney, and J. W. M. Bulte, *Magn. Reson. Med.* **51,** 273 (2003).

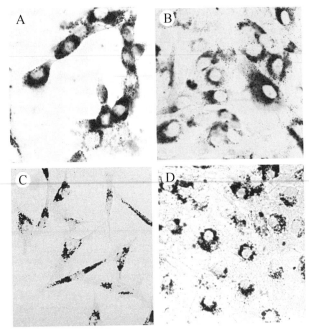

FIG. 3. Magnetic labeling of cells using magnetodendrimers (MD-100). Shown are DAB-enhanced Prussian Blue stains of labeled 3T3 mouse fibroblasts (A), C2C12 mouse muscle progenitors (B), HeLa human cervix carcinoma (C), and rat neural progenitors (D). Cells were labeled with 25 μg Fe/ml for 24–48 h, washed, and then replated on chamber slides for 24 h (A–C) or 5 days (D) to allow attachment. For details, see Bulte et al.[44] and Bulte et al.[42]

between MD-100 labeled and unlabeled cells in terms of proliferation and viability, the ability of neural stem cells to differentiate into neurons and to myelinate axons *in vivo*,[44] or the ability of muscle stem cells to differentiate into myotubes.[59]

Oligodendroglial progenitors have been derived from NSCs, labeled with MD-100, and transplanted in the ventricles of neonatal *les* rats.[44] Migration of labeled cells into the brain parenchyma could be observed at the earliest time points (2–3 weeks) throughout the latest time point of imaging (6 weeks) (Fig. 4). A good agreement was observed between the hypointense MR contrast and expression of the transfected *LacZ* gene, serving as a reporter gene for β-galactosidase. In these areas, new myelin was formed as demonstrated by antimyelin basic protein immunolabeling, proving that the cells were still functional following labeling. Labeled cells could also be readily identified when a clinical 1.5-Tesla (T) scanner was used. The contrast appeared to fade out somewhat at the later time points,

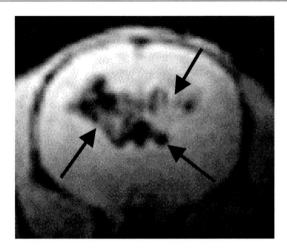

FIG. 4. Magnetic labeling of cells using magnetodendrimers (MD-100). Shown is a brain MR image (obtained at 313 μm resolution using a 4.7-T magnet) of a Long Evans shaker rat 6 weeks after transplantation of MD-100-labeled oligodendroglial progenitors. Arrows indicate the distribution of labeled cells that migrated from the ventricle (site of injection) into the brain parenchyma. For details, see Bulte *et al.*[44]

presumably through biodegradation of the iron oxide particles. A comparison of different pulse sequences demonstrated that gradient–echo-based techniques were most sensitive to the detection of label, due to the large magnetic susceptibility/T_2^* effects that these iron oxide particles induce.

In rats with experimental autoimmune encephalomyelitis (EAE), inflammatory cells can be found in the white matter, and this can induce targeted migration of transplanted neural precursor cells.[60] When MD-100-labeled neurospheres were transplanted into the ventricles of EAE rats at the peak of their disease, migration into white matter structures could be observed on the MR images[42] (Fig. 5). A good correlation was found between the histologic distribution of iron-labeled cells (Prussian Blue staining) and of bromodeoxyuridine (BrdU, immunostaining), indicating that the magnetic label was retained within labeled cells and not transferred to other cells *in vivo*.

Other MR tracking studies in the CNS include detection of transplanted, MD-100-labeled mouse neural progenitor cells in parkinsonian mice.[61] Mouse ES cells, labeled with MD-100 and differentiated into the

[60] T. Ben-Hur, O. Einstein, R. Mizrachi-Kol, O. Ben-Menachem, E. Reinhartz, D. Karussis, and O. Abramsky, *Glia* **41**, 73 (2003).
[61] J. M. Hakumaki, J. M. Savitt, J. D. Gearhart, T. M. Dawson, V. L. Dawson, and J. W. M. Bulte, *Dev. Brain Res.* **132**, A43 (2001).

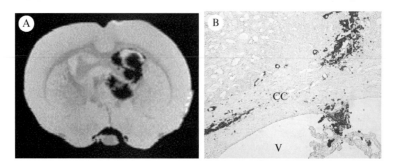

Fig. 5. Magnetic labeling of cells using magnetodendrimers (MD-100). Shown is a rat EAE brain 7 days after transplantation of MD-100-labeled neural precursor cells. (A) Microscopic MR image (obtained *ex vivo* at 104 μm resolution using a 4.7-T magnet) shows migration of labeled cells out of the ventricle (site of injection) into the corpus callosum. (B) DAB-enhanced Prussian Blue staining showing the corresponding areas of the ventricle (V) and corpus callosum (CC). For details, see Bulte *et al.*[42]

neural lineage using retonic acid, were transplanted into the spinal cords of rats that received a contusion injury.[58] Hypointense MRI contrast could be observed up to 1.3 cm away from the injection site. MD-100-labeled human small cell lung cancer cells also have been xenografted in a nude rat, and labeled cells could be seen to originate from the center, radiating out toward the periphery of the tumor.[57] In this way, intratumoral growth patterns of labeled tumor cells may be detected following implantation. Myoblasts have been labeled with MD-100 and transplanted into the muscle of *mdx* mice, a model of muscular dystropy.[59] The distribution of labeled cells on the MR images correlated well with the histologic staining for β-galactosidase, using *LacZ* as a reporter gene.

Use of Transfection Agent (TA)-Coated SPIOs

A disadvantage of the magnetodendrimers is that they have not yet been commercially developed (as of the present time) and are thus not widely available. Recently, however, a convenient magnetic labeling method that is based on the use of a commercially available (U)SPIO formulation, Feridex[62,63] or Sinerem,[64] has been developed. The method is based on the use of transfection agents (TAs), which are also commercially

[62] J. A. Frank, B. R. Miller, A. S. Arbab, H. A. Zywicke, E. K. Jordan, B. K. Lewis, L. H. Bryant, Jr., and J. W. Bulte, *Radiology* **228,** 480 (2003).

[63] J. A. Frank, H. A. Zywicke, E. K. Jordan, J. Mitchell, B. K. Lewis, B. Miller, L. H. Bryant, Jr., and J. W. M. Bulte, *Acad. Radiol.* **9,** S484 (2002).

[64] M. Hoehn, E. Kustermann, J. Blunk, D. Wiedermann, T. Trapp, S. Wecker, M. Focking, H. Arnold, J. Hescheler, B. K. Fleischmann, W. Schwindt, and C. Buhrle, *Proc. Natl. Acad. Sci. USA* **99,** 16267 (2002).

available. TAs are highly charged macromolecules and include classes such as dextrans, phosphonates, lipids, proteins, and dendrimers, and are used primarily to transfect DNA into cells. TAs complex to DNA via electrostatic interaction, and although the complexation increases the transfection efficiency of DNA into the cell, its major drawback in using TAs for DNA transfer is endosomal capture. Polycationic TAs such as poly-L-lysine (PLL), low and high molecular weight heat-inactivated dendrimers, and lipofectamine have zeta potentials that vary from 1 to 65 mV, whereas dextran-coated ferumoxides and MION-46L with carboxyl groups on the surface of the nanoparticle have zeta potentials of approximately -41 and 2 mV, respectively. Combining dextran-coated (U)SPIO and polycationic TA in a test tube in appropriate ratios resulted in complex formation through van der Waals interactions, resulting in a decrease in the MR relaxation rates of the (U)SPIO.[65] The alteration of the T_1 and T_2 relaxation rates indicated that the TAs were coating the (U)SPIO, preventing water protons from effectively shielding the normal water exchange with the paramagnetic sites on iron oxide nanoparticles for effective dipole–dipole T_1 relaxation. The increase in the T_2 relaxation rate indicates that there is an appreciable outer sphere relaxation effect presumably exerted by the TAs, clustering the nanoparticles and shielding the water from its surface.[62,63,65] These results indicate that under appropriate conditions, highly surface-charged macromolecular TAs complex through electrostatic interactions with the existing surface charges on dextran-coated (U)SPIO nanoparticles and then self-assemble into complexes. However, the shielding of the water from the surface of the iron dextran nanoparticles does not necessarily translate into cellular magnetic labeling efficiency, and results indicate that cellular labeling requires a balance between positive and negative charges on the surface of the nanoparticle.[62,63,66,67]

When the complexes are added to the cell culture, the TA effectively shuttles the (U)SPIO into the cell through formation of endosomes (Fig. 6). TAs that are suitable for this purpose include dendrimers (Superfect), PLL, lipofectamine, and FuGENE, but many other available TAs may be used (Fig. 7). Each combination of TA and (U)SPIO has to be carefully titrated and optimized, since lower concentrations may not lead to sufficient cellular uptake, whereas higher concentrations may induce precipitation of complexes or may be toxic to the cells. If excess iron oxides

[65] H. Kalish, A. S. Arbab, B. R. Miller, B. K. Lewis, H. A. Zywicke, J. W. M. Bulte, L. H. Bryant, Jr., and J. A. Frank, *Magn. Reson. Med.* **50**, 275 (2003).

[66] A. S. Arbab, L. A. Bashaw, B. R. Miller, E. K. Jordan, J. W. M. Bulte, and J. A. Frank, *Transplantation* **76**, 1123 (2003).

[67] A. S. Arbab, L. A. Bashaw, B. R. Miller, E. K. Jordan, B. K. Lewis, H. R. Kalish, and J. A. Frank, *Radiology* **229**, 838 (2003).

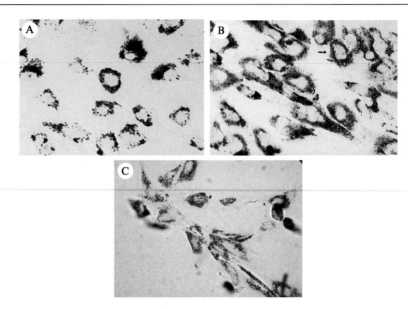

Fig. 6. Magnetic labeling of mesenchymal stem cells (MSCs) using iron oxide particles (Feridex) coated with a transfection agent (poly-L-lysine; PLL). Cells were labeled for 48 h (A, B) or 24 h with 25 μg Fe/ml Feridex and 375 ng/ml PLL. DAB-enhanced Prussian Blue stain of labeled porcine (A), human (B), and canine (C) MSCs shows an efficient intracellular uptake of particles into endosomes that is nonspecific across species.

are used, complexation to the TA may not occur or would form with an inappropriate size or charge distribution on the surface of the nanoparticle, decreasing the cellular labeling efficiency. The following provides the protocols used for labeling different cell types using PLL (Sigma, St. Louis, MO, catalog # P-1524) and Feridex.[66]

Preparation of Feridex–PLL Complex. Feridex at a concentration of 50 μg/ml is put into a mixing flask or tube containing respective complete or serum-free media (see the following), and then PLL is added to the solution at 1.5 μg/ml (for other than mononuclear hematopoietic-derived cells) or 3 μg/ml (for hematopoietic-derived cells such as CD 34[+] cells, AC 133[+] cells, lymphocytes, and B cells). The ferumoxides to PLL ratio ranges from 1:0.06 to 1:0.03. The solution containing ferumoxides and PLL is mixed in a rotator for 30–60 min. After 30–60 min, an equal volume of the solution containing Feridex–PLL complexes is added to the existing media in the cell culture. The final concentration of Feridex–PLL will be in a ratio of 25 μg/ml iron to PLL 0.75 μg/ml for adherent and rapidly dividing cells, and 25 μg/ml iron to PLL 1.5 μg/ml for hematopoietic-derived mononuclear cells.

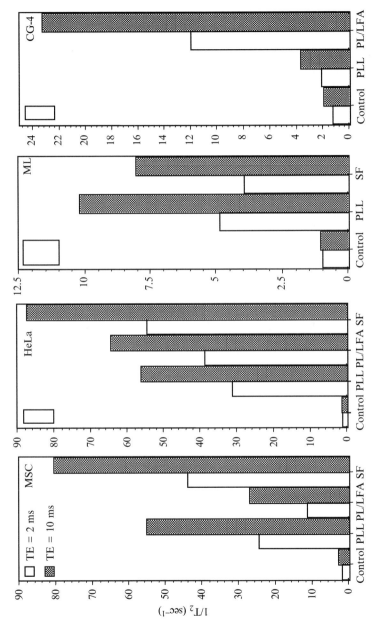

FIG. 7. Magnetic labeling of cells using iron oxide particles (Feridex) coated with various TAs. Bar graphs show $1/T_2$ values for TE = 2 and 10 ms at 1.0 T and room temperature for MSCs, HeLa tumor cells, mouse lymphocytes (ML), and CG-4 oligodendrocyte progenitor cells that were incubated with Feridex and various TAs. Control cells represent samples where no iron was added to the culture medium. The TE dependence of $1/T_2$ indicates clustering of the iron oxide nanoparticles within endosomes and cells. PLL, poly-L-lysine; PL/LFA, PLUS/lipofectamine; SF, Superfect. For details, see Frank et al.[62]

Labeling of Adherent Cells. Cells should be allowed to grow until 80–90% confluence of the surface area of the culture dish. After discarding old media, 50% of the media is replaced with fresh complete media, and 50% of the media contains Feridex–PLL complexes. Cells are then incubated overnight, for approximately 16–18 h. By using serum-free media for both the cells and Feridex–PLL complex formation, cells can be efficiently labeled within 2 h.

Labeling of Cells Grown in Suspension (Rapidly Growing Cells). On the day of cell labeling, cells are centrifuged and resuspended in fresh media and a specific volume of cell suspension containing approximately $1–2 \times 10^6$ cells/ml is placed into the culture dish. An equal volume of media containing Feridex–PLL complex is added to the cell suspension. The cells are then incubated overnight, for approximately 16–18 h. By using serum-free media for resuspension of cells and Feridex–PLL complex formation, cells can be efficiently labeled within 2 h.

Labeling of Hematopoietic Cells. Cell density should be controlled to maximize the incorporation of the Feridex–PLL complex into endosomes. On the day of cell labeling, cells are centrifuged and resuspended in old media and a specific volume of cell suspension containing approximately 4×10^6 cells/ml is placed into the culture dish before adding Feridex–PLL complexes. Feridex–PLL complex is made in serum-free media (containing all growth factors for the cells). The PLL should be added at 2.25–3 μg/ml (Feridex:PLL 1:0.045–0.06 μg/ml). An equal volume of media containing Feridex–PLL complex is added, and cells are incubated overnight, for approximately 12–18 h. Cells can be effectively labeled by 3 h using serum-free media for repeat centrifugation, resuspension, Feridex–PLL complex formation, and incubation.

High molecular weight PLL (Sigma P-1524, MW around 400 kDa) has now become the TA of choice in most recent studies,[62,68–71] since it is available from several different manufacturers at a low price and is not a proprietary TA. Following Feridex–PLL labeling, cell viability and proliferation capacity are unaffected, and the amount of cellular iron uptake, in the range of 10–20 pg Fe per cell,[62] is comparable to the results obtained with magnetodendrimer labeling.[44] Feridex–PLL labeled swine mesenchymal

[68] S. A. Anderson, J. Shukaliak-Quandt, E. K. Jordan, A. S. Arbab, R. Martin, H. F. McFarland, and J. A. Frank, *Proc. Intl. Soc. Magn. Reson. Med.* **11,** 763 (2003).

[69] K. S. Cahill, X. Silver, G. Gaidosh, B. J. Byrne, and G. A. Walter, *Proc. Intl. Soc. Magn. Reson. Med.* **11,** 368 (2003).

[70] D. L. Kraitchman, A. W. Heldman, E. Atalar, L. C. Amado, B. J. Martin, M. F. Pittenger, J. M. Hare, and J. W. M. Bulte, *Circulation* **107,** 2290 (2003).

[71] D. L. Kraitchman, P. Karmarkar, L. C. Amado, E. Atalar, L. V. Hofmann, A. W. Heldman, B. J. Martin, M. F. Pittenger, and J. W. M. Bulte, *Proc. Intl. Soc. Magn. Reson. Med.* **11,** 360 (2003).

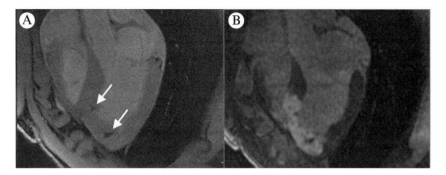

FIG. 8. Magnetic labeling of cells using Feridex–PLL. Magnetically labeled porcine mesenchymal stem cells (MSCs) were delivered intramyocardially under X-ray fluoroscopy in a swine myocardial infarct model. (A) Fast gradient echo (FGRE) MR image, obtained within 24 h following transplantation, shows the injection sites as hypointense lesions (arrows). (B) Delayed-enhanced (DE) image confirms that the cells were injected in the Gd-DTPA-enhanced area of the infarct. For details, see Kraitchman et al.[70]

stem cells (MSCs) have been injected intramyocardially under X-ray fluoroscopy in a swine model of closed-chest experimental myocardial infarct.[70] MR images of the heart obtained within 24 h after injection showed the injection sites as ovoid hypointense lesions with sharp borders (Fig. 8). At 1 week and 3 weeks after injection, the delineation of the borders was less clear. The size of the hypointense lesions were found to expand by 15% over 1 week, whereas the contrast decreased by 24%. In a dog myocardial infarct model, Feridex–PLL labeled canine MSCs have been delivered in the infarcted area under real-time MRI, using an active, MR-biocompatible catheter.[71] Sinerem–FuGENE-labeled embryonic cardiomyocytes have been implanted into the cryolesioned mouse heart, and their tissue location was visualized using high-resolution MRI.[72] Sinerem–FuGENE-labeled mouse ES cells have been implanted in the contralateral hemisphere of rats with experimental stroke (using a middle cerebral artery occlusion [MCAO] model).[64] In vivo imaging at high field (7 T) and high resolution showed a migration of labeled cells along the corpus callosum to the other hemisphere at the site of the injury (see Fig. 7). A good coregistration was achieved between the MR contrast and transfected green fluorescent protein (GFP) expression. It is believed that the local ischemia produces local inflammatory and chemotactic factors that attract cells to the injured site. For antigen-specific T lymphocytes, specific homing following adoptive transfer has been observed in the spinal cord of mice

[72] E. Kuestermann, W. Roell, S. Wecker, D. Wiedermann, C. Buehrle, A. Welz, J. Hescheler, B. K. Fleischmann, and M. Hoehn, Proc. Intl. Soc. Magn. Reson. Med. 11, 804 (2003).

with EAE.[68] Myoblasts have been labeled with Feridex–PLL and were transplanted into the muscle of *mdx* mice, a model of muscular dystropy.[69] The distribution of labeled cells on the MR images correlated well with the histologic staining for β-galactosidase, using *LacZ* as a reporter gene.

There are other approaches for (nonspecific) magnetic cell labeling that have recently been developed; these include the use of a biolistic gene gun[73,74] and the use of large, styrene–divinyl benzene-coated magnetic microspheres.[75–77] Using the gene gun, magnetically labeled neural progenitor cells were transplanted into the cisterna magna of rats that underwent experimental MCAO. Migration of labeled cells was visualized over time *in vivo* at 7 T, and cells were found to migrate throughout the ventricular system toward the ischemic brain parenchyma.[74] Labeled cells appeared to have differentiated into neurons and induced a significant improvement of neurologic function.

The distribution of magnetic microsphere-labeled porcine MSCs has been studied in a swine myocardial infarct model using MRI.[75,76] Homing of magnetically labeled rat MSCs, injected in the tail vein, to the infarcted rat heart could be detected using a 7-T magnet.[77] It remains to be seen if the microsphere iron oxide formulation is clinically applicable and can be used for labeling of nonphagocytic cells. These large iron oxides have the advantage, however, that a single particle can be detected by MRI.[78] Finally, anionic magnetic nanoparticles have been developed and used for intracellular labeling of macrophages.[79–81] The negative surface charge of the particles induces an uptake three magnitude orders higher as

[73] R. L. Zhang, L. Zhang, Z. G. Zhang, D. Morris, Q. Jiang, L. Wang, L. J. Zhang, and M. Chopp, *Neuroscience* **116**, 373 (2003).

[74] Z. G. Zhang, Q. Jiang, R. Zhang, L. Zhang, L. Wang, P. Arniego, K. L. Ho, and M. Chopp, *Ann. Neurol.* **53**, 259 (2003).

[75] A. J. Dick, M. A. Guttman, V. K. Raman, D. C. Peters, J. M. Hill, B. S. Pessanha, G. Scott, S. Smith, E. R. McVeigh, and R. J. Lederman, *Proc. Intl. Soc. Magn. Reson. Med.* **11**, 365 (2003).

[76] J. M. Hill, A. J. Dick, V. K. Raman, R. B. Thompson, B. S. Pessanha, Z. Yu, A. Hinds, T. R. Varney, C. E. Dunbar, E. R. McVeigh, and R. J. Lederman, *Proc. Intl. Soc. Magn. Reson. Med.* **11**, 361 (2003).

[77] J. M. Sorger, D. Despres, D. Schimel, E. R. McVeigh, and J. M. Hill, *Proc. Intl. Soc. Magn. Reson. Med.* **11**, 364 (2003).

[78] K. A. Hinds, J. M. Hill, E. M. Shapiro, M. O. Laukkanen, A. C. Silva, C. A. Combs, T. R. Varney, R. S. Balaban, A. P. Koretsky, and C. E. Dunbar, *Blood* **102**, 867 (2003).

[79] C. Billotey, C. Wilhelm, M. Devaud, J. C. Bacri, J. Bittoun, and F. Gazeau, *Magn. Reson. Med.* **49**, 646 (2003).

[80] C. Wilhelm, C. Billotey, J. Roger, J. N. Pons, J. C. Bacri, and F. Gazeau, *Biomaterials* **24**, 1001 (2003).

[81] C. Wilhelm, F. Gazeau, J. Roger, J. N. Pons, and J. C. Bacri, *Langmuir* **18**, 8148 (2002).

compared with conventional dextran-coated SPIOs. The particles also end up in endosomes, the shape of which can be deformed by applying an external magnet.[82]

Magnetically Labeled Cells: Methods of Analysis

From the previous sections, it is clear that several alternative methods have now been developed for a successful preparation of magnetically labeled cells. Time will tell us which method will be preferred by most investigators, and which method will be eventually implemented in a clinical setting. Regardless of the method employed, the analysis and evaluation of magnetic labeling will follow similar protocols. As mentioned earlier, an important qualitative method of analysis is Prussian Blue staining or Perls' staining. The cellular ferric iron forms a blue complex with the ferrocyanide salt at low pH, which can be further enhanced with diamino-benzidine (DAB). To this end, (4% glutaraldehyde) fixed cells are washed and incubated for 30 min with 2% potassium ferrocyanide in 6% HCl, washed, and counterstained with nuclear fast red. For DAB-enhanced Prussian Blue staining, slides are reacted with unactivated and activated (containing 0.03% H_2O_2) 14% w/v DAB for 15 min each and can be counterstained either with nuclear fast rad or hematoxylin. Examination of labeled cells by light microscopy at high power will reveal if the magnetic label is located intracellular (in endosomes), or remains mostly bound on the outer cell membrane. Labeling can also be assessed by transmission electron microscopy (TEM), due to the inherent electrodense properties of the nanoparticles. The appearance of the particles on TEM is very similar to that of ferritin, which is used as a TEM marker in immunoelectron microscopy. In short, 4% glutaraldehyde-fixed cells are also embedded in 2% agar, and stained with 0.1% osmium tetroxide for 30 min and with 0.5% uranyl acetate overnight. Ultrathin sections are then further stained with lead citrate and examined by TEM.

A quantitative evaluation of magnetic labeling determines the amount of cellular iron, usually expressed as pg of iron per cell. As an alternative to ^{59}Fe-labeled cells that allows direct quantification through measurement of radioactivity, two alternative methods have been described[44] for assessing the cellular iron content (i.e., a relaxometric and a spectrophotometric method). For both methods, the labeled, washed cells are counted using a hemocytometer and resuspended in 500 μl 4% w/w gelatin at a density of approximately 1×10^7 cells. The cell suspensions are dried at 110°

[82] C. Wilhelm, A. Cebers, J. C. Bacri, and F. Gazeau, *Eur. Biophys. J.* **32,** 655 (2003).

for 16 h, and then completely digested in acid. To each sample, 375 μl 70% ultrapure perchloric acid (containing <2 ppb Fe), and 125 μl 100% ultrapure nitric acid (<0.6 ppb Fe) are added, and samples are digested for 3 h at 60° using a heating block. On these 500-μl samples, the two methods of iron quantification can be applied. For the relaxometric method, the $1/T_2$ is measured at room temperature and a fixed magnetic field (i.e., 0.47 T). Calibration standards of ferrous chloride containing 0.2–10.0 mM Fe in the same acid mixture need to be included. The relationship between the relaxation rates and Fe concentration will be linear for the entire Fe range, with a slope of about 12–13 s^{-1}/mM Fe). For the spectrophotometric method, the iron content is determined by a Ferrozine-based assay,[83] with modification as described.[84] The method has a sensitivity of about 3 μg Fe/500 μl acid digested cell suspension. Triplicate 50-μl samples from the acid-digested cell suspensions are used, as well as iron calibration standards (ammonium iron [II] sulfate hexahydrate), adjusted to contain between 0.1 and 1.0 mM Fe. To the 50-μl sample or standard in 1.5-ml polypropylene tubes, 50 μl 60% perchloric acid and 50 μl 30% hydrogen peroxide are added, and the tubes are boiled in a waterbath for 30 min to denature protein. After cooling down on ice, the tubes are centrifuged at 1000g for 2 min in an Eppendorf minicentrifuge. To each tube, 100 μl hydroxylamine is added slowly, followed by incubation for 30 min at room temperature (RT). Hydroxylamine, 50 μl, is added again, and the tubes are incubated for an additional 10 min at RT. Following this reaction, 500 μl Ferrozine and 500 μl pyridine ("Photrex" reagent) are added and mixed, and the 562-nm absorbance line is measured using a spectrophotometer.

Functional tests comparing labeled cells with unlabeled cells are essential to ensure that the followed magnetic labeling procedure does not interfere with cell proliferation or differentiation. A simple but robust viability assay is the use of trypan blue as an exclusion dye. For further assessment of toxicity and normal proliferation of labeled cells, cells can be tested in an MTT (3-[4,5-dimethylthiazol-2-yl]-2,5-diphenyl tetrazolium bromid) mitochondrial assimilation assay. To this end, cells are seeded in flat-bottom well plates at 1.5×10^4 cells per well in 90 μl medium, and the magnetic label is added at the appropriate concentration(s). During the final 4 h of labeling, 10 μl of MTT is added at a final concentration of 0.5 mg/ml medium. The reaction is then initiated by adding 100 μl of solubilization solution (Roche Molecular Biochemicals, Indianapolis, IN), and incubated

[83] L. L. Stookey, *Anal. Chem.* **42,** 779 (1970).
[84] J. W. M. Bulte, G. F. Miller, J. Vymazal, R. A. Brooks, and J. A. Frank, *Magn. Reson. Med.* **37,** 530 (1997).

overnight at 37°. The absorbance of the formazon product is then measured at a wavelength of 570 nm with 750 nm as (subtracted) reference.

The introduction of SPIO complexes into cells may increase the formation of reactive iron species and hydroxyl free radicals following the release of ionic iron into the cytoplasm from the endosome. In turn, this may alter cell metabolism or increase the rate of apoptosis, or cell death.[85,86] For assessment of reactive oxygen species (ROS), cells are grown in 25-cm^2 flasks to confluence. After media change, cells are incubated with media containing Feridex–PLL complex or media alone (control). Cells are then washed twice with phosphate buffered saline (PBS) at different time points and incubated in fresh culture media. Feridex–PLL-labeled and unlabeled cells are maintained under identical conditions, and ROS assay and apoptosis rates are determined at various time intervals. For the ROS assay, cells are collected after trypsinization, washed twice with PBS, and resuspended in 1 ml of PBS at 1×10^6 cells/ml. The intracellular formation of ROS is detected by using the fluorescent probe CM-H$_2$DCFDA (Molecular Probes Inc., Eugene, OR). CM-H$_2$DCFDA is added at a final concentration of 10 μM, and cells are incubated for 60 min at 37°. CM-H$_2$DCFDA is a nonfluorescent agent, which forms fluorescent esters when reacted with ROS inside cells. The intensity of the fluorescence is measured using an LS-55 Fluorescent Plate Reader (PerkinElmer Life Sciences, Norwalk, CT) using a 490–500-nm wavelength for excitation and 525 nm for emission. Control cells are run side by side to determine the percent increased ROS production in the labeled cells (Fig. 9). A transient increase in ROS production can be observed for the labeled cells.

To determine the number of cells undergoing apoptosis, cells are collected after trypsinization, washed twice with ice-cold PBS, and resuspended in 1 ml of annexin media (Vybrant apoptosis assay kit #2, Molecular Probes, Inc.) at 1×10^6 cells/ml. Ten μl of fluorescent-labeled annexin V is added to 100 μl of cell suspension, which is then kept at room temperature for 15–20 min. Cells are washed twice with annexin media and put into fluorescent plates for fluorescent reading. The percentage of dead cells and cells undergoing apoptosis can also be determined by a flow cytometer. Labeled and unlabeled cells are incubated with 10 μl of annexin V and 2 μl propidium iodide (PI) at room temperature for 15–20 min, and flow cytometry is performed using a fluorescent-activated cell sorter.

It is also important to confirm that labeled cells retain their functional and differential capacity. There are many methods to determine the cell-specific functional or differential capacity. An example is to determine

---

[85] J. Emerit, C. Beaumont, and F. Trivin, *Biomed. Pharmacother.* **55**, 333 (2001).
[86] J. M. Gutteridge, D. A. Rowley, and B. Halliwell, *Biochem. J.* **206**, 605 (1982).

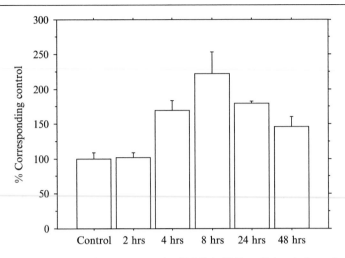

FIG. 9. Production of reactive oxygen species (ROS) in HeLa cells in relation to incubation time with Feridex–PLL complexes. The increased ROS production might be due to increased iron concentration in the cells following longer incubation time. However, published results show that the increased ROS in the cells following iron labeling is transient. For details, see Arbab et al.[67]

the differential capacity of MSCs. To assess if MSCs retain the ability to differentiate following labeling with Feridex–PLL, both unlabeled controls and labeled MSCs are subjected to stimulation for adipogenic differentiation according to the cell supplier's recommendations (Bio Whittaker, Rockland, ME). Alternate induction and maintenance cycles are performed for 2 weeks. After 2 weeks, both unlabeled and labeled cells are fixed with 10% formalin and stained with Oil Red O (for fat globules) and counterstained with hematoxylin (Fig. 10).

Magnetically Labeled Cells: Detection Limits

A number of studies have indicated that MRI cell tracking of small numbers of labeled cells or even single labeled cells may be feasible. At higher magnetic fields, single T cells were detected with a good correlation of the corresponding fluorescent images.[87] Single magnetodendrimer-labeled oocytes, which are large mammalian cells, were imaged at 11.7 T, and clear distinction between the nuclear and cellular compartments was

---

[87] S. J. Dodd, M. Williams, J. P. Suhan, D. S. Williams, A. P. Koretsky, and C. Ho, *Biophys. J.* **76,** 103 (1999).

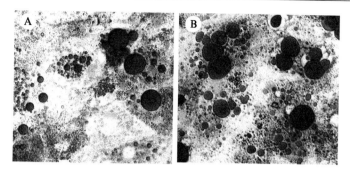

FIG. 10. Adipogenic differentiation of human MSCs stained with Oil Red O. (A) Unlabeled cells and (B) Feridex–PLL labeled cells. The circular dark spots represent the fat globules.

achieved.[88] At a clinical field strength of 1.5 T, using customized gradient coil inserts, single magnetically labeled macrophages could be identified with a matching overlay to the corresponding fluorescent microscopic images. Calculations of the minimum amount of iron needed to detect a single cell at $100 \times 100 \times 200$ μm resolution using the FIESTA sequence indicated a range of 1.4–3.0 pg of iron per cell,[89] a level that is easily achieved when labeling nonphagocytic cells with magnetodendrimers or (U)SPIO–TAs. With a symmetric cell division and even dilution of magnetic label among progenitor cells, it can be estimated that an individual magnetically labeled cell with an initial iron content of 25 pg can be detected up to at least 4 cell divisions. Taken together, the excellent anatomic resolution of MRI, allowing detection of small numbers of cells, will make MRI cell tracking an important tool in biology and medicine.

Acknowledgments

J.W.M.B. is supported by grants RO1 NS045062 (NIH) and PP0922 (Multiple Sclerosis Society).

[88] P. D. Majors, E. J. Ackerman, R. A. Wind, T. Douglas, and J. W. M. Bulte, *Proc. Intl. Soc. Magn. Reson. Med.* **10,** 2571 (2002).
[89] C. Heyn, C. V. Bowen, B. K. Rutt, and P. J. Gareau, *Proc. Intl. Soc. Magn. Reson. Med.* **11,** 805 (2003).

Section III

General Methods

[14] The Application of Magnetic Resonance Imaging and Spectroscopy to Gene Therapy

By KISHORE K. BHAKOO, JIMMY D. BELL, I. JANE COX, and SIMON D. TAYLOR-ROBINSON

Introduction

The central aim of gene therapy is to augment or substitute the function of mutated genes with the correct genetic code. Gene therapy can theoretically modify specific genes to correct the underlying cause of the disease, whereas normal management of an altered disease phenotype requires agents that interact with gene products, or are themselves gene products. Successful gene therapy is often limited by the inefficient delivery of genes because of short *in vivo* half-lives, lack of cell-specific targeting, and low transfection efficiencies. A number of gene delivery systems have been developed for subsequent application in humans. These systems include replicant deficient viral vectors, such as adenovirus,[1,2] retrovirus,[3,4] and herpes simplex virus,[5–7] synthetic nonviral vectors [8–12] (such as liposomes, poly-L-lysine, dendrimers, and molecular conjugates)[13]; physical methods (such as the gene gun and electroporation)[14]; naked DNA[15]; and combinations of these various technologies.[16,17] Although less efficient than viral

[1] S. C. Ko, A. Gotoh, G. N. Thalmann, H. E. Zhau, D. A. Johnston, W. W. Zhang, C. Kao, and L. W. Chung, *Hum. Gene Ther.* **7**, 1683 (1996).

[2] D. M. Nguyen, S. A. Wiehle, P. E. Koch, C. Branch, N. Yen, J. A. Roth, and R. J. Cristiano, *Cancer Gene Ther.* **4**, 191 (1997).

[3] M. Wiznerowicz, A. Z. C. Fong, A. Mackiewicz, and R. G. Hawley, *Gene Ther.* **4**, 1061 (1997).

[4] E. G. Hanania and A. B. Deisseroth, *Cancer Gene Ther.* **1**, 21 (1994).

[5] R. Blasberg and J. Tjuvajev, *Q. J. Nucl. Med.* **43**, 163 (1999).

[6] D. Ross, B. Kim, and B. Davidson, *Clin. Cancer Res.* **1**, 651 (1995).

[7] D. Schellingerhout, A. Bogdanov, E. Marecos, M. Spear, X. Breakefield, and R. Weissleder, *Hum. Gene Ther.* **9**, 1543 (1998).

[8] P. Y. P. Kuo and W. M. Saltzman, *Crit. Rev. Eukaryot. Gene Expr.* **6**, 59 (1996).

[9] B. Abdallah, L. Sachs, and B. A. Demeneix, *Biol. Cell* **85**, 1 (1995).

[10] P. L. Felgner, *Sci. Am.* **276**, 102 (1997).

[11] R. G. Crystal, *Nat. Med.* **1**, 15 (1995).

[12] F. Ledely, *Hum. Gene Ther.* **6**, 1129 (1995).

[13] M. Zenke, P. Steinlein, E. Wagner, M. Cotton, H. Beug, and M. L. Birnstiel, *Proc. Natl. Acad. Sci. USA* **87**, 3655 (1990).

[14] T. Suzuki, B. C. Shin, K. Fujikura, T. Matsuzaki, and K. Takata, *FEBS Lett.* **425**, 436 (1998).

[15] J. P. Yang and L. Huang, *Gene Ther.* **3**, 542 (1996).

Copyright 2004, Elsevier Inc.
All rights reserved.
0076-6879/04 $35.00

vectors in terms of the amount of gene required for cell transfection, nonviral vectors have a reduced risk of eliciting an immune response. However, although many of these strategies are relatively effective within the confines of a laboratory environment, clinical trials have been less promising.

In an attempt to overcome some of these problems, it has been necessary to develop multimodality protocols that can be readily applicable in the biologic and clinical setting. Moreover, because of the possibility that only transient expression[18] of the therapeutic gene may be achieved, these methods need to be noninvasive, nontoxic, and capable of determining temporal events. Furthermore, because of the large volume of tissue that is necessary to target in humans, methods will also be required to assess the degree of gene transfer achieved. Thus an ideal *in vivo* reporter system, essential for assessing the effectiveness of any gene therapy strategy, must therefore satisfy a number of criteria. First, it must provide a unique signal so that transgene expression can be monitored *in vivo* against a background of natively expressed genes. Second, it must be small enough so that it can be expressed with therapeutic genes in bicistronic constructs and serve as a direct marker for the entire delivery system. Third, it must not interfere with normal tissue function.

At present, a number of imaging methods are currently under development that will allow gene delivery, uptake, expression, and subsequent therapeutic changes to be mapped out *in vivo,* with a view to tailoring gene therapies to individual patient requirements. These include clinical imaging techniques such as ultrasound, computed tomography (CT), and magnetic resonance imaging (MRI), as well as other existing modalities that have until now largely remained a preserve of the research environment, principally positron emission tomography (PET) and single photon emission tomography (SPECT).[19] New technologies such as electric source imaging (using electroencephalographic or electrocardiographic techniques),[20,21] electrical impedance mapping,[22] magnetic field gradient measurements

[16] E. Wagner, K. Zatloukal, M. Cotten, H. Kirlappos, K. Mechtler, D. T. Curiel, and M. L. Birnstiel, *Proc. Natl. Acad. Sci. USA* **89,** 6099 (1992).

[17] B. Bonnekoh, D. A. Greenhalgh, D. S. Bundman, K. Kosai, S. H. Chen, M. J. Finegold, T. Krieg, S. L. Woo, and R. Roop, *J. Invest. Dermatol.* **106,** 1163 (1996).

[18] X. Ye, V. M. Rivera, P. Zoltick, F. Cerasoli, Jr., M. A. Schnell, G. Gao, J. V. Hughes, M. Gilman, and J. M. Wilson, *Science* **283,** 88 (1999).

[19] W. H. Theodore, A. V. Delgado-Escueta, and R. J. Porter, *Adv. Neurol.* **79,** 865 (1999).

[20] C. Grimm, A. Schreiber, R. Kristeva-Feige, T. Mergner, J. Hennig, and C. H. Lucking, *Electroencephalogr. Clin. Neurophysiol.* **106,** 22 (1998).

[21] H. J. Huppertz, M. Otte, C. Grimm, R. Kristeva-Feige, T. Mergner, and C. H. Lucking, *Electroencephalogr. Clin. Neurophysiol.* **106,** 409 (1998).

[22] J. P. Morucci and B. Rigaud, *Crit. Rev. Biomed. Eng.* **24,** 655 (1996).

(magnetoencephalography or magnetocardiography),[23,24] microwave scattering tomography,[25] and bioluminescence techniques[26] are also under assessment for use in humans. However, uniquely, magnetic resonance (MR) techniques bridge the gap between technologies that have gained everyday acceptance in the clinical arena (MRI) and those that still are research tools (magnetic resonance spectroscopy [MRS]). Here we review the application of MRI and MRS to the study of gene therapy and their potential role in defining and refining future gene therapy development strategies.

Magnetic Resonance Techniques

The last 2 decades have seen an increasing interest in the use of nuclear magnetic resonance (NMR) techniques as applied to biologic and clinical problems. Indeed, since its first human applications in the early 1980s, MRI has become one of the main clinical imaging modalities, whereas MRS, which provides direct biochemical information on the function of animal and human organs *in vivo* and *in vitro*, has become the technique of choice for much clinical and biologic *in vivo* research.

A great advantage of both MRI and MRS is that they are noninvasive, safe, and reproducible modalities and can be repeated to allow serial information to be collected, or in the case of MRS, dynamic biochemical changes to be observed. Furthermore, MR techniques can encompass both the laboratory (high-resolution MR in animal studies or *in vitro*) and clinical needs (whole body MR scanning).

In vivo MR studies are normally carried out using large, whole body superconducting magnets. The strength of magnetic fields is usually measured in Tesla (T), with most clinical MRI systems operating between 0.5 and 3 T. The usual field strength of magnets employed for whole body (*in vivo*) MRS is 1.5 T, although there are clinical research systems that operate at 4.7 T and some at 8 T. With increasing magnetic field strengths, improved sensitivity and resolution is achieved, but the cost of high-field whole body systems (>3 T) can be prohibitive. Animal systems range from 2–11.7 T, with most modern systems ranging from 4.7–9.4 T. This allows high-resolution images of murine models in relatively short periods of time. *In vitro* MRS of cell suspensions and tissue samples or detailed MRI at the

[23] T. P. Roberts, D. Poeppel, and H. A. Rowley, *Neuropsychiatry Neuropsychol. Behav. Neurol.* **11,** 49 (1998).
[24] Y. Nakaya and H. Mori, *Clin. Phys. Physiol. Meas.* **13,** 191 (1992).
[25] S. Caorsi, G. L. Gragnani, and M. Pastorino, *IEEE Trans. Biomed. Eng.* **41,** 406 (1994).
[26] P. Pasini, M. Musiani, C. Russo, P. Valenti, G. Aicardi, J. E. Crabtree, M. Baraldini, and A. Roda, *J. Pharm. Biomed. Anal.* **18,** 555 (1998).

cellular level are often performed at up to 11.7 T, well beyond the range of most clinical systems.

MR Overview

There are a number of MR-sensitive nuclei available for biologic and clinical investigation, including hydrogen-1 (^{1}H), carbon-13 (^{13}C), nitrogen-15 (^{15}N), fluorine-19 (^{19}F), sodium-23 (^{23}Na), and phosphorus-31 (^{31}P). These nuclei allow the production of either an image, providing anatomic information for MRI (principally proton, although ^{31}P images can be obtained under some very specific conditions), or a frequency spectrum, providing biochemical information (MRS). Biochemical and pharmacologic information can thus be obtained from endogenous and/or exogenous low molecular weight metabolites, containing ^{1}H (signal from amino acids, organic acids, lipids, sugars, and cell membrane components), ^{31}P (signal from nucleoside triphosphates such as adenosine triphosphate [ATP] and from phosphocreatine, inorganic phosphate [Pi] and phosphorylated intermediates on the cell membrane synthetic [phosphomonoesters] and degradation pathways [phosphodiesters]), ^{13}C (glycogen, sugars, amino acids, and lipids), and ^{19}F (fluorinated drug products).

MR Imaging (MRI)

The ability to generate images of intact living objects noninvasively and nondestructively has made MRI one of the leading clinical and biologic imaging modalities. In general, MR images consist of high-resolution maps of the intracellular tissue water. The contrast normally observed in an MR image therefore corresponds to differences in content and/or MR characteristics of the water molecules within different regions of a given organ or tissue. The clinical information (anatomic and/or dynamic), obtained from MRI examinations can be modified either by tailoring the MR pulse sequences to highlight or nullify signal from different water-containing or fat compartments in the body, or by the use of MR contrast agents. Thus excellent tissue contrast, and therefore discrimination of areas of interest, can be obtained by proper choice of experimental parameters. Among many applications, MRI techniques are being used to determine organ development and/or atrophy, angiography, functional imaging (fMRI), dynamic changes in arterial blood flow, or biliary drainage, for example.

Recent advances in high-field magnets have permitted an expansion of the MRI modality into previously unforeseen areas of research.[27] This area of application, which has led to the imaging of isolates within cells, is termed "MRI microscopy." The resolution is about 1 μm, and in combination with contrast agents, which highlight physiologic changes in the cell, has opened up exciting possibilities.

MR Contrast Agents

The use of intravenously injected contrast agents is a well-established method for improving the conspicuity of certain body tissues and therefore for improving diagnostic yield in the clinical and research arenas.[28] MR contrast agents are usually biologically inert chelates of gadolinium (Gd) or other paramagnetic elements such as iron and exert their action by altering MR relaxation times.[29,30] Thus the presence of Gd-based contrast agent, such as Gd-DTPA, will normally lead to areas of hyperintensity on MRI, whereas agents such as superparamagnetic iron oxides (SPIOs) cause hypointense areas on MRI.[31] However, the ability of a particular contrast agent to highlight a tissue depends on its specificity and molecular weight, if it is not freely diffusible.[32]

Newer contrast agents have been developed that have particular affinity for particular body compartments such as the blood, bile, the hepatic parenchyma, or the reticuloendothelial system.[33] For example, in the liver, Gd-DTPA and manganese dipyridoxyl diphosphate are concentrated in hepatic parenchymal tissue and subsequently in the bile, producing a brightening of the liver, whereas SPIOs are taken up by Kupffer cells and other reticuloendothelial cells and produce hepatic darkening.[34] This is important in improving the conspicuity of focal lesions, especially in the context of a cirrhotic liver, where tumors develop on the background of nodular regeneration.[35]

[27] R. W. Bowtell, A. Peters, J. C. Sharp, P. Mansfield, E. W. Hsu, N. Aiken, A. Horsman, and S. J. Blackband, *Magn. Reson. Med.* **33,** 790 (1995).

[28] R. Mathur de Vre and M. Lemort, *Br. J. Radiol.* **68,** 225 (1995).

[29] J. E. Kirsch, *Top Magn. Reson. Imaging* **2,** 1 (1991).

[30] M. F. Tweedle, J. J. Hagan, K. Kumar, S. Mantha, and C. A. Chang, *Magn. Reson. Imaging* **9,** 409 (1991).

[31] A. D. Watson, S. M. Rocklage, and M. J. Carvlin, *in* "Magnetic Resonance Imaging" (D. Stark and W. C. Bradley, eds.), p. 372. Mosby, St Louis, 1992.

[32] T. S. Desser, D. L. Rubin, H. H. Muller, F. Qing, S. Khodor, G. Zanazzi, S. W. Young, D. L. Ladd, J. A. Wellons, K. E. Kellar *et al., J. Magn. Reson. Imaging* **4,** 467 (1994).

[33] J. P. Earls and D. A. Bluemke, *Magn. Reson. Imaging Clin. N. Am.* **7,** 255 (1999).

[34] O. Clement, N. Siauve, M. Lewin, E. de Kerviler, C. A. Cuenod, and G. Frija, *Biomed. Pharmacother.* **52,** 51 (1998).

[35] R. N. Low, *J. Magn. Reson. Imaging* **7,** 56 (1997).

Monitoring Gene Delivery by MRI

Thus far, MR strategies to monitor gene expression have relied on receptor-mediated or probe activation strategies,[36–38] xenografts,[39] and the expression of an endogenous gene.[40] Noninvasive imaging techniques are used increasingly to follow the various stages, from gene delivery[41] to observation of transgene expression[42–45] and therapeutic modifications.[46] Molecular imaging, based on highly complex specific contrast agents, gives insights into molecular abnormalities or molecular modification in living cells and opens up the possibility of detection of gene delivery[38] and gene expression.[47,48]

MRI has been used to image gene delivery by conjugating paramagnetic contrast agents to the gene delivery vector. Kayyem and colleagues[49] conjugated human transferrin to poly-L-lysine (PLL), a cationic polymer that can form ionic complexes with negatively charged DNA. Poly-D-lysine (PDL) was attached to the paramagnetic contrast agent gadolinium diethylenetriaminepentaacetic acid (Gd-DTPA). Particles were then complexed by adding varying amounts of conjugated PLL to plasmid DNA, followed by the addition of Gd-DTPA–PDL to neutralize the negative DNA charge. K562 leukemia cells were transfected *in vitro* with the polymer–DNA complexes and analyzed using MRI T_1-weighted images of the

[36] A. Bogdanov, Jr. and R. Weissleder, *Trends Biotechnol.* **16,** 5 (1998).

[37] A. Moore, J. P. Basilion, E. A. Chiocca, and R. Weissleder, *Biochim. Biophys. Acta* **1402,** 239 (1998).

[38] G. de Marco, A. Bogdanov, E. Marecos, A. Moore, M. Simonova, and R. Weissleder, *Radiology* **208,** 65 (1998).

[39] L. D. Stegman, A. Themetulla, B. Beattie, E. Kievit, T. S. Lawrence, R. G. Blasberg, J. G. Tjuvajev, and B. D. Ross, *Proc. Natl. Acad. Sci. USA* **96,** 9821 (1999).

[40] A. P. Koretsky, M. J. Bronsan, L. H. Chen, J. D. Chem, and T. Van Dyke, *Proc. Natl. Acad. Sci. USA* **87,** 3112 (1990).

[41] X. Yang, E. Atalar, D. Li, J.-M. Serfaty, D. Wang, A. Kumar, and L. Cheng, *Circulation* **104,** 1588 (2001).

[42] C. Nichol and E. E. Kim, *J. Nucl. Med.* **42,** 1368 (2001).

[43] C. Bremer and R. Weissleder, *Acad. Radiol.* **8,** 15 (2001).

[44] D. C. MacLaren, T. Toyokuni, S. R. Cherry, J. R. Barrio, M. E. Phelps, H. R. Herschman, and S. S. Gambhir, *Biol. Psychiatry.* **48,** 337 (2000).

[45] A. Honigman, E. Zeira, P. Ohana, R. Abramovitz, E. Taylor, I. Bar, Y. Zilberman, R. Rabinovsky, D. Gazit, A. Joseph, A. Panet, E. Shai, A. Palmon, M. Laster, and E. Galun, *Mol. Ther.* **4,** 239 (2001).

[46] V. Allamand and K. P. Campbell, *Hum. Mol. Genet.* **9,** 2459 (2000).

[47] A. Y. Louie, M. M. Huber, E. T. Ahrens, U. Rothbacher, R. Moats, R. E. Jacobs, S. E. Fraser, and T. J. Meade, *Nat. Biotechnol.* **18,** 321 (2000).

[48] R. Weissleder, A. Moore, U. Mahmood, R. Bhorade, H. Benveniste, E. A. Chiocca, and J. P. Basilion, *Nat. Med.* **6,** 351 (2000).

[49] J. Kayyem, R. Kumar, S. Fraser, and T. Meade, *Chem. Biol.* **2,** 615 (1995).

transfected cells. The authors found that the cells transfected with the Gd-containing particles showed MRI contrast enhancement and that MRI could be used to noninvasively monitor the delivery of the polymeric gene delivery system.

MRI has also been used to image other nonviral vector for gene delivery. In a study by de Marco et al.,[38] dextran-PLL-iron oxide–DNA particles were evaluated in vitro using a 293 embryonic kidney cell line and in vivo using adult Sprague–Dawley rats. They showed that MRI can be used to image gene delivery vectors. However, these studies imaged the delivery system rather than the expressed protein.

More recently, Gd-chelating cationic lipids have been synthesized to obtain lipoplexes with MRI contrast properties. These compounds were designed to follow the biodistribution of synthetic DNA for gene delivery by nuclear MRI. In vivo intratumoral administration of MCO-I-68-Gd–DNA lipoplexes to tumor model led to an increase of the NMR signal.[50]

Clearly these studies are beginning to take advantage of the power of noninvasive imaging modalities to monitor an important aspect of gene therapy, principally the vector delivery system.

Imaging Gene Expression

A number of strategies have been developed to assess in vivo gene expression by MRI. Weissleder and colleagues[51] developed an amplification strategy with a second reporter system, using a pcDNA3tyr plasmid that encodes for human tyrosinase. The tyrosinase enzyme is central in the formation of melanin, which can bind with paramagnetic metals. A mouse fibroblast L929 cell line and a human embryonic kidney 293 cell line were transfected with the tyrosinase gene. Both nontransfected and mock-transfected cells were used as controls. A Fontana stain with silver nitrate established the melanin production capability of these transfected cells. ^{111}In binding studies were performed to determine the metal-binding capacity of the transfected cells. The transfected 929 cells were found to have a significantly higher binding affinity (up to 35% by weight), particularly of iron, than either of the control cells, and the binding was dependent on the dose of DNA. MR studies were performed on 293 transfected and mock-transfected cells that were incubated in media with iron sulfate. T_1-weighted images were obtained of the cells in culture, where higher signal intensity corresponded to higher expression of tyrosinase, which produced

[50] F. Leclercq, M. Cohen-Ohana, N. Mignet, A. Sbarbati, J. Herscovici, D. Scherman, and G. Byk, *Bioconjug. Chem.* **14,** 112 (2003).
[51] R. Weissleder, M. Simonova, A. Bogdanova, S. Bredow, W. Enochs, and A. Bogdanov, *Radiology* **204,** 425 (1997).

the iron-binding melanin.[52] This study demonstrated the feasibility of imaging for gene expression.

More recently, a similar system has been used to evaluate gene expression *in vitro*.[53] The authors transfected a reporter tyrosinase gene into a hepG2 hepatocyte cell line. The plasmid, pcDNA3tyr, carrying the full-length cDNA of tyrosinase gene, was transfected into HepG2 cells, subsequently giving rise to its product, melanin. The transfected HepG2 cell synthesized large amounts of melanin, enough to be detected by MR, which appeared as high signal in T_1-weighted MRI. Furthermore, the signal intensity correlated with the amount of melanin granules, confirmed in HepG2 cell using Fontana staining and transfected plasmid for tyrosinase gene by reverse taqman polymerase chain reaction (RT-PCR).

Another amplification system, based on the cellular internalization of (super)-paramagnetic probes, such as monocrystalline iron oxide nanoparticles (MION), has also been developed. Moore and co-workers[37] used a gene that encoded for human transferrin receptor (hTfR) to transfect rat 9L gliosarcoma cells with three forms of hTfR. A protected iron containing a magnetic hTfR probe was used to show that the receptor expression could be visualized using MRI.

Paramagnetic chelates that change magnetic properties upon enzymatic hydrolysis ("smart" contrast agents) have been used recently to image gene expression. In one example, a paramagnetic galactopyranoside–galactosidase system was injected into cells and used to image β-galactosidase activity in *Xenopus* embryos.[47] Galactosidase activity was revealed by measuring enzyme-mediated T_1 changes.

MRI and Gene Therapy Efficacy

MRI has been included in the diagnostic armory of outcome measures in assessing the efficacy of gene therapy. For example, MR angiography provided qualitative evidence of improved distal flow in a study investigating the safety and efficacy of intramuscular gene therapy with vascular endothelial growth factor (VEGF) in patients with chronic critical leg ischemia.[54] In a study of the efficacy of adv/tk/GCV gene therapy in a syngeneic BT4C rat malignant glioma model, MRI was one of the measures used to assess treatment effect and tissue responses by comparison of signal intensities and tumor volume.[55]

[52] W. Enochs, P. Petherick, A. Bogdanova, U. Mohr, and R. Weissleder, *Radiology* **204,** 417 (1997).

[53] J. P. Yuan, B. L. Liang, B. K. Xie, and J. L. Zhong, *Ai Zheng* **22,** 156 (2003).

[54] K. G. Shyu, H. Chang, B. W. Wang, and P. Kuan, *Am. J. Med.* **114,** 85 (2003).

MR Spectroscopy (MRS)

Information from MRS is normally presented in the form of a spectrum. Whatever the chosen nucleus under MRS examination, the nuclei from individual metabolites resonate at a given frequency, depending on the chemical environment of each nucleus. This phenomenon is known as chemical shift. The intensity of each metabolite signal is related to the concentration of the metabolite. Analysis of the MR spectrum allows noninvasive insight into metabolite concentrations, intracellular pH, and the metabolic state of pathologic tissue, as well as dynamic changes in the metabolism of living tissue in both health and disease. However, only compounds present at millimolar concentrations are normally detectable utilizing the clinical MR systems currently available.[56]

MRS and Gene Expression

MRS could play an important role in gene therapy, both as a tool to assess gene expression and as a noninvasive method to determine the efficacy of gene therapies. MRS can provide biochemical and metabolic information on intact cells, isolated organs, or different areas of the body *in vivo*.[57–59] Moreover, its ability to detect signal noninvasively has been exploited to assess drug metabolism *in vivo* and *in vitro*. Changes in a MR spectrum brought about by delivery of exogenous (fluorinated, phosphorylated, or ^{13}C-enriched) or naturally occurring compounds may be detected using this technique. Today this ability is being harnessed to develop MRS as a noninvasive tool to detect gene expression, albeit indirectly.

However, regardless of the approach, it is important that the accumulation of the "marker" substrate reflects gene expression and be proportional to the marker gene and the therapeutic gene. Thus the marker metabolite needs to be nontoxic, not be metabolized by the host tissue (unless expressing the marker gene), and must accumulate within the transduced cells.

The ability of MRS to distinguish signals from chemically distinct compounds also offers the potential to measure gene expression. In one study, tumor cell lines expressing the cDNA encoding for cytosine deaminase

[55] K. Tyynela, A. M. Sandmair, M. Turunen, R. Vanninen, P. Vainio, R. Kauppinen, R. Johansson, M. Vapalahti, and S. Yla-Herttuala, *Cancer Gene Ther.* **9,** 917 (2002).

[56] E. B. Cady, "Clinical Magnetic Resonance Spectroscopy." Plenum Press, New York, 1990.

[57] R. Jalan, S. D. Taylor-Robinson, and H. J. F. Hodgson, *J. Hepatol.* **25,** 414 (1996).

[58] D. K. Menon, J. Sargentoni, S. D. Taylor-Robinson, J. D. Bell, I. J. Cox, D. J. Bryant, G. A. Coutts, K. Rolles, A. K. Burroughs, and M. Y. Morgan, *Hepatology* **21,** 417 (1995).

[59] S. D. Taylor-Robinson, J. Sargentoni, J. D. Bell, N. Saeed, K. K. Changani, B. R. Davidson, K. Rolles, A. K. Burroughs, H. J. Hodgson, C. S. Foster, and I. J. Cox, *Liver* **17,** 198 (1997).

from yeast (yCD) were grown in animals. Conversion of the nontoxic pro-drug 5-fluorocytosine to the antimetabolite 5-fluorouracil by yCD could be observed and quantified *in vivo* using ^{19}F MRS.[39] This study demonstrated local conversion of 5-fluorocytosine, validating the concept of localizing chemotherapy with enzyme–prodrug gene therapy.

Walter and co-authors[60] have developed a noninvasive detection method for expression of viral-mediated gene transfer using a recombinant adenovirus constructed using the gene for arginine kinase (AK), which is the invertebrate correlate to the vertebrate ATP-buffering enzyme, crea-tine kinase. Gene expression was noninvasively monitored using ^{31}P MRS. The product of the AK enzyme, phosphoarginine (PArg), served as an MRS-visible reporter of AK expression. The recombinant adenovirus coding for arginine kinase (rAdCMVAK) was injected into the right hind-limbs of neonatal mice. Two weeks after injection of rAdCMVAK, a unique ^{31}P-MRS resonance was observed. It was observable in all rAdCM-VAK-injected hindlimbs and was not present in the contralateral control or the vehicle-injected limb. AK activity was demonstrated *in vivo* by moni-toring the decreases in PArg and ATP resonances during prolonged ische-mia. AK activity persisted for at least 8 months after injection, providing a useful reporter gene that allows noninvasive and repeated monitoring of gene expression after viral-mediated gene transfer to muscle.

Finally, a fast, although less specific, method to detect transgene expres-sion could be achieved by the use of *in vitro* MRS.[61,62] This would be ac-complished by measuring the level of "marker metabolites" in body fluids such as urine, blood, cerebrospinal fluid (CSF), and bile, instead of transfected cells. This method would be limited to a binomial form of detecting gene expression, [on/off], and would depend on the ability of the "marker metabolites" to cross the cell membrane and accumulate in body fluids. Nevertheless, this method could be relatively effective as a broad way of detecting transgene expression. One could envisage it as a fast and inexpensive first-pass approach to detect gene expression in human subjects, before more sophisticated *in vivo* methods are applied.

MRS and Gene Therapy Efficacy

When the proteins expressed as a result of gene transfection alter the concentration of cellular metabolites, then MRS provides a unique mea-sure for monitoring gene therapy efficiency. For example, in a study of the

[60] G. Walter, E. R. Barton, and H. L. Sweeney, *Proc. Natl. Acad. Sci. USA* **97**, 5151 (2000).
[61] J. D. Bell, N. E. Preece, and H. G. Parkes, *in* "NMR in Physiology and Biomedicine" (R. J. Gillies, ed.), p. 221. Academic Press, London, 1994.
[62] J. K. Nicholson, J. C. Lindon, and E. Holmes, *Xenobiotica* **29**, 1181 (1999).

antitumor efficacy of the gene-encoding herpes simplex virus thymidine kinase (HSVtk) activating the prodrug ganciclovir (GCV) in a model of rat prostate cancer MRS studies, serial *in vivo* MRS of the tk-transfected MATLyLu tumors demonstrated a decreased ATP–Pi ratio during growth and an increase in the ATP–Pi ratio during regression initiated by treatment with GCV. Also significant differences were found in the phosphomonoester (PME) to total phosphate ratios in treated compared with untreated tumors.[63]

Conclusions

Analysis of gene delivery and expression in the clinical setting currently involves molecular assays of histologic biopsy material, but MR techniques hold the promise of noninvasive and dynamic assessment of the location, magnitude, and duration of transgene expression. The development of "smart" contrast agents may allow the visualization of the effects of gene expression through physiologic changes within the cell that cause an increase in MRI-detectable signal, whereas the delivery of marker genes that encode for MR visible metabolites means that MRS has exciting prospects in monitoring the effects of therapeutic intervention. Furthermore, both MR modalities are uniquely equipped to extend the monitoring role from the laboratory bench into the patient.

Acknowledgments

Some of the work described in this chapter has been supported by the British Medical Research Council.

[63] J. D. Eaton, M. J. Perry, S. M. Todryk, R. A. Mazucco, R. S. Kirby, J. R. Griffiths, and A. G. Dalgleish, *Gene Ther.* **8,** 557 (2001).

[15] Voxelation Methods for Genome Scale Imaging of Brain Gene Expression

By DANIEL M. SFORZA and DESMOND J. SMITH

Introduction

Imaging techniques have granted access to significant results on structure and function in the central nervous system. In addition, the completion of the human genome sequence promises to accelerate even further the already extraordinary developments of molecular biology. Therefore it is a natural step to seek an integration of both fields for a better understanding of the molecular basis of the brain. One strategy employs reporter genes, and this has been a major advance in molecular imaging. However, although this approach has important advantages, such as *in vivo* imaging, it does not provide convenient access to genome-wide information but only the examination of one, or at most, a few genes at a time.[1–5] One recently developed approach for genome-scale acquisition of brain gene expression patterns employs microarray analysis of spatially registered voxels (cubes) harvested from the brain.[6,7] This method is called voxelation and provides a mapping of gene expression analogous to the images reconstructed in biomedical imaging techniques, such as X-ray computerized tomography (CT), positron emission tomography (PET), and magnetic resonance imaging (MRI).

Voxelation has permitted high-throughput gene expression reconstructions in human and rodent brains.[8–11] One study using human brains compared gene expression between brains from normal individuals and brains

[1] S. S. Gambhir, J. R. Barrio, H. R. Herschman, and M. E. Phelps, *Nucl. Med. Biol.* **26,** 481 (1999).

[2] S. S. Gambhir, J. R. Barrio, H. R. Herschman, and M. E. Phelps, *J. Nucl. Cardiol.* **6,** 219 (1999).

[3] H. R. Herschman, D. C. MacLaren, M. Iyer, M. Namavari, K. Bobinski, L. A. Green, L. Wu, A. J. Berk, T. Toyokuni, J. R. Barrio, S. R. Cherry, M. E. Phelps, E. P. Sandgren, and S. S. Gambhir, *J. Neurosci. Res.* **59,** 699 (2000).

[4] A. Y. Louie, M. M. Huber, E. T. Ahrens, U. Rothbacher, R. Moats, R. E. Jacobs, S. E. Fraser, and T. J. Meade, *Nat. Biotechnol.* **18,** 321 (2000).

[5] D. A. Zacharias, G. S. Baird, and R. Y. Tsien, *Curr. Opin. Neurobiol.* **10,** 416 (2000).

[6] D. Liu and D. J. Smith, *Methods* **31,** 317 (2003).

[7] R. P. Singh and D. J. Smith, *Biol. Psychiatry* **53,** 1069 (2003).

[8] V. M. Brown, A. Ossadtchi, A. H. Khan, S. R. Cherry, R. M. Leahy, and D. J. Smith, *Genome Res.* **12,** 244 (2002).

[9] V. M. Brown, A. Ossadtchi, A. H. Khan, S. Yee, G. Lacan, W. P. Melega, S. R. Cherry, R. M. Leahy, and D. J. Smith, *Genome Res.* **12,** 868 (2002).

Copyright 2004, Elsevier Inc.
All rights reserved.
0076-6879/04 $35.00

from individuals with Alzheimer's disease.[8] The mouse studies investigated global gene expression patterns in brains of a Parkinson's disease model induced by toxic doses of methamphetamine compared with controls.[9]

The previously mentioned voxelation studies used fresh tissue slabs, but a weakness of this approach is the deformation of the tissue during the process of physical voxelation, which can make problematic the registration of gene expression analysis results to the neuroanatomy. Although this is not a major problem for low-resolution studies, it becomes a significant factor at higher resolution. Therefore we have developed a new voxelation protocol to overcome this drawback and make feasible higher-resolution studies.[12] The new procedure involves the use of a fixation process to avoid tissue deformation and to allow histologic staining,[13] as well as cryoprotection to allow harvesting of slabs from frozen specimens.[14] These protocols have been shown to preserve RNA quality and yield at excellent levels, and allow the use of quality demanding downstream applications such as real-time polymerase chain reaction (PCR) or DNA microarrays.[12]

In the following sections we review the complete protocols and methods for high-resolution voxelation, from samples preparation to data analysis and image reconstruction.

Methods

The following protocol can be applied to rodent or monkey studies and requires the use of intercardiac perfusion. Here we provide the times and volumes of the solutions used for adult C57BL/6J male mice (10 to 24 weeks, 25 to 31 g) for which the protocol was developed,[12] but it can be easily adapted to the other cases. For all perfusions, a small mechanical pump was used that allowed a flux of 1.5 ml min^{-1}. To easily switch among the different solutions and avoid the formation of bubbles, a serial connection of three-way stopcocks with swivel male luer lock was used (MX231-1L; Medex, Hilliard, OH) and carefully filled with the solutions before starting the procedure.

[10] A. Ossadtchi, V. M. Brown, A. H. Khan, S. R. Cherry, T. E. Nichols, R. M. Leahy, and D. J. Smith, *Neurochem. Res.* **27,** 1113 (2002).

[11] R. P. Singh, V. M. Brown, A. Chaudhari, A. H. Khan, A. Ossadtchi, D. M. Sforza, A. K. Meadors, S. R. Cherry, R. M. Leahy, and D. J. Smith, *J. Neurosci. Methods* **125,** 93 (2003).

[12] D. M. Sforza, J. Annese, D. Liu, S. Levy, A. W. Toga, and D. J. Smith, *Neurochem. Res.* **29,** 1291 (2004).

[13] J. Annese and A. W. Toga, *in* "Brain Mapping: The Methods" (A. W. Toga and J. C. Mazziotta, eds.), p. 537. Academic Press, New York, 2002.

[14] D. L. Rosene and K. J. Rhodes, *in* "Methods in Neuroscience, Vol. 3: Quantitative and Qualitative Microscopy" (P. M. Conn, ed.), p. 360. Academic Press, New York, 1990.

Sample Preparation

1. Mice must undergo intracardiac perfusion under deep halothane anesthesia. For each animal the vasculature is cleared with 11–22 ml of ice-cold phosphate buffered saline solution (PBS).
2. Fixation: 67 ml of 4% paraformaldehyde in PBS (pH 7.4) is perfused for a period of 45 min.
3. Cryoprotection: mice are additionally perfused with 67 ml of 10%, 20%, and 30% ice-cold phosphate buffered sucrose solutions (pH 7.4). The sucrose solutions are each perfused for 45 min.
4. After perfusion, brains are removed from the skull and immediately frozen in chilled isopentane or liquid nitrogen. Tissue blocks are kept at $-70°$ until RNA testing and histologic processing.

This complete protocol lasts approximately 190 minutes.

Histologic Staining

As voxelation studies are pushed to higher resolution, registration between the harvested voxels and the neuroanatomy becomes increasingly important. At lower levels of resolution, visual inspection and digital photographs of fresh sections are sufficient for registration. However, the inherent low contrast of fresh specimens makes it difficult to identify boundaries between structures of small size or subtly different microscopic anatomy, and this hinders image reconstruction at high resolution. Proper use of the described fixation and cryoprotection protocol allows the use of histologic stains to improve contrast in the voxelated section and facilitate image registration. There are methods available for histologic staining of tissue compatible with the recovery of good quality RNA for gene expression studies.[15]

Physical Voxelation

The equipment required for voxelation depends on the specimen under study and on the desired resolution. A detailed description of voxelation devices and procedures for human and rodent species can be found in Liu and Smith.[6] Furthermore, supplementary information on the design and construction of these pieces of equipment (including blueprints) is given at our web site (http://pharmacology.ucla.edu/smithlab/). The devices essentially consist of two-dimensional arrays of blades, together with

[15] P. Bonaventure, H. Guo, B. Tian, X. Liu, A. Bittner, B. Roland, R. Salunga, X. J. Ma, F. Kamme, B. Meurers, M. Bakker, M. Jurzak, J. E. Leysen, and M. G. Erlander, *Brain Res.* **943,** 38 (2002).

supporting and guiding mechanisms to facilitate registered harvesting of voxels. Voxelation devices have been constructed that harvest square voxels of size 3.3 mm (used in the human studies, a 425-voxel template, consisting of 25 voxels in length and 17 voxels in width) to 1 mm (used in the rodent studies, a 400-voxel template, 20 × 20 voxels).[11]

The selection of the actual section for voxelation obviously depends on the nature of the investigation. At the moment all voxelation studies have investigated coronal sections. Human studies on Alzheimer's disease focused on left coronal hemisections that included the hippocampus, corresponding to section 17 of the University of Maryland Brain and Tissue Bank protocol, method 2 (http://medschool.umaryland.edu/BTBank/).[8] The thickness of the slab is also an important factor in order to obtain enough RNA for microarray hybridization, as well as for the accuracy of the regions identified on the section. In the human studies, the slice thickness was 8 mm. In the study of the mouse model of Parkinson's disease,[9] a different strategy was applied. The whole brain was divided into slices of thickness 1 mm using a brain matrix (Harvard Apparatus, Inc., Holliston, MA). Each slice was then divided into four voxels, giving a total of 40 voxels, and an average volumetric resolution of 7.5 μl. The existence of voxelation instruments for the rodent brain will allow acquisition of expression images at 1 mm (1 μl) resolution.[11]

It is important to obtain digital pictures of the selected slab before and after voxelation in order to evaluate possible deformation induced by the procedure, as we already mentioned, this is more likely to be a problem as resolution is increased. In addition, digital pictures are required for image registration and reconstruction (see last section).

RNA Isolation and Quality Check

There are many commercially available RNA isolation kits. Lately we have obtained excellent yields and quality using the RNeasy Lipid Tissue Mini Kit column (Qiagen, Hilden, Germany). Frozen samples are placed in QIAzol Lysis Reagent (Qiagen) and homogenized using a Tissue-Tearor homogenizer (BioSpec Products, Bartlesville, OK). Subsequent steps are done according to the instructions provided by the manufacturer of the RNeasy kit. Total RNA quality is assessed by microcapillary electrophoresis using the Agilent BioAnalyzer 2100 (Agilent Technologies, Palo Alto, CA). Evaluation of 18-s and 28-s RNA peaks and background noise are employed for this purpose.

The cryoprotection step is very important to achieve good RNA quality and recovery. It is well known that paraformaldehyde fixation results in an apparent decrease in both the recovery and quality of RNA extracted from brain.[16–18] However, we have shown that sucrose perfusion following fixation helps preserve the RNA.[12]

The long time required for the complete protocol might raise doubts about RNA quality, but these are unfounded. Our experience is consistent with published work in the sense that the quality of RNA from human brains depends on brain pH, a measure of premortem condition, rather than postmortem interval.[19]

Microarrays, RNA Labeling, and Hybridization

High-resolution voxelation requires the use of a large number of microarrays, so the use of spotted cDNA arrays is the natural option due to their relatively low cost. The list of clones used in published voxelation investigations can be obtained at our web site (http://www.pharmacology.ucla.edu/smithlab/). To produce targets for hybridization to the cDNA microarrays, 10 μg of total RNA is labeled with fluorescent nucleotides by chemical coupling following reverse transcription. The total RNA is mixed with 6 μg of anchored oligo-dT (5'-TTTTTTTTTTTTTTTTTT VN-3') and hybridization achieved by incubation at 70° for 10 min, followed by 10 min at 4° in a total volume of 18 μl. The annealed RNA is then reverse-transcribed in a 30-μl reaction mix containing reaction buffer (50 mM Tris–HCl, 75 mM KCl, 3 mM MgCl$_2$, pH 8.3); 10 mM dithiothreitol; 200 μM dATP, dGTP, and dCTP; 51 μM dTTP; 149 μM aminoallyl-UTP (Sigma, St. Louis, MO); and 200U SuperScript II reverse transcriptase (Life Technologies, Gaithersburg, MD). The reactions are incubated at 42° for 2 h. Following incubation, 10 μl of 1 M NaOH and 10 μl of 0.5 M EDTA are added to degrade the template RNA, and the samples are incubated at 70° for 10 min. The reaction is neutralized by the addition of 10 μl of 1 M HCl, followed by 300 μl of Qiagen Buffer PB (PCR Cleanup kit; Qiagen, Valencia, CA). Reactions are purified using Qiagen's PCR cleanup kit following the manufacturer's directions with the substitution of 80% ethanol for Qiagen's PE buffer and water for Qiagen's EB buffer.

Following purification, individual samples are desiccated, resuspended in 7 μl 0.1 M sodium bicarbonate buffer, and chemically coupled to

[16] S. M. Goldsworthy, P. S. Stockton, C. S. Trempus, J. F. Foley, and R. R. Maronpot, *Mol. Carcinog.* **25,** 86 (1999).

[17] R. Parlato, A. Rosica, V. Cuccurullo, L. Mansi, P. Macchia, J. D. Owens, J. F. Mushinski, M. De Felice, R. F. Bonner, and R. Di Lauro, *Anal. Biochem.* **300,** 139 (2002).

[18] S. J. Scheidl, S. Nilsson, M. Kalen, M. Hellstrom, M. Takemoto, J. Hakansson, and P. Lindahl, *Am. J. Pathol.* **160,** 801 (2002).

[19] P. J. Harrison, P. R. Heath, S. L. Eastwood, P. W. Burnet, B. McDonald, and R. C. Pearson, *Neurosci. Lett.* **200,** 151 (1995).

monofunctional reactive cyanine dyes (Amersham Biosciences, Bucking-hamshire, England). Unbound cyanine dyes are removed by purification with Qiagen's PCR cleanup kit as described earlier. Following purification, the labeled samples are combined in Cy3–Cy5 pairs, desiccated, and resus-pended in 100 μl hybridization buffer (25% formamide, 5 × SSC, 0.1% sodium dodecyl sulfate [SDS], 10 μg yeast tRNA, 10 μg poly-A-RNA, 1 μg human COT-1 DNA). Immediately prior to hybridization, the micro-arrays are prehybridized in a solution of 5 × SSC, 1% SDS, and 1% bovine serum albumin (BSA) at 55° for 45 min. Following prehybridization, the arrays are vigorously washed in ddH$_2$O to remove all traces of the prehybridization solution, rinsed in isopropanol, and dried at room temperature (RT).

The labeled RNA samples are denatured at 95° for 2 min and placed on the prehybridized arrays, covered with Lifterslip coverslips (25 mm × 60 mm; Erie Scientific, Inc., Portsmouth, NH), placed in a hybridization chamber (Corning, Acton, MA) with 30 μl hybridization buffer to maintain humidity, and incubated at 42° for 16 h in a standard hybridization oven. After hybridization, the arrays are washed in 2 × SSC to remove the cover-slips, placed in 2 × SSC, 0.1% SDS at 55° with agitation for 5 min then washed with two successive 5 min washes with 1 × SSC and 0.1 × SSC at RT. The washed arrays are dried by centrifugation at 50g for 5 min and immediately scanned using a GenePix 4000A microarray scanner (Axon Instruments, Union City, CA).

Data Analysis

Voxelation studies yield large amounts of data, and a variety of differ-ent analytical techniques can be employed for data mining. There are low- and high-level analyses of the microarray data, described in the next two subsections. Also important is how to display the large amount of informa-tion retrieved from microarrays in a meaningful and appealing manner. This is discussed in the last subsection. Unless otherwise specified, all com-putational algorithms described are written in Matlab scripting language (Matlab 5.3 or later required) and can be retrieved from the lab web site (http://www.pharmacology.ucla.edu/smithlab/). The algorithms are run on a personal computer and are not highly demanding of resources, so a workstation is not required.

Low-Level Analysis

Low-level analysis refers to image processing of scanned microarray images and normalization (within array and interarrays). Images are usu-ally processed by the software bundle included with the scanner, and they allow automatic spot segmentation and acquisition of intensity values in both channels. In our case the microarray scanner is a GenePix 4000A (Axon Instruments) with GenePix Pro 3.0 software. An important step is the normalization of the data to correct the multiple sources of systematic variation in cDNA microarray experiments that affect measured gene expression levels.[20] The normalization procedure we have applied in our studies consists of the removal of spatial trends due to array printing by a nonlinear transformation of the data set and the compensation for differences in the labeling of Cy3 and Cy5 dyes by aligning the histograms of the dye signals both within, as well as between, microarrays. The normalization method relies on the assumption that most of the genes do not change significantly between different experimental conditions.

High-Level Analysis

After the data have been properly normalized, calculated gene expression values can be subjected to further analysis in the search of relevant biologic information. The most common first step in the analysis of the data is the identification of significantly changing expression values; this singles out genes that are upregulated or downregulated in the comparison between control and experimental conditions. Even more interesting is the identification of clusters of coregulated and antiregulated genes, which can potentially identify regulatory networks.[8,9]

There are also exploratory techniques that do not rely on a priori hypotheses or assumptions concerning the data, such as singular value decomposition. This method can identify clusters of genes (gene "vectors") that most efficiently explain the variance in the data. It has been shown that these gene vectors display interesting regional patterns of expression, and their associated genes may play an important role in differentiation of the mouse and human brain.[8,9] In addition, a gene vector has been identified that shows a significant spatial shift away from the striatum in the normal mouse brain toward the hippocampus in the Parkinson's brain.[9] These results suggest that high-throughput acquisition of gene expression patterns in combination with singular value decomposition has the potential to identify functionally abnormal neuroanatomic regions in neurologic disease states. This is especially relevant given the fact that for many neuropsychiatric disorders, such as schizophrenia, Down syndrome, and autism, the location of functionally abnormal brain regions associated with the diseases

[20] Y. H. Yang, S. Dudoit, P. Luu, D. M. Lin, V. Peng, J. Ngai, and T. P. Speed, *Nucleic Acids Res.* **30,** e15 (2002).

remains enigmatic. The higher-resolution gene expression images that can be obtained following the protocols described here would potentially allow the precise identification of such important regions, given that they exist.

Another technique that has proven powerful in modeling microarray data from voxelation is the analysis of variance (ANOVA).[10,21,22] With microarray data analysis, there is always an underlying question: how to calculate valid estimates of gene expression values that take into account potential sources of variation, both from experimental design and error. ANOVA can provide corrected estimates of change in gene expression with the estimation of potential confounding effects. The use of ANOVA to analyze voxelation data was found to produce results consistent with those from singular value decomposition.[10]

Image Reconstruction

High-resolution voxelation can result in demanding data sets for image reconstruction. The addition of a staining step in the protocol can help greatly, especially at high resolution, as we have already discussed. The basic problem is to align an actual voxelated slab to a corresponding atlas image. The final image represents the overlapping of an anatomic image and gene expression levels converted into pseudocolor. The registration between the picture of the actual voxelated section and the atlas image is accomplished using an implementation of the thin-plate splines warping method.[23] This method interpolates surfaces between scattered fiducial landmarks, and its name refers to a physical analogy involving the bending of a thin sheet of metal subject to point constraints (the landmarks that relate both images). The manual selection of corresponding fiducial landmarks on both images allows the algorithm to warp the desired image onto the atlas reference. Gene expression images are further improved using interpolation and smoothing functions. For the high-resolution human expression images, the smoothing used about 2–4 voxels and for the high-resolution rodent images, only 1 voxel. Examples of final images obtained from voxelation are displayed in Fig. 1.

[21] M. K. Kerr and G. A. Churchill, *Genet. Res.* **77,** 123 (2001).
[22] M. K. Kerr, M. Martin, and G. A. Churchill, *J. Comput. Biol.* **7,** 819 (2000).
[23] F. L. Bookstein, *IEEE Trans. Patt. Anal. Mach. Intell.* **11,** 567 (1989).

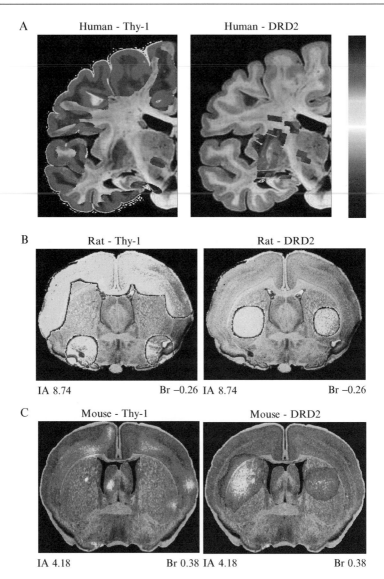

A   Human - Thy-1   Human - DRD2

B   Rat - Thy-1   Rat - DRD2

IA 8.74   Br −0.26 IA 8.74   Br −0.26

C   Mouse - Thy-1   Mouse - DRD2

IA 4.18   Br 0.38 IA 4.18   Br 0.38

FIG. 1. Human and rodent brain voxelation gene expression images for Thy-1 and DRD2 (dopamine D2 receptor) genes. Images are the result of the methods explained in the text. Gene expression patterns are shown in pseudocolor and smoothed across voxels. Thy-1 is expressed in the cortex, DRD2 in the striatum (caudate/putamen). (A) Human brain, atlas section from Virtual Hospital: The Human Brain (http://www.vh.org). The voxel size was 3.3 mm. (B) Rat brain, coronal atlas section from Paxinos and Watson.[24] The voxel size was 1 mm. Coordinates in mm: IA, interacural; Br, bregma. (C) Mouse brain, coronal atlas section from the Mouse Brain Library (http://www.mbl.org).[25] The voxel size was 1 mm. (See color insert.)

Acknowledgments

Supported by the NIH/NIDA (DA015802, DA05010) NARSAD Young Investigator Award, Tobacco-Related Disease Research Program (11RT-0172), and Alzheimer's Association (IIRG-02–3609).

[24] G. Paxinos and C. Watson, "The Rat Brain in Stereotaxic Coordinates." Academic Press, Orlando, FL, 1986.
[25] R. W. Williams, *Results Probl. Cell Differ.* **30**, 21 (2000).

[16] MR–Intracranial Compliance and Pressure: A Method for Noninvasive Measurement of Important Neurophysiologic Parameters

By NOAM ALPERIN

Introduction

Advances in magnetic resonance imaging (MRI) technology over the last decade resulted with dramatic decrease in MR image acquisition time from minutes to few milliseconds. With short acquisition time it is now possible to capture dynamic biologic processes such as the beating of the heart or changes in cerebral blood oxygenation level following neural activation. The strictly anatomic information provided by MRI early on is now augmented with functional information. MRI, being a noninvasive imaging modality, therefore offers new capabilities for visualization and quantitation of biologic processes in humans.

A recently developed methodology to study cerebral physiology through dynamic measurements of blood and cerebrospinal fluid (CSF) flow is described in this chapter. The MRI-based method provides measurements of the following physiologic parameters: total cerebral blood flow (TCBF), intracranial compliance (the inverse of elastance), and intracranial pressure (ICP). In addition to the overall compliance of the intracranial compartment, the method provides information on the biomechanical state of sub-compartments of the intracranial space such as the cerebral vasculature (e.g., the cerebral vascular compliance). These physiologic parameters are important for characterization of the hemodynamic and hydrodynamic state of the craniospinal system (i.e., the craniospinal biomechanical state). The biomechanical state of the craniospinal system is not constant; it changes with normal activities (e.g., change in body posture), due to aging and by head trauma and diseases (e.g., intracranial tumors, hydrocephalous,

Copyright 2004, Elsevier Inc.
All rights reserved.
0076-6879/04 $35.00

Arnold-Chiari malformations). Therefore characterization and quantitation of the cerebral biomechanics may improve our understanding of the healthy cerebral physiology and pathophysiology of the disease state.

The second part of this chapter describes the application of the method to study and quantify the effect of posture on the craniospinal biomechanical state in healthy subjects. The cerebral biomechanical state is different in the upright and supine postures. The application of the MR–ICP methodology to study healthy subjects in supine and upright postures provides an insight into the coupling that exists between CSF and blood flow—that is, hemodynamic and hydrodynamic coupling—and the way in which intracranial compliance and pressure affect the CSF and blood flow dynamics.

The MRI-Based Method for Measurement of Intracranial Compliance and Pressure

The Neurophysiology Basis of the MR–ICP Method

The MRI-based method integrates human neurophysiology and fluid dynamics principles with dynamic MRI techniques to measure intracranial elastance and ICP. The method takes advantage of the pulsatile nature of blood flow, which causes small fluctuation in intracranial volume and ICP with each cardiac cycle. The intracranial elastance is derived from the ratio of these changes during the cardiac cycle. A mean ICP value is then derived from the known relationship between intracranial pressure and elastance.

In a closed system such as the cranium, the pressure and volume are related. This relation has been studied extensively using invasive techniques in animals and in humans. The change in pressure due to volume change is determined by the overall mechanical elastance of the system. Ryder et al.[1] and others[2-4] studied the pressure–volume relationship by injection of fluid into the CSF space. Measurements of pressure change resulting from a given volume change permitted characterization of the intracranial pressure–volume dependence. Marmarou et al.[2] proposed the following expression for the pressure–volume curve:

[1] H. W. Ryder, F. F. Espey, F. D. Kimbel, E. J. Penka, A. Rosenauer, and J. P. Evans, J. Lab. Clin. Med. **41**, 428 (1953).

[2] A. Marmarou, K. Shulman, and J. La Morgese, J. Neurosurg. **43**, 523 (1975).

[3] F. H. Sklar and I. Elashvili, J. Neurosurg. **47**, 670 (1977).

[4] J. Szewczykowski, S. Sliwka, A. Kunicki, P. Dyko, and J. Korsak-Sliwaka, J. Neurosurg. **47**, 19 (1977).

$$P = P_1 e^{E_1 V} \tag{1}$$

where E_1 is a constant elastance coefficient (units–volume^{-1}), P_1 is a pressure coefficient related to the exponential shape of the curve, and V is the intracranial volume (CSF + blood + brain tissue). The monoexponential elastance curve, shown in Fig. 1, illustrates the dependence of the ratio of pressure to volume change during the cardiac cycle, dP/dV, on the mean pressure. At a low pressure this ratio is small, whereas at a high pressure (larger volume) the ratio is large. This ratio (i.e., the change in pressure for a unit change in volume) is defined as elastance and is a linear function of ICP as shown in Eq. (2):

$$\frac{dP}{dV} = P_1 E_1 \times e^{E_1 \times V} = E_1 \times P \tag{2}$$

The pressure–volume relationship was not determined from a direct measurement of the volume of the intracranial space; there was no *in vivo* means of measuring that volume. Instead, it was inferred from the linear relationship between elastance and pressure.[2,3] The elastance was measured with a volume–pressure response test. In this test, shown in Fig. 2 (from Marmarou *et al.*[2]), the total intracranial volume is rapidly increased by injecting a known amount of fluid into the ventricles. The elastance is then derived from the ratio of the resultant pressure change (Q_2–Q_1) to the amount of injected volume, ΔV.

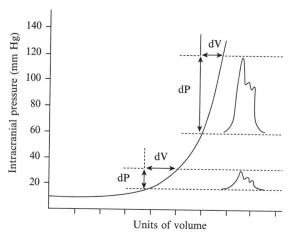

FIG. 1. The monoexponential curve depicting the relation between the intracranial pressure and volume. Note that at low pressures the ratio of dP/dV (elastance) is small, whereas at high pressures the ratio is large.

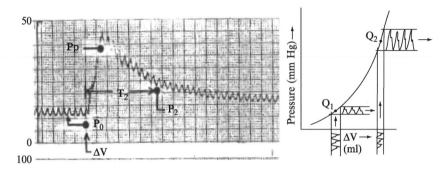

FIG. 2. The volume–pressure response test for invasive measurement of elastance (i.e., inverse of compliance). *Left:* The pressure trace showing the increase in pressure following injection of a volume of fluid, ΔV. *Right:* Q_1 is the initial mean ICP and Q_2 is the mean ICP following the injection. The elastance is then derived from the ratio of the resultant pressure change (Q_2–Q_1) to the amount of injected volume, ΔV. From Marmarou et al.[2]

Similarly, the MR–ICP method provides a measure of elastance from the ratio of pressure and volume changes that occurs naturally with each cardiac cycle. Intracranial pressure then is derived through the linear relationship between elastance and pressure. The small volume change that occurs naturally with each cardiac cycle is analogous to the injected volume used in the volume–pressure response test. Volume and pressure changes occurs because of the pulsatile nature of blood flow. During systole, more volume flows into the cranial vault (arterial inflow exceeds venous and CSF outflows), whereas during diastole, the outflow is larger. The MRI method measures the pulsatile arterial, venous, and CSF flows into and out from the cranial vault. The small (on the order of 1 ml) volume change is derived from the differences between inflow and outflow at different time points in the cardiac cycle. The pressure change is derived from the CSF velocities using fluid dynamics principles. The derivation of these quantities with MRI is described in following sections of this chapter.

The proportionality factor between elastance and pressure, the elastance coefficient constant, was estimated by Szewczykowski et al.[4] from the slope of the linear relationship between ICP change over the cardiac cycle and mean ICP. Intracranial pressure was manipulated by a continuous injection of fluid into the CSF space. The intracranial volume change (ICVC) during the cardiac cycle was not measured, but rather assumed to vary randomly around the same value for all patients. A relatively small variability (SD; standard deviation) in the measured slopes was found (0.329 +/− 0.084 SD). It is possible that even less variability in the elastance coefficient constant would have been found if the actual value of

ICVC had been measured for each patient. It is unknown to what extent the elastance coefficient is affected by pathology. An open cranial vault or variation in the elasticity of the dura mater might alter this value.

Intracranial Volume Change During the Cardiac Cycle

In a compartment with inlets and outlets, *a change* in the volume of that compartment can be determined from the instantaneous difference between volumetric flows into and out of the compartment as long as the flowing fluid and the content inside are incompressible. The change in the volume of the intracranial space at each time point in the cardiac cycle is measured from the difference between blood and CSF inflows and outflows on the basis that brain tissue, blood, and CSF are incompressible. The Monro–Kellie doctrine makes this assumption.[5] This doctrine also assumes that the volume of the intracranial space is constant. Today, we know that a small periodic change in the intracranial volume occurs with each cardiac cycle. Sensitive ultrasonic methods measure slight changes in the distance between the lateral temporal walls of the cranium during the cardiac cycle.[6] The mean intracranial volume, however, is constant when averaged over a cardiac cycle at steady state. The change in the intracranial volume during the cardiac cycle can be calculated directly from the net volumetric flow rates into and out of the cranium as described in Eq. (3) and from the condition described by Eq. (4)

$$\text{ICVC}(t) = [Q_A(t) - Q_V(t) - Q_{CSF}(t)]\Delta t$$
$$0 < t < T \tag{3}$$

$$\text{ICVC}(T) = \Sigma[Q_A(t) - Q_V(t) - Q_{CSF}(t)]\Delta t = 0 \tag{4}$$
$$\text{Cardiac cycle}$$

Where $Q_A(t)$ is the total arterial volumetric flow rate, $Q_V(t)$ is the total venous volumetric outflow rate, $Q_{CSF}(t)$ is the rates of CSF volumetric outflows through the foramen magnum, and T is the time period of the cardiac cycle. Equation (4) is equivalent to the Monro–Kellie doctrine assumption of constant mean intracranial volume.

The MRI method measures the small intracranial volume change (ICVC) during the cardiac cycle using a previously described craniospinal flow–volume–pressure model.[7,8] This model, shown in Fig. 3, includes

[5] F. Magendie, "Recherches Physiologiques et Cliniques sur le Liquide Céphalorachidien ou Cérébro-Spinal," p. 66. Librairie Medicale de Mequigenon-Marvis Files, Paris, 1842.

[6] T. Ueno, L. M. Shuer, W. T. Yost, and A. R. Hargens, *Biol. Sci. Space* **12**, 270 (1998).

[7] N. Alperin, E. M. Vikingstad, B. Gomez-Anson, and D. N. Levin, *Magn. Reson. Med.* **35**, 741 (1996).

[8] N. Alperin and C. Stelzig, *Proc. Intl. Soc. Magn. Reson. Med.* **3**, 2011 (1999).

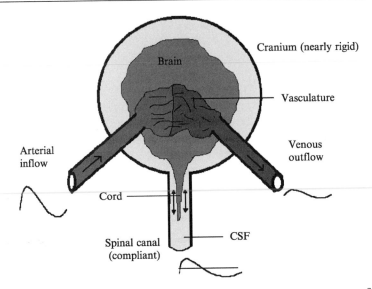

FIG. 3. The craniospinal flow–volume–pressure model proposed by Alperin *et al.*[7,8] This model shows the arterial inflow, the venous outflow, and the CSF flow that oscillates between the cranium and spinal canal. During systole, arterial blood inflow is greater than the venous outflow, which is accomodated by the CSF movement to the spinal canal. The CSF moves to the cranium during diastole as the venous outflow becomes greater. (See color insert.)

intracranial and spinal canal compartments, as well as the following inputs and outputs: arterial inflow, venous outflow, and CSF oscillatory flow between the cranium and the spinal canal. During the systolic phase of the cardiac cycle, arterial blood inflow is greater than venous outflow; thus the intracranial blood volume increases. This produces a rise in pressure that in turn forces CSF to flow from the cranium into the spinal canal. During the diastolic phase, blood volume decreases and the CSF flow is reversed. Motion-sensitive MRI techniques allow for direct imaging and quantitation of these flow dynamics.

MRI Measurements of CSF Flow, Total Cerebral Blood Flow, and Volume Change During the Cardiac Cycle

Quantitation of flow measurements employs a motion-sensitive technique, dynamic velocity-encoded phase contrast (PC) MRI.[9] The reported accuracy of the MR velocity and volumetric flow measurements in

[9] N. J. Pelc, R. Herfkens, A. Shimakawa, and D. R. Enzman, *Magn. Reson. Med.* **7**, 229 (1991).

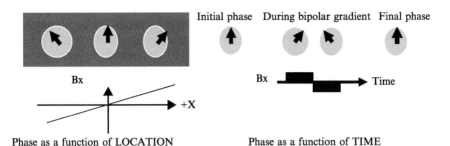

Phase as a function of LOCATION   Phase as a function of TIME

Fig. 4. A schematic representation of the effect a magnetic field gradient has on stationary protons. *Left:* Protons at different locations along the x-axis accumulate a different phase when a magnetic field gradient is applied. *Right:* Time evolution of the phase of a stationary proton located at +X that is experiencing a bipolar field gradient. The positive phase shift caused by the positive lobe is canceled after the application of the negative lobe, resulting with no phase shift for a stationary proton. A moving proton will accumulate a phase shift that is proportional to its velocity. (See color insert.)

nonsteady (pulsatile) flow are within 7.5% and 2.8%, respectively,[10] making cine PC MRI one of the most reliable and accurate techniques for noninvasive *in vivo* measurement of volumetric flow rate. This technique provides a series of images with values of picture elements proportional to the velocity value at that location (velocity maps). The PC MRI technique is based on the principle that the precession frequency of the protons is proportional to the magnetic field strength. Therefore velocity can be phased–encoded by application of magnetic field gradient. Figure 4 demonstrates the principle of the velocity-encoding PC technique. When a gradient field is applied along an axis for a short time, the proton's phase will change based on its location along that axis. When a bipolar (positive and then negative) gradient field is applied, the phase of the stationary protons (located at +X) will increase during the positive portion (lobe) of the bipolar gradient and then will decrease during the negative lobe. If the lobes were of equal area, no net phase change would occur. However, moving protons, such as those in the blood or CSF, will experience different field strength during each lobe as they change their location; this will result in a net phase change proportional to the proton velocity.

Examples of MRI PC images of CSF and blood flow obtained from a healthy volunteer are shown in Figs. 5 and 6, respectively. The oscillatory CSF flow between the cranial and the spinal compartments is

[10] R. Frayne, D. A. Steinman, C. R. Ethier, and B. K. Rutt, *J. Magn. Reson. Imaging* **5**, 428 (1995).

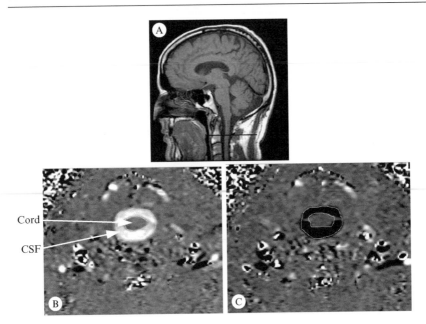

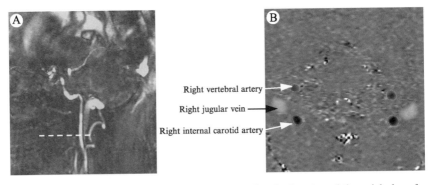

FIG. 5. (A) Anatomic midsagittal T_1-weighted MR image showing the location of the axial plane used for CSF flow measurement (dark line). (B) and (C) Phase contrast (PC) MRI images of CSF flow in the spinal canal. (B) CSF flow during systole. (C) CSF flow during diastole. The pixel values in these images are proportional to velocities in a direction perpendicular to the image plane. Gray, static tissue; white, outward flow (caudal direction); black, inward flow (cranial direction).

FIG. 6. (A) A blood vessel MRI scout image showing the location of the axial plane for blood flow measurement (dashed line). (B) A phase contrast (PC) MRI image of blood flow through that location. Black pixels indicate arterial inflow, and white are venous outflow.

visualized in images taken in a transverse (an axial) anatomic orientation through the upper cervical spinal canal. The location of this plane is indicated on a midsagittal T_1-weighted MR image shown in Fig. 5A; Fig. 5B depicts outflow (white pixels) during the systolic phase and Fig. 5C depicts inflow (black pixels) during the diastolic phase. The CSF flow was imaged with a velocity-encoding value of 7 cm/sec. A much higher velocity-encoding value of 70 cm/sec is used to image the faster blood flow through the neck arteries and veins. The location of the imaging plane used for blood flow measurement is shown in Fig. 6A and a velocity-encoded image of blood flow is shown in Fig. 6B.

Volumetric flow rates are obtained by integration of the velocities throughout a lumen cross-sectional area defined by its boundary. It has been shown that the use of an automated technique for delineation of the lumen boundary improves measurements' reproducibility and accuracy.[11] Our lab has recently developed an automated method for delineation of the lumen boundaries, Pulsatility-Based Segmentation.[12] The method incorporates the *temporal* information of the pulsatile flows to enhance the robustness of the lumen segmentation process. The lumen segmentation is based on differences in temporal information between lumen pixels and background pixels. Velocity waveforms sampled inside the lumen are similar to each other, whereas those sampled outside the lumen are different. This difference is used for the lumen boundary identification process. Pixels with velocity waveform similar to a reference velocity waveform selected from inside a lumen of interest are segmented and identified as the lumen region. The degree of similarity between waveforms is used to differentiate between lumen and background pixels. The method incorporates computation of an unbiased threshold to increase measurement reproducibility. Evaluation of the method performance with phantom data and with arterial, venous, and CSF flow data from human subjects demonstrated significant improvements in measurement accuracy and reproducibility compared with manual delineation of the lumens. For example, in flow phantom data, average lumen area measurement error of about 18%, obtained manually by a skilled observer, was reduced to 2.9%. Analysis of 15 sets of CSF and blood flow measurements used for derivation of the elastance and pressure demonstrated an average of 4-fold reduction in interobserver measurement variability.

The total volumetric arterial flow rate—that is, total cerebral blood flow (TCBF)—is calculated directly from the sum of the volumetric flow through the four vessels carrying blood to the brain (internal carotid and

[11] P. Summers, "Proceedings of the 9th Meeting of the ISMRM," p. 110. Glasgow, UK, 2001.
[12] N. Alperin and S. H. Lee, *Magn. Reson. Med.* **49,** 934 (2003).

vertebral arteries) over one cardiac cycle. An example of the volumetric flow waveforms for CSF, arterial inflow, and venous outflow measured in a healthy volunteer is shown in Fig. 7.

The venous blood outflow is derived by summation of the flow through the jugular veins, and epidural and vertebral veins when present. Flows through these veins do not always represent the entire venous outflow. We found that the measured total venous flow can be lower by 0–15% than the total measured arterial inflow. This may be due to secondary venous outflow through other channels (e.g., ophthalmic veins). The constraint of no mean net change in the intracranial volume over one cardiac cycle as formulated in Eq. (4) is used to account for the unmeasured venous flow. The measured venous flow is increased by the ratio of the sum of the arterial inflow to the measured venous outflow.

The rate of the time-varying volume change (net transcranial volumetric flow rate) is obtained by subtracting outflow rates from inflow rates at each time point. The intracranial volume change (delta of volume from a given reference point) is obtained by integrating that waveform with respect to time. The net transcranial volumetric flow rate and the ICVC waveforms obtained from the CSF and blood flow waveforms shown in Fig. 7 are shown in Fig. 8A and 8B respectively. The maximum volume change in this example is close to 0.5 ml.

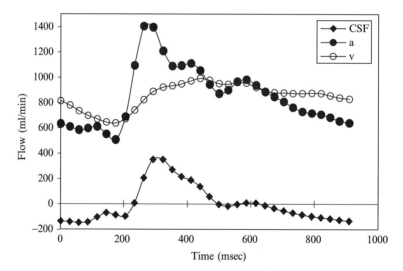

FIG. 7. The volumetric flows into and out of the cranial vault during the cardiac cycle derived from the MRI phase contrast scans. Total arterial inflow (filled circles), venous outflow (open circles), and the cranial-to-spinal CSF volumetric flow rate (diamonds) during the cardiac cycle. Note that arterial inflow is greater than venous outflow during systole.

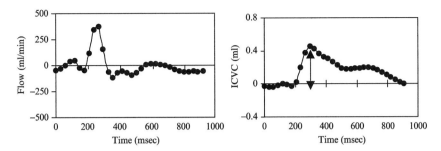

FIG. 8. *Left:* The MRI-derived net transcranial volumetric flow rate waveform. *Right:* The intracranial volume change (ICVC) during the cardiac cycle derived by integrating the net transcranial volumetric flow waveform on the left. Note that the maximal volume change in this subject is 0.5 ml.

MRI Measurements of the ICP Change During the Cardiac Cycle

The change in ICP during the cardiac cycle is derived from the change in the CSF pressure gradient. A method to measure *pressure gradient* of pulsatile flow in tubes with MRI was reported by Urchuk and Plewes.[13] Pulsatile pressure gradients are derived from the MRI velocity-encoded PC images using the Navier–Stokes relationship between pressure gradient and temporal and spatial derivatives of the fluid velocity for incompressible fluid in a right tube.[14] This relationship is shown in Eq. (5):

$$\nabla P = -\rho \left(\frac{\partial V}{\partial t} + V \cdot \nabla V \right) + \mu \nabla^2 V \tag{5}$$

Where ρ is the fluid density, μ is the fluid viscosity, and V and P are the velocity vector and pressure, respectively. This equation is essentially a momentum balance, which states that a small volume element moving with the fluid is accelerated because of the forces acting upon it. The first and second terms on the right-hand side represent momentum changes due to transient (pulsatile) and convective accelerations, respectively. Transient accelerations are those related to velocity changes during a period of time (i.e., blood accelerates during systolic phase of the cardiac cycle). Convective accelerations are those related to velocity changes with a change in position (i.e., fluid with constant flow rate accelerates as the conduit's cross-sectional area decreases). The third term is momentum loss due

[13] S. N. Urchuk and D. B. Plewes, *J. Magn. Reson. Imaging* **4**, 829 (1994).

[14] R. B. Bird, W. E. Stewart, and E. N. Lightfoot, "Transport Phenomena." John Wiley & Sons, New York, 1960.

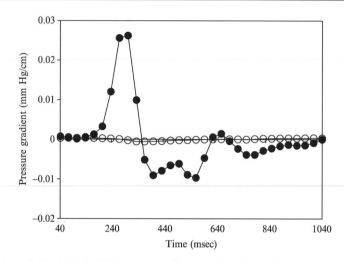

FIG. 9. The MRI-derived CSF pressure gradient waveform from a healthy volunteer (filled circles). The open circles show the small contribution of the viscous term in the Navier–Stokes equation (Eq. 1).

to viscous effects. Our lab has adapted this method to calculate CSF pulsatile pressure gradients in the upper cervical spine.[15] The assumptions of a rigid conduit and incompressible fluid are justified in the case of CSF flow in the cervical spinal canal.

The CSF pressure gradient waveform is calculated from the velocity-enclosed MRI images of the CSF pulsatile flow. The inertia component of the pressure gradient is approximated by the first-order central difference templates of the time series images. The shear (viscous) component is derived by using a pair of second-order central difference operators. The shear and inertia components are summed to obtain a pressure gradient value for each time point of the cardiac cycle. Figure 9 shows the MRI-derived CSF pressure gradient waveform from a healthy subject. The open circle curve shows the contribution of the viscous term to the total pressure gradient. The CSF flow dynamics in the upper cervical spine is clearly dominated by inertia, where the contribution of shear term is negligible. A detailed analysis of the hydrodynamics of CSF motion in the cervical spine can be found in Loth et al.[16]

[15] N. J. Alperin, S. H. Lee, F. Loth, P. B. Raksin, and T. Lichtor, *Radiology* **217,** 877 (2000).
[16] F. Loth, M. A. Yardimci, and N. Alperin, *J. Biomechan. Eng.* **123,** 71 (2001).

Validation Studies

Validation Accuracy and Reproducibility of the ICVC Measurement

The volume of the entire intracranial space is on the order of 1500 ml; the change in that volume during the cardiac cycle is on the order of 1 ml, less than 0.1%. Therefore reproducible and accurate ICVC measurement poses a challenge. The validity of the ICVC measurement was assessed from studies in patients with pathology that affect the ICVC in a predicted manner, and the inherent accuracy was assessed with a specially designed craniospinal flow phantom.[17] These studies were necessary, since currently there is no other *in vivo* method that could be used as a reference. The reproducibility was established from repeated measurements in healthy volunteers.[15,18]

Measurements before and after corrective surgical procedure (cranioplasty) in patients with a skull defect offered an opportunity to test the validity of the ICVC measurement by MRI because a larger volume change is expected during the "open skull" state. In all three patients studied, a larger maximum ICVC was measured before cranioplasty, as expected. The ICVC waveforms measured before and after cranioplasty in one of these patients are shown in Fig. 10. The maximum ICVC values before and after surgery were 1.01 and 0.52, respectively.

Measurement results obtained in the cranioplasty patients supported the validity of the method but could not be used for assessment of the method accuracy, since the true ICVC value at the two states is unknown. The inherent accuracy of the method was determined using a specially built craniospinal flow phantom where the ICVC value can be determined independently of the MRI-based method. A flow phantom composed of a nearly rigid container and tubes leading to and from the container was constructed to simulate the intracranial space and the arterial and venous flow channels.[17] A low-friction glass syringe attached to the container simulated the CSF channel and the spinal canal. A pulsatile flow pump was used to drive the system at a frequency of 1 Hz. Measurements were repeated five times with high velocity-encoding (VENC = 125 cm/sec) used to quantify the rapid blood flow and low velocity-encoding (VENC = 6 cm/sec) used to quantify the slower CSF flow. Pressure waveforms were recorded from within the container using a tip transducer sensor (Codman, Rayham, MA). The experimental setup is shown in Fig. 11. A picture of the flow

---

[17] N. Alperin, Y. Kadkhodayan, F. Loth, and R. Yedavalli, *Proc. Intl. Soc. Magn. Reson. Med.* **3,** 1981 (2001).

[18] H. Dhoondia and N. Alperin, *Proc. Intl. Soc. Magn. Reson. Med.* **1,** 793 (2003).

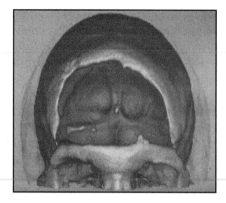

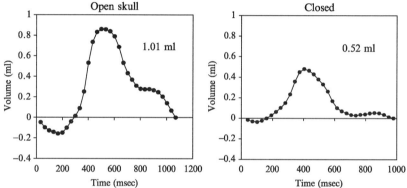

FIG. 10. Intracranial volume change (ICVC) waveforms in a patient with a skull defect that was later corrected. The maximum volume change at each state is indicated on the graph. *Top:* CT-rendered image of the open skull state. *Left:* ICVC during that state. *Right:* ICVC after skull closure. As expected, the maximum volume change is reduced after cranioplasty (skull closure) surgery.

phantom and the MRI velocity image of this phantom are shown in Fig. 12A and B, respectively.

The average mean inflow volume, outflow volume, ICVC, and oscillatory CSF volume (the volume that moves back and forth between the cranium and the spinal canal) were calculated and compared with independent measurements of these quantities. The intracranial volume change was measured independently of the MRI measurement by blocking the arterial and venous channels and deflecting the syringe to produce a pressure value equal to the pulse pressure recorded while the channels are open. Measurements results are summarized in Table I. The average maximum

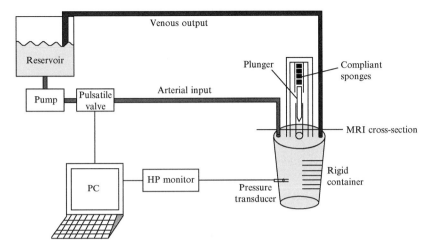

FIG. 11. The experimental setup of the flow phantom study to quantify the inherent accuracy and reproducibility of the method for ICVC measurement. The rigid container represents the cranial compartment, and the glass syringe attached to the container simulates the CSF channel and spinal canal. A pulsatile flow pump was used to drive the arterial input (red) through the tube to the cranial compartment. The venous outflow (blue) was drained to a reservoir. (See color insert.)

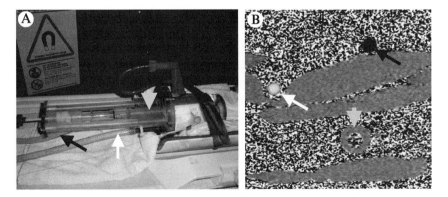

FIG. 12. (A) A picture of the flow phantom described in the experimental setup shown in Fig. 11. (B) The MRI velocity image for the phantom obtained at the cross section indicated in Fig. 11. The arterial inflow (black arrow), the venous outflow (white arrow), and CSF flow channel (short arrow) are indicated on the physical phantom, as well as on the MRI velocity image. (See color insert.)

TABLE I
SUMMARY OF RESULTS OBTAINED FROM THE FLOW PHANTOM STUDY[a]

	Mean value (ml/cardiac cycle)	SD (ml)	SD (%)	Independent measurement (ml)
Volume of arterial inflow	17.21	0.71	4	17 +/− 0.1
Volume of venous outflow	16.89	0.58	3	17 +/− 0.1
Intracranial volume change	1.94	0.09	5	2.0 +/− 0.1
Oscillatory CSF volume	1.41	0.11	8	1.35 +/− 0.1

[a] Under controlled conditions, the reproducibility of the intracranial volume change measurement is within 5% of the actual value.

ICVC value measured by MRI was within 5% of the ICVC value measured independently. Furthermore, the SD of the ICVC measurements, 0.09 ml, is much smaller than the SDs of the mean inflow and outflow volumes, 0.71 and 0.58 ml, respectively. This seeming inconsistency is explained by the fact that the main source of error in the mean inflow and outflow volumes is a baseline phase shift in the phase-encoding MR images. The mean inflow volume should be equal to the mean outflow volume. However, this baseline phase shift causes the mean inflow volume to be larger and the mean outflow volume to be smaller than the actual volume. The effect of this baseline shift on the ICVC measurement is removed by the use of Eq. (4) to ensure a zero mean volume change over a cardiac cycle. The high reliability by which ICVC is measured by MRI is attributed to the excellent temporal response of nonsteady flow measurement with the cine PC technique.

The phantom experiments demonstrated high intrinsic accuracy and reproducibility of the method. In contrast to the phantom studies, early evaluation of measurement reproducibility in human subjects demonstrated a much larger measurement variability of 18%.[15] This rather large variability needed to be improved before the method could become clinically viable. The source of this variability was found to be a change in the subject mean heart rate that often occurred between the two separate cine PC scans used for imaging of the fast blood flow and the slow CSF flow. Two separate scans were used because scanning parameters needed to be adjusted for the measured velocity range. This problem was solved by the development of a new MRI data acquisition scheme, the dual velocity-encoding scheme, which allows for simultaneous acquisition of fast and slow flow in an interleaved fashion.[18] With this technique, the CSF and the blood flow measurements are always synchronized—regardless of a change in the patient's heart rate. A study comparing the ICVC measurement variability in human subjects with the new and the previous MRI data acquisition techniques demonstrated overall improvement of 2.7 times in the reproducibility of the ICVC measurement with the dual VENC MRI acquisition technique. The current ICVC measurement reproducibility of approximately 7% in humans is close to the intrinsic reproducibility of the technique established by the phantom study.

Validation Accuracy and Reproducibility of the ICP Change Measurement

The intracranial pressure change during the cardiac cycle is derived from change in the CSF pressure gradient. The relationship between time-varying change in pressure and in pressure gradient was evaluated experimentally with a nonhuman primate[15] and theoretically with computational fluid dynamics (CFD).[16] The experimental validation required the use of a large nonhuman primate (baboon) because hydrodynamics is scale-dependent and important fluid dynamic parameters, such as the size of the spinal canal and heart rate in baboons, are similar to those of humans. MRI image of the cervical spine anatomy of a baboon demonstrating this similarity is shown in Fig. 13.

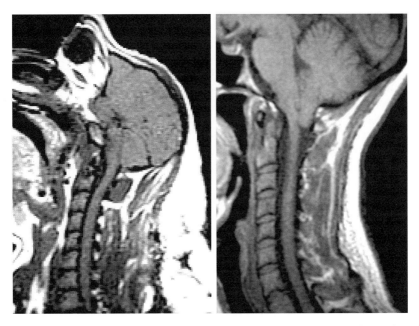

FIG. 13. Anatomic sagittal 2D T_1-weighted MR images of the cervical spine region of a baboon *(left)* and a human *(right)*. The similarity between the cervical spinal anatomy of the baboon and the human is demonstrated.

In the baboon validation study, pressures in the cranium and in the spinal canal were measured invasively using tip transducer catheters (Model V420, Camino Laboratories, San Diego, CA). Pressure and CSF flow velocities were measured at three different ICP values (baseline, elevated, and reduced pressure) achieved by either addition or withdrawal of fluid from the CSF space. The relationship between the peak-to-peak (PTP) CSF pressure gradient measured by MRI and invasively measured PTP intracranial pressure change in a baboon at the three different values of absolute ICP was found to be a linear relationship.

A linear relationship between PTP changes in pulsatile pressure and pressure gradient was also verified with CFD simulation.[16] In these simulations the spinal canal geometry was represented with two concentric cylinders, with the diameters of the inner (spinal cord) and outer (spinal canal) cylinders being 15 and 6 mm, respectively. Calculations were based on the finite volume method to solve the Navier–Stokes equations using a commercial software package (Star-CD 3.0; Adapco, Inc., New York, NY). The walls were assumed to be rigid, and the fluid was assumed to behave as a Newtonian fluid with a viscosity of 1.1 cP. A linear relationship was demonstrated between the PTP pressure and pressure gradient calculated for two different boundary conditions, high and low flow rates which correspond to high and low PTP pressure values.

The ICP change measurement reproducibility is therefore determined by the reproducibility of the PTP CSF pressure gradient measurement. The CSF pressure gradient measurement reproducibility was assessed from repeated MRI scans of healthy subjects. A measurements variability of 8% was found in these studies.[15] The MRI-derived intracranial compliance and pressure (MR–ICP) are derived from the ratio of the pressure and volume changes. Therefore the overall MR–ICP measurement reproducibility is the square root of the sum of the square of individual fractional standard deviations, that is, the fractional SD of the volume and the pressure change measurements, which are currently approximately 8% each. Therefore the overall MR–ICP measurement variability is approximately 10%. This measurement reproducibility is comparable or better than that of invasive techniques. Furthermore, a 10% difference in the ICP value is not clinically significant.

Early Assessment of the MR–ICP Method Performance in Humans

Direct comparison between MRI-derived elastance index (the ratio of the pressure change to volume change) and mean ICP value measured invasively at the time of the MRI study in five patients demonstrated a linear relationship, as expected, with high degree of correlation ($r^2 = 0.96$).[15] This linear relationship allows for expression of the MRI-derived elastance index in units of absolute ICP value. A large number of MRI measurements were acquired in human subjects (healthy volunteers as well as patients) for whom no invasive ICP values were available. Over 100 MR studies were performed in 19 healthy volunteers with no neurologic problems or a previous head trauma of any kind (18 males, 1 female, mean age of 27 +/− 8.3 years). MRI-derived ICP values ranged from 3.8 mm Hg to 18.8 mm Hg; most measurements were between 8 and 10 mm Hg. Wide distribution of ICP values measured invasively in normal human subjects has been previously reported: from 2 to 16 mm Hg in one study[19] and up to 18 mm Hg in another.[20] These reports also mentioned that invasively measured ICP values are affected by the lumen size of the needle used. Lower values were measured with wider needles. Thus it is reasonable to assume that the small volume of CSF that leaks from the intracranial space during these measurements could bias the reading toward lower ICP values. Distribution of ICP values measured invasively in 1033 healthy subjects and the distribution of the MRI-derived measurements are shown in Fig. 14A and B, respectively. Three of the 110 MRI measurements (two from the same subject) were between 16 mm Hg and 18.7 mm Hg, above the upper range found in the 1937 study. However, it is not possible to determine whether this represents a normal variation of ICP, an MRI measurement error, or the true ICP value, because no invasive values are available for comparison. Based on the current view that ICP value of 20 mm Hg is a critical threshold for elevated ICP,[21] no false positives were measured with MRI. The fact that 96% of the MRI measurements were between 4 to 16 mm Hg and no values were found above 20 mm Hg is strong evidence for the method's ability to correctly identify normal ICP (i.e., a zero-false-positive rate).

The distribution of MRI-derived TCBF measurements in the same subjects is shown in Fig. 15. It is interesting to note that the TCBF spans a much narrower range relative to the ICP. It is known that the blood flow to the brain is highly regulated; a narrower distribution may reflect that degree of regulation.

[19] H. H. Merritt and F. Fremont-Smith, "The Cerebrospinal Fluid." WB Saunders, Philadelphia, 1937.

[20] J. B. Ayer, *Assn. Res. Nerv. Mental Dis.* **4,** 159 (1924).

[21] A. Marmarou, H. M. Eisenberg, and M. A. Foulkes, *J. Neurosurg.* **75,** s59 (1991).

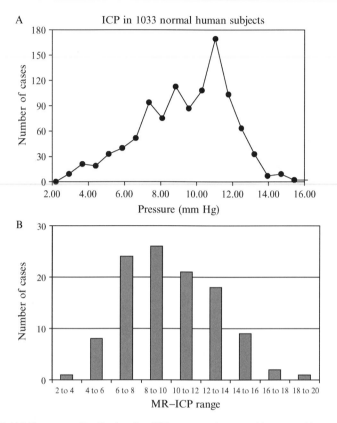

Fig. 14. (A) Frequency distribution for CSF pressures in normal human subjects measured invasively; original ICP values were given in cm H$_2$O (from Merritt and Fremont-Smith,[19] and (B) frequency of 110 measurements of MR–ICP in 19 healthy volunteers.

Application of MR–ICP to Study the Effect of Posture on Cerebral Hemodynamic Hydrodynamic Coupling

Studying the effect of posture on the craniospinal system with MR–ICP offers new insight into the coupling that exists between cerebral hemodynamics and intracranial hydrodynamics. The study of normal neurophysiology under different conditions has been limited by the invasiveness of methods currently used to quantify cerebral hemodynamics and hydrodynamics parameters. For example, blood flow is usually measured with xenon computed tomography (CT), which involves radiation, and ICP measurement requires insertion of a pressure transducer catheter into the CNS. The MR–ICP provided means for comprehensive characterization of

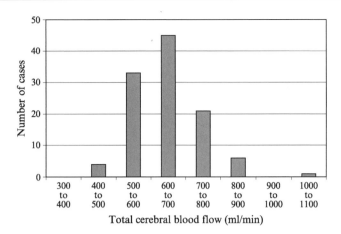

FIG. 15. Frequency distribution of MRI-derived total cerebral blood flow (TCBF) measurements obtained in the same 19 healthy volunteers.

the normal hemodynamic and hydrodynamic noninvasively. Body posture affects both the cardiovascular and the craniospinal system; the steady state of the craniospinal system in the sitting position is different than that in the supine position. These two normal but different biomechanical states were characterized and quantified in healthy subjects by application of the MR–ICP method to study healthy subjects in both upright and supine positions. The effect of posture on (-1) the distribution of the venous drainage, (-2) the TCBF, (-3) the volume of CSF that flows back and forth through the foramen magnum during the cardiac cycle, (-4) the cerebral vascular compliance, (-5) the intracranial compliance, and (-6) the intracranial pressure was imaged and quantified using the MR–ICP methodology.[22]

Seven healthy volunteers with no history of neurologic problems or head trauma were imaged in both upright and supine positions in a vertical gap MRI scanner (Signa SP/i 0.5 T; GE Medical Systems, Milwaukee, WI). The vertical gap allows the subject to be positioned seated upright. The MR–ICP scanning protocol was applied in both postures. An MR angiography technique (2D Time of Flight) was added to the MR–ICP protocol to visualize the cerebral venous drainage at the two postures. In addition to measurements of the overall intracranial compliance in the supine and upright positions, the change in the compliance of the cerebrovascular subcompartment between the two postures was estimated as well from

[22] N. Alperin, S. G. Hushek, A. Sivaramakrishnan, S. Surapaneni, R. Moser, and N. Hoerter, *Proc. Intl. Soc. Magn. Reson. Med.* **2**(Suppl.), 5 (2003).

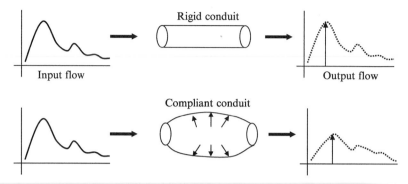

FIG. 16. A schematic representation of the modulation of a pulsatile input flow to a rigid conduit *(top)* and to a compliant conduit *(bottom)*. The output flow from a compliant conduit is less pulsatile than the output from the rigid conduit.

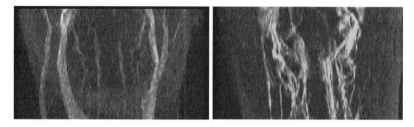

FIG. 17. The 2DTOF MR images of the veins (MRV) in the neck from one of the healthy subjects obtained in the supine *(left)* and upright *(right)* postures. Both jugular veins are visualized in the supine position, whereas in the upright posture these veins are collapsed. The redistribution of the venous drainage in the upright posture to the epidural and vertebral veins is clearly demonstrated.

the relationships between the venous outflow and the arterial inflow waveforms.[23] Estimation of the cerebrovascular compliance from inflow outflow dynamics is based on the observation that a compliant conduit tends to smooth pulsatile flow more than a rigid conduit, as illustrated in Fig. 16.

Representative 3D MR images of the neck veins at the two postures are shown in Fig. 17. These images demonstrate different patterns of venous drainage between the supine and sitting postures. In the supine posture the flow is primarily through the internal jugular veins (IJV), whereas in the upright posture the IJV collapse and the main pathway for venous

[23] N. Alperin, B. Varadarajalu, C. Fisher, and T. Lichtor, *Proc. Intl. Soc. Magn. Reson. Med.* **2,** 1281 (2002).

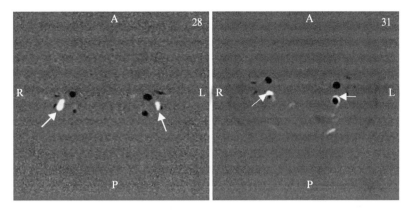

FIG. 18. One of the 32 high velocity-encoding PC MR images obtained in the supine *(left)* and upright *(right)* postures. Black pixels indicate inflow into the cranium (arteries), and white pixels indicate outflow (veins). In the supine posture, the dominant venous drainage is through the jugular veins (arrowheads), and in the upright posture, the dominant flow is through the vertebral veins (white arrows).

outflow is the venous plexus. The corresponding high velocity-encoding PC images from the same subject are shown in Fig. 18. In these images, white pixels represent velocity in the cranio-caudal direction (i.e., venous flow), and black pixels flow toward the brain (i.e., arterial flow). In the supine posture, high flow velocities are seen mainly in the IJV, whereas in the sitting posture, high flow velocities are seen mainly in the vertebral veins.

Plots of the total arterial inflow and venous outflow waveforms from one of the subjects are shown in Fig. 19. Arterial inflow (open circles) and venous outflow (filled circles) during the cardiac cycle in the supine position are shown on the left and in the sitting position on the right. Although the arterial volumetric flow rates are slightly lower in the sitting posture, the overall shape (dynamics) and amplitude (the difference between the maximum and the minimum) in the two postures are similar. In contrast, the shape and the amplitude of the venous flow were different between the two postures; the venous flow is much less pulsatile in the upright position.

An average of a 2-fold decrease in the pulsatility of the venous outflow was found. The reduced venous pulsatility implies a more compliant cerebral vascular compartment. Because the vascular compartment can be regarded as the conduit for the blood flow through the brain, an increase in the compliance of the vascular compartment will result in a larger attenuation of the arterial pulsatility (i.e., less pulsatile venous outflow).

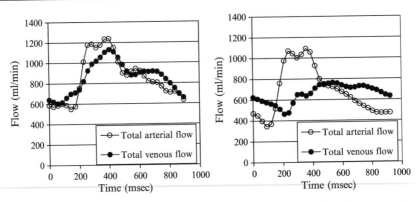

Fig. 19. MRI-based measurements of the arterial inflow (open circles) and venous outflow (filled circles) volumetric flow waveforms during the cardiac cycle in supine *(left)* and sitting *(right)* postures. Note the slightly lower arterial flow rates in the sitting position. Although the amplitude of the arterial waveform at the two postures is similar, the amplitude of the venous flow is significantly smaller (less pulsatile flow) at the upright posture, suggestive of higher cerebral vascular compliance.

The most likely cause for the increased compliance is a reduction in the volume of blood that resides in the cerebral venous vasculature.

Changes in resistance to flow could have modulated the pulsatility of the venous outflow as well. The relative small change in TCBF (about 13%), however, suggests that postural changes have no or only a small effect on the arterial subcompartment of the cerebral vasculature, where resistance to flow is regulated. Unchanged cerebrovascular resistance with head elevation in injured patients[24] further supports this assumption. Therefore the 2-fold reduction in the pulsatility of the venous outflow is mainly due to an increase in the vascular compliance, for which the venous subcompartment (e.g., venules) contributes the most.

Plots of the measured CSF flow waveforms at the two postures are shown in Fig. 20. The CSF flow waveforms are plotted together with the net transcranial blood flow (arterial–venous; A–V) to demonstrate how the net blood flow drives the CSF flow. Note that in the supine position, the CSF flow waveform follows the A–V flow waveform "more closely"— an indication of a lower intracranial compliance. The differences in the hydrodynamic parameters between the two postures are also statistically significant. They include a 2-fold increase in the ICVC, a 2-fold reduction in the CSF volume that moves in and out of the cranium, and a large

[24] Z. Feldman, M. J. Kanter, C. S. Robertson, C. F. Contant, C. Hayes, M. A. Sheinberg, C. A. Villareal, R. K. Narayan, and R. G. Grossman, *J. Neurosurg.* **76**, 207 (1992).

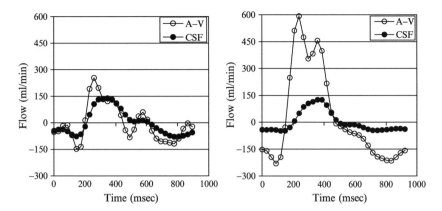

FIG. 20. Graphs depicting the A–V and CSF flow waveforms in the supine *(left)* and upright *(right)* postures. Note that the CSF follows the A–V less closely in the upright posture compared with the supine posture.

increase (3.6-fold) in the intracranial compliance with a corresponding decrease in pressure. All these changes are consistent with a more compliant intracranial compartment caused by a smaller mean volume of blood and CSF residing in the intracranial space in the sitting position.

Another parameter strongly affected by posture is the net transcranial blood flow (A–V). The amplitude of the net transcranial blood flow is much larger in the sitting position than in the supine position (Fig. 20). This is the result of the reduced pulsatility of the venous outflow together with a relatively unchanged arterial inflow dynamics. A larger net transcranial blood flow results in a larger volume of blood entering the cranium during systole. However, because the overall mean intracranial volume is smaller in the sitting position, the blood entering the cranium can be easily accommodated even with less CSF being displaced into the spinal canal. The combination of smaller amount of CSF that leaves the cranium with each cardiac cycle, and larger net transcranial inflow explains the larger ICVC measured in the sitting position.

A commonly accepted explanation for a lower ICP in an upright posture argues that venous flow is increased in that position and this avoids compression of the jugular veins.[25] Because in steady state, mean venous outflow equals mean arterial inflow, total venous outflow, like TCBF, is relatively unchanged in the two postures. The lower ICP is therefore most

[25] B. R. Nathan, *in* "Textbook of Clinical Neurology" (C. G. Goetz and E. J. Pappert, eds.), p. 489. WB Saunders, Philadelphia, 1999.

likely the result of the increased intracranial compliance due to a lower mean volume of the intracranial compartment caused by the reduced volume of CSF and blood. The shift of blood and CSF volume out of the cranium occurs during the transition from the supine to the upright posture.

Potential Clinical Role of the MR–ICP Method

Physiologic parameters measured by the MRI-based method (e.g., TCBF, intracranial compliance and pressure) are important for many neurologic problems. They are especially important for traumatic brain injury (TBI) patients because it has been shown that TCBF and ICP are important markers for progression of head trauma to secondary brain injury.[26,27] The MRI method provides these parameters noninvasively. The role of MRI-based ICP measurement may be different from that of the invasive technique. Although invasive monitoring provides continuous ICP measurements, the MRI study provides a single time point measurement. There are several clinical settings in which a "snapshot" of ICP may be beneficial. Placement of an invasive monitor to track ICP continuously is recommended for patients with severe head injuries, defined as Glasgow Coma Scale (GCS) score = 3–8. However, the necessity of ICP monitoring for patients with intermediate GCS scores (9–12)—and in particular, those with a normal CT scan on presentation—has been the subject of debate.[28,29] Between 10–30% of these patients will develop increased ICP over the first few days after injury.[30–32] Invasive monitoring techniques are not routinely employed in this population. Yet, at the same time, it is not clear which of these patients will progress to an increased pressure state. Likewise, patients with diffuse axonal injury (DAI) may demonstrate a GCS score in the severe injury range but are found to have normal ICP upon placement of an invasive monitor. In these settings, noninvasive MR–ICP measurement would provide a means of objective assessment of the need for an invasive monitor without incurring the potential morbidities.

MR–ICP may also play an important role in diagnosis and serial evaluation of several chronic disorders that may be associated with changes in

[26] T. A. Gennarelli, in "Head Injury" (P. R. Cooper, ed.), p. 137. Williams and Wilkins, New York, 1993.

[27] J. H. Siegal, D. R. Gens, T. Marmentov et al., Crit. Care. Med. **19,** 1252 (1991).

[28] B. Schmidt, J. Klingelhöfer, J. J. Schwarze, D. Sander, and I. Wittich, Stroke **28,** 2465 (1997).

[29] B. G. H. Schoser, N. Riemenschneider, and H. C. Hansen, J. Neurosurg. **91,** 744 (1999).

[30] H. M. Eisenberg, H. E. Gary, Jr., E. F. Aldrich et al., J. Neurosurg. **73,** 688 (1990).

[31] R. D. Lobato, R. Sarabia, J. J. Rivas et al., J. Neurosurg. **65,** 784 (1986).

[32] M. G. O'Sullivan, P. F. Stratham, P. A. Jones et al., J. Neurosurg. **80,** 46 (1994).

ICP. Such processes include hydrocephalus, pseudotumor cerebri, intracranial mass lesions, and toxic–metabolic encephalopathy (where a depressed level of consciousness may or may not correspond to an increased ICP). MR–ICP measurement may prevent unnecessary invasive monitoring in an increased risk setting. Single time measurement of ICP may also be helpful in the evaluation of patients with possible ventriculoperitoneal shunt malfunction—particularly young children who present with nonspecific complaints and/or who are unable to communicate their symptoms. The MRI-based technique would limit exposure to harmful radiation from repeated CT scans. Also, a finding of normal ICP by the MRI-based method might prevent unnecessary shunt exploration.

If proven accurate and reliable, noninvasive, MRI-based measurement of intracranial compliance and pressure and TCBF may provide a valuable adjunct to the evaluation of patients with various neurologic problems. In the setting of TBI, a finding of increased ICP may provide an early warning signal for a physician to adjust therapy. This tool also may prove valuable in situations where external factors confound clinical evaluation (i.e., drug intoxication), where "coma" does not necessarily reflect an increased pressure state (i.e., diffuse axonal injury), or where coexisting medical problems make placement of an invasive monitor too dangerous (i.e., bleeding disorders). In each case, noninvasive MR–ICP measurement would enhance the array of currently available evaluation modalities, providing a novel adjunct to patient diagnosis and serial monitoring with limited expense and morbidity.

[17] Near-Infrared Spectroscopy and Imaging of Tumor Vascular Oxygenation

By Hanli Liu, Yueqing Gu, Jae G. Kim, and Ralph P. Mason

Introduction

Current Development in Optical Methods for Vascular Hemoglobin Oxygenation

In recent years, a large number of investigations have been conducted in both laboratory and clinical settings to noninvasively monitor tissue vascular oxygenation using near-infrared (NIR) spectroscopy (NIRS) and imaging. Although the NIR imaging techniques are limited by their spatial resolutions, they have a great potential to be developed as a new imaging

Copyright 2004, Elsevier Inc.
All rights reserved.
0076-6879/04 $35.00

modality because of their capabilities to provide functional images. NIR spectroscopy and imaging research has been mainly focused on two organs: (1) the brain and (2) the breast.

The NIR studies of the brain include detection of brain injury or trauma,[1] determination of cerebrovascular hemodynamics and oxygenation,[2,3] and functional brain imaging in response to a variety of neurologic activations.[4-7] NIR functional brain imaging increasingly has become of great interest in studying hemodynamic response to brain activation.[8] This is mainly because the optical signals of the NIR techniques are able to non-invasively penetrate through the scalp and skull of an adult human and are sensitive to changes in the concentration of oxygenated (HbO) and deoxygenated hemoglobin (Hb). Although it is difficult to obtain very accurate quantification of cerebral HbO and Hb concentrations from the NIR imaging techniques due to rigorous requirements of theory and boundary conditions,[9] the techniques can offer relatively accurate measurements of changes of HbO and Hb, thus providing quantitative changes in total cerebral blood volume (CBV, which is assumed to be proportional to total hemoglobin concentration, HbT, where HbT = HbO + Hb).

The objective of NIR breast imaging is to develop a novel functional imaging modality for early breast cancer detection and diagnosis beyond currently available techniques. Various efforts by several research groups[10-15] have been made in either laboratory or clinical studies. For example, the research groups of Paulsen and colleagues[12] and of Jiang[15]

[1] S. Gopinath, C. S. Robertson, R. G. Grossman, and B. Chance, *J. Neurosurg.* **79,** 43 (1993).

[2] C. Cheung, J. P. Culver, K. Takahashi, J. H. Greenberg, and A. G. Yodh, *Phys. Med. Biol.* **46,** 2053 (2001).

[3] G. Zhang, A. Katz, R. R. Alfano, A. D. Kofinas, P. G. Stubblefield, W. Rosenfeld, D. Beyer, D. Maulik, and M. R. Stankovic, *Phys. Med. Biol.* **45,** 3143 (2000).

[4] M. Fabiani, G. Gratton, and P. M. Corballis, *J. Biomedi. Opti.* **1,** 387 (1996).

[5] R. Wenzel, H. Obrig, J. Ruben, K. Villringer, A. Thiel, J. Bernarding, U. Dirnagl, and A. Villringer, *J. Biomed. Opt.* **1,** 399 (1996).

[6] B. Chance, E. Anday, S. Nioka, S. Zhou, L. Hong, K. Worden, C. Li, T. Murray, Y. Ovetsky, D. Pidikiti, and R. Thomas, *Opt. Express.* **2,** 411 (1998).

[7] D. A. Boas, T. Gaudette, G. Strangman, X. Cheng, J. J. A. Marota, and J. B. Mandeville, *Neuroimage* **13,** 76 (2001).

[8] D. A. Boas, G. Jasdzewski, G. Strangman, J. P. Culver, and R. Poldrack, "OSA Biomedical Topical Meetings, Technical Digest," p. 307. Miami, FL, April 7–10, 2002.

[9] J. C. Hebden, E. M. C. Hillman, A. Gibson, N. Everdell, R. M. D. Yusof, D. T. Delpy, S. R. Arridge, T. Austin, and J. H. Meek, "OSA Biomedical Topical Meetings, Technical Digest," p. 587. Miami, FL, April 7–10, 2002.

[10] B. Tromberg, N. Shah, R. Lanning, A. Cerussi, J. Espinoza, T. Pham, L. Svaasand, and J. Butler, *Neoplasia* **2,** 26 (2000).

[11] S. Fantini, S. A. Walker, M. A. Franceschini, M. Kaschke, P. M. Schlag, and K. T. Moesta, *Appl. Opt.* **37,** 1982 (1998).

have developed a frequency-domain (FD) 16-source, 16-detector breast imager and have reported their *in vivo* results of optical properties of abnormalities from female volunteers and patients. The research group led by Chance has employed a time-domain (TD) 32-channel imaging system in conjunction with magnetic resonance imaging (MRI) to increase specificity and sensitivity of breast cancer detection.[13] Because of simplicity and low cost in comparison to the FD and TD imaging systems, continuous-wave (CW) NIR breast imaging systems also have been developed by the groups of Barbour *et al.*[14] and Chance.[16] These systems are currently under clinical tests for better breast cancer detection and diagnosis.

Although Hb, HbO, and HbT concentrations and light-scattering properties of tumors may be different from those of surrounding tissues, the optical contrast between tumor and surrounding tissue is about 2–3-fold at most in absorption, and much less in light scattering.[17] Thus much effort on increasing the optical contrast between tumor and healthy surrounding tissues also has been made using fluorescence imaging[18] or molecular beacons[16] to detect and diagnose cancer or tumor with improved sensitivity and specificity.

However, efforts in using NIR techniques to monitor tumor responses to therapeutic interventions[19–21] and therapy, such as to chemotherapy[22] or photodynamic therapy,[23] are very limited and preliminary. Moreover,

[12] T. O. McBride, B. W. Pogue, S. Jiang, U. L. Österberg, and K. D. Paulsen, *Opt. Lett.* **26,** 822 (2001).

[13] V. Ntziachristos and B. Chance, "SPIE Proceedings of Optical Tomography and Spectroscopy of Tissue III," Vol. 3597, p. 565, 1999.

[14] R. L. Barbour, C. H. Schmitz, H. L. Graber, and Y. Pei, "OSA Biomedical Topical Meetings, Technical Digest," p. 456. Miami, FL, April 7–10, 2002.

[15] X. Gu, N. Iftimia, Y. Xu, H. Jiang, and L. L. Fajardo, "OSA Biomedical Topical Meetings, Technical Digest," p. 468. Miami, FL, April 7–10, 2002.

[16] B. Chance, "OSA Biomedical Topical Meetings, Technical Digest" p. 450. Miami, FL, April 7–10, 2002.

[17] J. B. Fishkin, O. Coquoz, E. R. Anderson, M. Brenner, and B. J. Tromberg, *Appl. Opt.* **36,** 10 (1997).

[18] R. H. Mayer, J. S. Reynolds, and E. M. Sevick-Muraca, *Appl. Opt.* **38,** 4930 (1999).

[19] H. Liu, Y. Song, K. L. Worden, X. Jiang, A. Constantinescu, and R. P. Mason, *Appl. Opt.* **39,** 5231 (2000).

[20] J. G. Kim, Y. Song, D. Zhao, A. Constantinescu, R. P. Mason, and H. Liu, *J. Biomed. Opt.* **8,** 53 (2003).

[21] E. L. Hull, D. L. Conover, and T. Foster, *Br. J. Cancer* **79,** 1709 (1999).

[22] D. B. Jakubowski, A. E. Cerussi, F. Bevilacqua, N. Shah, and B. J. Tromberg, "OSA Biomedical Topical Meetings, Technical Digest," p. 456. Miami, FL, April 7–10, 2002.

[23] H. Wang, T. C. Zhu, M. Solonenko, S. M. Hahn, J. Metz, A. Dimofte, J. Mile, and A. G. Yodh, "OSA Biomedical Topical Meetings, Technical Digest." Miami, FL, April 7–10, 2002.

few reports have been found on using NIR spectroscopy and imaging as a prognostic tool for therapy planning and optimization, or for tumor prognosis. Based on the existing knowledge and development of NIR spectroscopy and imaging for the brain and breast, we were motivated a few years ago by the possibility of using NIRS as an efficient, real-time, noninvasive means to monitor tumor vascular oxygenation during respiratory interventions. The recent results in our study were obtained from animal breast and prostate tumors in vivo with a one-channel NIR spectrometer. The data have demonstrated that the NIR techniques could have applications as a prognostic means accompanying cancer therapy.[19,20,24]

Tumor Oxygenation as an Indicator for Tumor Response to Interventions

The presence and significance of tumor hypoxia has been recognized since the 1950s. Hypoxic cells in vitro and in animal tumors in vivo are documented to be three times more resistant to radiation-induced killing compared with aerobic cells.[25] Recent studies have shown that tumor hypoxia is a possible prognostic indicator and is related to the aggressiveness of a tumor, the clinical stage, and poor clinical outcome.[26–30] To improve the efficacy of oxygen-dependent treatment modalities, one possible strategy is the reduction of tumor hypoxia by raising the arterial oxygen partial pressure, P_aO_2 to overcome diffusion-limited hypoxia. Raising the arterial pO_2 by breathing hyperoxic or hyperbaric gas mixtures could be effective.[31,32] However, attempts to apply increased oxygen breathing in the clinic have not always been successful, and this may be attributed to the inability to identify those patients with hypoxic tumors.[33] More specifically, it may be

[24] Y. Gu, V. Bourke, J. G. Kim, A. Constantinescu, R. P. Mason, and H. Liu, Appl. Opt. **42**, 2960 (2003).

[25] S. M. Evans, W. T. Jenkins, M. Shapiro, and C. J. Koch, Adv. Exp. Med. Biol. **411**, 215 (1997).

[26] M. Höckel, K. Schlenger, B. Aral, M. Mitze, U. Schäffer, and P. Vaupel, Cancer Res. **56**, 4509 (1996).

[27] E. K. Rofstad and T. Danielsen, Br. J. Cancer **80**, 1697 (1999).

[28] A. W. Fyles, M. Milosevic, R. Wong, M. C. Kavanagh, M. Pintile, A. Sun, W. Chapman, W. Levin, L. Manchul, T. J. Keane, and R. P. Hill, Radiother. Oncol. **48**, 149 (1998).

[29] D. M. Brizel, S. P. Scully, J. M. Harrelson, L. J. Layfield, J. M. Bean, L. R. Prosnitz, and M. W. Dewhirst, Cancer Res. **56**, 941 (1996).

[30] J. H. A. M. Kaanders, K. I. E. M. Wijffels, H. A. M. Marres, A. S. E. Ljungkvist, L. A. M. Pop, F. J. A. van den Hoogen, P. C. M. de Wilde, J. Bussink, J. A. Raleigh, and A. J. van der Kogel, Cancer Res. **62**, 7066 (2002).

[31] S. Dische, Br. J. Radiol. **51**, 888 (1979).

[32] O. Thews, D. K. Kelleher, and P. Vaupel, Radiother. Oncol. **62**, 77 (2002).

attributed to the inability to identify those patients who would benefit from such interventions. Therefore evaluation of tumor oxygenation distributions and their changes during various stages of tumor growth, during therapeutic interventions, and during therapy is needed. Such evaluation can provide a better understanding of tumor development and tumor response to therapy, potentially allowing therapy to be planned to individual characteristics.

Numerous studies on tumor oxygen tension (pO_2) measurements have been conducted in recent years using a variety of methods,[34–36] such as microelectrodes,[37] optical reflectance,[38] electron spin resonance (EPR),[39,40] or MRI.[36,41–45] The latter two methods have advantages of making repeated measurements of pO_2 noninvasively, and MRI has the further advantage of providing dynamic maps of pO_2, which can reveal tumor heterogeneity.[46] Although NIRS does not quantify pO_2, it can indicate dynamic changes in vascular oxygenation and has the advantage of being entirely noninvasive, providing real-time measurements, and being cost-effective and portable. Our recent studies have revealed the need for NIR imaging of tumor vasculatures to study tumor heterogeneous response to therapeutic interventions and therapy. In the following sections, we briefly introduce basic algorithms used to quantify tumor hemoglobin oxygenation, followed by the NIR instrument description, system calibration, and experimental methods. Then, we provide several representative results taken

[33] J. Overgaard, in "Progress in Radio-Oncology V" (H. D. Kogelnik, ed.), p. 469. Monduzzi Editore, Bologna, Italy, 1995.

[34] H. B. Stone, J. M. Brown, T. Phillips, and R. M. Sutherland, Radiat. Res. 136, 422 (1993).

[35] R. P. Mason, S. Ran, and P. E. Thorpe, J. Cell. Biochem. 87, 45 (2002).

[36] D. Zhao, L. Jiang, and R. P. Mason, Methods Enzymol. 386, 378 (2004).

[37] O. Thews, D. K. Kelleher, B. Lecher, and P. Vaupel, Adv. Exp. Med. Biol. 428, 123 (1996).

[38] F. Steinberg, H. J. Röhrborn, K. M. Scheufler, S. Asgari, H. A. Trost, V. Seifert, D. Stolke, and C. Streffer, Adv. Exp. Med. Biol. 428, 553 (1996).

[39] P. E. James, J. A. O'Hara, O. Y. Grinberg, T. Panz, and H. M. Swartz, Adv. Exp. Med. Biol. 428, 97 (1996).

[40] J. A. O'Hara, F. Goda, J. F. Dunn, and H. M. Swartz, Adv. Exp. Med. Biol. 411, 233 (1996).

[41] D. Le, R. P. Mason, A. Constantinescu, B. R. Barker, and P. P. Antich, Magn. Reson. Imaging 15, 971 (1997).

[42] S. Hunjan, R. P. Mason, A. Constantinescu, P. Peschke, E. W. Hahn, and P. P. Antich, Int. J. Radiat. Oncol. Biol. Phys. 41, 161 (1998).

[43] D. Zhao, A. Constantinescu, L. Jiang, E. W. Hahn, and R. P. Mason, Am. J. Clin. Oncol. 24, 462 (2001).

[44] D. Zhao, A. Constantinescu, C.-H. Chang, E. W. Hahn, and R. P. Mason, Radiat. Res. 159, 621 (2003).

[45] D. Zhao, S. Ran, A. Constantinescu, E. W. Hahn, and R. P. Mason, Neoplasia 5, 308 (2003).

[46] R. P. Mason, A. Constantinescu, S. Hunjan, D. Le, E. W. Hahn, P. P. Antich, C. Blum, and P. Peschke, Radiat. Res. 152, 239 (1999).

from both prostate and breast tumors under hyperoxic respiratory interventions, using both one-channel and multichannel NIR systems, as well as one-channel and three-channel pO_2 fiber optic needle probes. At the end, we wish to demonstrate that the NIR techniques are complementary with tumor pO_2 readings and can be used as a new prognostic means for cancer therapy prognosis and therapy planning.

Theory and Algorithms

It is well known that hemoglobin concentrations and oxygen saturation in tissue vasculature can be determined using NIRS, since light absorptions of HbO and Hb in the NIR range are distinct. As with our previous work,[19,20] we assumed that HbO and Hb are the major significant absorbing species in tissue vasculature, including tumors, within the selected NIR range of 700–900 nm. Although diffusion theory has been a well-accepted theoretical approach to mathematically quantify light–tissue interaction,[6,7,10,47] no analytical solution has been available for solid tumors because of their finite size and high heterogeneity. While our future work will use the finite element method to obtain numerical solutions for the diffusion equation, our current approach is based on modified Beer–Lambert's law to account for light scattering in tumor tissue.

For a nonscattering medium, Beer–Lambert's law gives rise to the following expressions for the relationships between the absorption coefficient, μ_a, and the extinction coefficient for deoxyhemoglobin (ε_{Hb}) and oxyhemoglobin (ε_{HbO}):

$$\mu_a^{758} = 2.3\{\varepsilon_{Hb}^{758}[Hb] + \varepsilon_{HbO}^{758}[HbO]\} \tag{1}$$

$$\mu_a^{785} = 2.3\{\varepsilon_{Hb}^{785}[Hb] + \varepsilon_{HbO}^{785}[HbO]\} \tag{2}$$

where [HbO] and [Hb] are concentrations of HbO and Hb, respectively, the factor of 2.3 results from the different definitions of μ_a and ε in relation to the incident and detected optical intensities. The conventional definitions for μ_a and ε are $I = I_0 e^{-\mu_a L}$ and $I = I_0 10^{-\varepsilon C L}$, respectively, where I_0 and I are the incident and detected optical intensities in transmission measurement from a nonscattering medium, C is the concentration of hemoglobin measured in mM/L, and L is the optical pathlength through the medium in cm. By eliminating I and I_0 in these two expressions, we arrived at a relationship of $\mu_a = 2.3 \, \varepsilon C$.

[47] M. S. Patterson, B. Chance, and B. C. Wilson, *Appl. Opt.* **28,** 2331 (1989).

Because of light scattering in tissue, Beer–Lambert's law cannot be applied directly to biologic tissue, such as tumors. By taking an empiric approach to modify Beer–Lambert's law, we arrived at $\mu_a = 2.3 \varepsilon C = 2.3/L_S \log(I_0/I)$, where L_S represents the optical pathlength in a scattering medium and is no longer equal to the physical distance between the source and detector, d. In a highly light-scattering medium, L_S is much longer than d. Specifically, in the tumor study, we use I_0 and I to represent the incident and detected light intensities, when the tissue sample is without and with light absorption from [HbO] and/or [Hb]. Then, changes in absorption coefficient of the tumor, $\Delta\mu_a$, between baseline and transient conditions under respiratory intervention can be expressed as

$$\Delta\mu_a = \mu_{aT} - \mu_{aB} = 2.3\log(I_B/I_T)/L_S \tag{3}$$

where I_B and I_T are baseline and transient light intensities of the measured optical signals, respectively.

Using the transmitted light intensities at $\lambda_1 = 758$ nm and $\lambda_2 = 785$ nm and manipulating Eqs. (1)–(3), we can quantify the changes of tumor [HbO] and [Hb] due to an intervention as follows:

$$\Delta[\text{HbO}] = -11.73 \times \frac{\log(I_B/I_T)^{\lambda 1}}{L_S^{\lambda 1}} + 14.97 \times \frac{\log(I_B/I_T)^{\lambda 2}}{L_S^{\lambda 2}} \tag{4}$$

$$\Delta[\text{Hb}] = 8.09 \times \frac{\log(I_B/L_T)^{\lambda 1}}{L_S^{\lambda 1}} - 6.73 \times \frac{\log(I_B/I_T)^{\lambda 2}}{L_S^{\lambda 2}} \tag{5}$$

where $L_S^{\lambda 1}$ and $L_S^{\lambda 2}$ are optical path lengths between the source and detector at the respective wavelengths. The units of $\Delta[\text{HbO}]$ and $\Delta[\text{Hb}]$ in Eqs. (4) and (5) are mM. The constants given in the equations were computed with the extinction coefficients for oxygenated and deoxygenated hemoglobin at the two wavelengths used.[48]

In principle, $L_S^{\lambda 1}$ and $L_S^{\lambda 2}$ in Eqs. (4) and (5) are variables, depending on the actual separation of source and detector, as well as the optical properties of tumor. Previously, a differential pathlength factor (DPF), (i.e., $L_S = d \times DPF$) has been introduced to associate L_S with d.[49] The DPF values of blood-perfused tissues have been studied intensively for muscles[50] and

[48] W. G. Zijlstra, A. Buursma, and W. P. Meeuwsen-van der Roest, *Clin. Chem.* **37,** 1633 (1991).

[49] D. T. Delpy, M. Cope, P. van der Zee, S. Arridge, S. Wray, and J. Wyatt, *Phys. Med. Biol.* **33,** 1433 (1988).

[50] M. Ferrari, Q. Wei, L. Carraresi, R. A. De Blasi, and G. Zaccanti, *J. Photochem. Photobiol.* **16,** 141 (1992).

brains[51] with approximate values of 4–6 and 5–6, respectively. However, little is known about DPF for tumors, although a DPF value of 2.5 has been used by others.[52] In our approach, we made two assumptions: (1) DPF is the same for both $\Delta[\text{HbO}]$ and $\Delta[\text{Hb}]$ at 785 nm (i.e., $DPF = DPF_{\text{HbO}}^{785} = DPF_{\text{Hb}}^{785}$). This assumption was based on the fact that the absorption difference between oxygenated and deoxygenated blood at 785 nm is much smaller than that at 758 nm. The maximal relative error caused by this assumption in tumor oxygen interventions was estimated to be less than 12%, and detailed justification and discussion were given by Liu et al.[19] Furthermore, we associated L_S to μ_a by $L = \sqrt{3}/2d\sqrt{\mu_s'/\mu_a}$, where μ_s' is the reduced scattering coefficient, according to Sevick-Muraca et al.[53] and Liu.[54] The second assumption was $\mu_s'(758 \text{ nm}) \cong \mu_s'(785 \text{ nm})$ because of weak wavelength dependence of light scattering from tissue. With all the conditions mentioned and further derivation, as given in details by Kim et al.[20] we arrived at the final expressions for $\Delta[\text{HbO}]$ and $\Delta[\text{Hb}]$:

$$\Delta[\text{HbO}] = \frac{-10.63 \cdot \log\left(\frac{A_B}{A_T}\right)^{758} + 14.97 \cdot \log\left(\frac{A_B}{A_T}\right)^{785}}{d} \tag{6}$$

$$\Delta[\text{Hb}] = \frac{8.95 \cdot \log\left(\frac{A_B}{A_T}\right)^{758} - 6.73 \cdot \log\left(\frac{A_B}{A_T}\right)^{785}}{d} \tag{7}$$

where d is the direct physical distance between source and detector. $\Delta[\text{Hb}]_{\text{total}}$ is followed by adding Eqs. (6) and (7), as

$$\Delta[\text{Hb}]_{\text{total}} = \Delta[\text{HbO}] + \Delta[\text{Hb}] \tag{8}$$

Equations (6) to (8) will be used in calculating $\Delta[\text{HbO}]$, $\Delta[\text{Hb}]$, and $\Delta[\text{Hb}]_{\text{total}}$ in tissue phantoms and tumors during gas interventions in later sections.

The units for $\Delta[\text{HbO}]$, $\Delta[\text{Hb}]$, and $\Delta[\text{Hb}]_{\text{total}}$ in Eqs. (6) to (8) are mM/ DPF, which still depends on the optical properties of tumor at a particular wavelength. Because our focus is on dynamic changes in [HbO] due to respiratory challenges, in some cases, we normalize $\Delta[\text{HbO}]$ at its maximal

[51] P. van der Zee, M. Cope, S. R. Arridge, M. Essenpreis, L. A. Potter, A. D. Edwards, J. S. Wyatt, D. C. McCormick, S. C. Roth, E. O. R. Reynolds, and D. T. Delpy, *Adv. Exp. Med. Biol.* **316**, 143 (1992).

[52] R. G. Steen, K. Kitagishi, and K. Morgan, *J. Neurooncol.* **22**, 209 (1994).

[53] E. M. Sevick-Muraca, B. Chance, J. Leigh, S. Nokia, and M. Maris, *Anal. Biochem.* **195**, 330 (1991).

[54] H. Liu, *Appl. Opt.* **40**, 1742 (2001).

value (i.e., $\Delta[HbO]/\Delta[HbO]_{max}$) to eliminate the effect of DPF on our results.

Experimental Methods

One-Channel and Multichannel NIR System

A dual-wavelength (at 758 and 785 nm), one-channel NIR system (NIM, Inc., Philadelphia, PA) uses an in-phase and quadrature-phase chip (IQ chip). As shown in Fig. 1, the NIR system starts with a radiofrequency (RF) source to modulate the light intensities of two laser diodes (LD_1 and LD_2) at 140 MHz through a time-sharing system. The light passes through a bifurcated fiber optic probe, is transmitted through the tumor tissue, and is collected by a second fiber bundle. The light is then amplified by a photomultiplier tube (PMT), demodulated by the IQ circuit, and filtered by a lowpass filter (LPF) for passing only the direct current (DC) components. The signals are digitized by an analog-to-digital converter (ADC) and stored in a laptop computer. The measured DC signals at the IQ branches

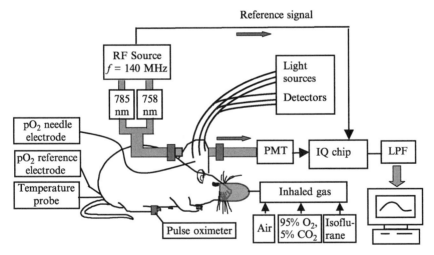

Fig. 1. Experimental setup for simultaneous tumor oximetry, using an NIRS system with either a needle electrode or a three-channel fiber-optic FOXY system. The NIR system consists of two laser diodes modulated at 140 MHz, two fiber bundles for light delivery and collection, a photomultiplier tube (PMT), an in-phase and quadrature-phase demodulator (IQ chip), and lowpass filter (LPF) for retrieving amplitude and phase information. The FOXY system comprises three fiber-optic, oxygen-sensing probes that are inserted into different regions of the tumor. Alternatively, a needle electrode probe can be inserted into the tumor for single-channel pO_2 reading.

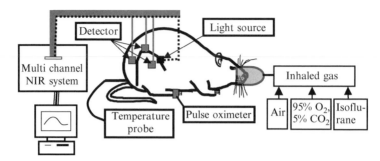

FIG. 2. Schematic experimental setup for three-channel NIRS experiments. One of the three detectors was placed opposite to the light source in order to detect light in transmission mode, and the other two detectors were set in the semireflection mode on the tumor surface.

at the measured wavelengths, $I_{bc}(\lambda)$ and $Q_{bc}(\lambda)$, lead to the quantification of optical amplitudes, $A(\lambda)$, and phase, $\theta(\lambda)$, that have passed through the tumor tissues[55]:

$$A(\lambda) = \sqrt{I(\lambda)_{DC}^2 + Q(\lambda)_{DC}^2} \tag{9}$$

$$\theta(\lambda) = \arctan\left(\frac{Q(\lambda)_{DC}^2}{I(\lambda)_{DC}^2}\right) \tag{10}$$

where λ represents the respective wavelengths used in the NIR system. Then, changes in light intensity through the tumor caused by hyperoxic interventions are used to compute changes in tumor vascular [HbO] and [Hb]. Whether the readings of $\theta(\lambda)$ can be linked to tumor physiology remains to be further explored.

Besides a one-channel NIR system, we have also used a multichannel NIR system (Fig. 2) to explore and reveal intratumoral vascular heterogeneity by having simultaneous readings at several separations of the source and detectors. Unlike the single-channel system, the multichannel system uses two laser diodes at 730 and 850 nm with constant light intensities (i.e., DC light) and three photo detectors. These detectors are placed along the circumference together with the light source, with three different source–detector separations in order to detect different depths within the tumor volume, as demonstrated in Fig. 3.

Following the same procedures as those described in the previous section, we arrived at the expression of $\Delta[HbO]$ for the multichannel system with the two wavelengths of 730 and 830 nm, as follows:

[55] Y. Yang, H. Liu, X. Li, and B. Chance, $Opt.\ Eng.$ **36,** 1562 (1997).

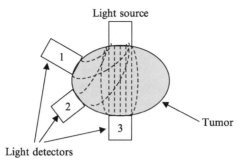

FIG. 3. A schematic diagram showing the locations of the three detectors and possible internal tumor volumes interrogated by the different detectors.

$$\Delta[\text{HbO}] = \frac{-0.674 \cdot \log\left(\frac{A_B}{A_T}\right)^{730} + 1.117 \cdot \log\left(\frac{A_B}{A_T}\right)^{830}}{d} \tag{11}$$

where d is the direct source–detector separation in cm, and the unit of $\Delta[\text{HbO}]$ is mM/DPF.

System Testing and Calibration

It is important to characterize and verify linearity of the NIRS system before any animal experiment starts. We have performed two critical system tests: (1) to quantify the output voltage range, within which the NIR system has a linear response, and (2) to perform liquid phantom measurements to confirm the physiologic readings derived from the optical NIRS system.

Linearity Testing Associated with Crosstalk Between the Amplitude and Phase. In principle, changes in amplitude of an FD photon migration system should be independent of its changes in phase (i.e., the crosstalk between amplitude and phase should be minimal for an ideal NIR system): However, in reality, it is difficult to have such a perfect condition for an actual instrument, so it is necessary to conduct the crosstalk test. We altered the optical densities (O.D.) of optical filters in front of the two light sources (758 and 785 nm) to attenuate the detected amplitude, so we could see whether changes in amplitude and phase were correlated. The test showed that phase–amplitude crosstalk exists, if the measured output intensity is either too small (<50 mV) or too large (>380 mV), as shown in Fig. 4A and 4B for 758 nm and 785 nm, respectively. We also plotted O.D. values of the filters versus log (amplitude of the output signal) to see the electrical attenuation of the system output in response to the optical

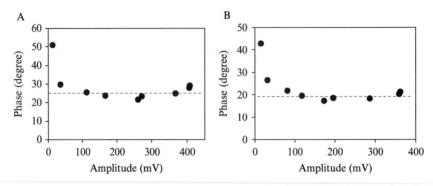

FIG. 4. Relationships between the output amplitude (mV) and phase (in degree) of the NIR single-channel system at (A) 758 nm and (B) 785 nm.

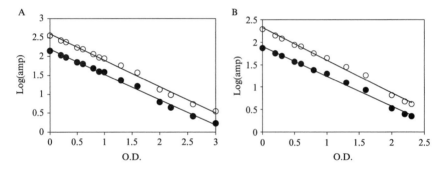

FIG. 5. Linearity tests for the NIR system between the optical input attenuation and electrical output signals at 758 nm (open circles) and 785 nm (solid circles). The X variable is the optical density (O.D. = $\log[I_0/I]$) of the filters, which were inserted before the detector to attenuate the light intensity. Cases (A) and (B) correspond to a higher (350 mV for 758 nm) and lower (193 mV for 758 nm) output voltage, respectively. Both 350 and 193 mV were obtained when the measurement was performed without any optical attenuation.

attenuation. It was found that as long as the output is in the range 100–350 mV, good linearity is achieved for the system response, as shown in Fig. 5A and 5B.

Blood Phantom Study. To calibrate the NIRS system for quantifying hemoglobin concentrations, we used human blood for tissue phantom experiments. Two packets of Sigma (St. Louis, MO) P-3813 phosphate buffered saline (PBS) (pH 7.4) powder was used to make a 2-liter buffer, and 100 ml of 20% intralipid was added into the buffer for a 1% intralipid solution, as shown in Fig. 6A. Then 14 g of baking yeast was mixed with tap

A

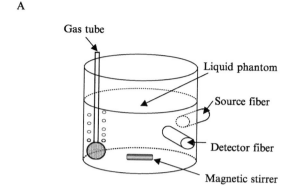

B

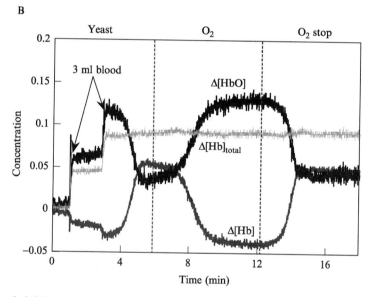

Fig. 6. (A) Experimental setup for system calibration using a liquid phantom with 1% Intralipid and baker's yeast in saline buffer. The NIRS probes were placed in reflectance geometry, while the gas bubbler was placed opposite to minimize liquid movement effects. (B) Simultaneous dynamic changes of $\Delta[HbO]$, $\Delta[Hb]$, and $\Delta[Hb]_{total}$ in the phantom solution measured using the NIR single-channel system. The dark solid curve is for $\Delta[HbO]$, the lighter solid line is for $\Delta[Hb]_{total}$, and the gray solid line shows $\Delta[Hb]$ in the phantom solution. Blood was deoxygenated by the yeast's oxygen consumption and oxygenated by bubbling O_2. During the oxygenation and deoxygenation process, $\Delta[Hb]_{total}$ was maintained as a constant.[20]

water before being added into the 2-liter buffer solution to deoxygenate the tissue phantom solution. Amount of blood for each addition to the solution was 3 ml for two additions after the yeast was well mixed in the solution.

When the blood was fully deoxygenated by yeast, pure oxygen was introduced in the solution to oxygenate the blood. After the blood was fully oxygenated, oxygen blowing was stopped in order to re-deoxygenate the solution by yeast again. The gas tube for oxygen delivery was placed far from the NIRS probes to minimize any liquid movement effects, as shown in Fig. 6A. Source and detector probes for the NIRS system were placed in reflection geometry with a direct separation of $\sim$3 cm. The solution was stirred constantly with a magnetic stirrer to maintain homogeneity.

After getting the raw amplitude data from the experiment, we used Eqs. (6) to (8) to calculate $\Delta[\text{HbO}]$, $\Delta[\text{Hb}]$, and $\Delta[\text{Hb}]_{\text{total}}$. As shown in Fig. 6B, $\Delta[\text{Hb}]_{\text{total}}$ remained constant, as expected, whereas $\Delta[\text{HbO}]$ and $\Delta[\text{Hb}]$ altered in opposite directions during the oxygenation and deoxygenation cycle. This demonstrates that the two assumptions used to derive Eqs. (6) and (8) are correct and necessary to compensate the differences in DPF caused by two different wavelengths.[20] Notice that there appears to be a delay between the gas switch and responses of $\Delta[\text{HbO}]$ and $\Delta[\text{Hb}]$ as shown in Fig. 6B. This delay is possibly caused by the entire process of transporting oxygen gas from the tube to the solution, oxygenating/deoxygenating the solution, and reaching a homogeneous state near the detectors.

Animal Models

In the prostate tumor study, Dunning prostate R3327-HI rat tumors (originally obtained from Dr. Peter Peschke, DKFZ Heidelberg) were implanted in pedicles on the foreback of adult male Copenhagen rats.[56] In the breast tumor study, rat mammary adenocarcinomas 13762NF (originally obtained from the Division of Cancer Therapeutics, National Cancer Institute [DCT NCI]) grown in the hindlimb of adult female Fisher 344 rats ($\sim$200 g) were used. Once the tumors reached approximately 1–2 cm in diameter, the rats were anesthetized with 0.2 ml ketamine hydrochloride (100 mg/ml; Aveco, Fort Dodge, IA) and maintained under general gaseous anesthesia with isoflurane in air (1.3% isoflurane at 1 dm^3/min air) through a mask placed over the mouth and nose. Tumors were shaved to improve optical contact for transmitting light. Body temperature was maintained by a warm water blanket and was monitored by a rectally

---

[56] E. W. Hahn, P. Peschke, R. P. Mason, E. E. Babcock, and P. P. Antich, *Magn. Reson. Imaging* **11**, 1007 (1993).

inserted thermal probe (see Fig. 1) connected to a digital thermometer (Digi-Sense, Model 91100–50; Cole-Parmer Instrument Company, Vernon Hills, IL). A pulse oximeter (Model 8600; Nonin, Inc., Plymouth, MN) was placed on the hindfoot to monitor arterial oxygenation (S_aO_2). Tumor volume V (in cm^3) was estimated as $V = 4\pi/3(L + W + H/6)^3$, where L, W, and H are the three respective orthogonal dimensions.

In general, the source–detector fiber separation was about 1–1.5 cm in transmittance geometry, and thus the tumor volume interrogated by NIR light can be estimated as follows. The radius of probe is 0.4 cm, which makes the volume of the cylinder 0.5–0.75 cm^3. By diffusion approximation, the optical penetration depth is about one third of the direct distance between the source and detector, which makes an additional spherical volume with a radius from 0.5–0.75 cm. Therefore the total tumor volume interrogated by NIR light can be estimated in the range of 1.0–2.5 cm^3, depending on the actual source–detector separation.[20]

Measurements of pO_2 Using Oxygen Needle Electrode and Fiber-Optic Fluorescence-Quenched Probe

Needle Electrode. We used a needle-type oxygen electrode for local pO_2 readings in several tumors. Linear two-point calibrations were performed with air (21% O_2) and pure nitrogen (0% O_2) saturated saline buffer solutions before the animal experiments, and we estimated an instrumental precision of 2–3 mmHg. After calibration, the needle electrode was inserted carefully into the tumor with the reference electrode being placed rectally, as described in detail previously.[20] Both the sample and reference electrodes were connected to a picoammeter (Chemical Microsensor; Diamond Electro-Tech, Inc., Ann Arbor, MI) and polarized at −0.75 V. Measurement points of pO_2 were manually recorded, whereas the NIRS data were acquired automatically. Measurements of pO_2 and NIRS were initiated, while rats breathed air for ~10 min to demonstrate a stable baseline. The inhaled gas was then switched to carbogen (5% CO_2 and 95% O_2) or oxygen for 15 min and switched back to air.

Fiber-Optic Fluorescence-Quenched Probe. For multiple pO_2 readings, we used a multichannel, fiber-optic, oxygen-sensing system (FOXY; Ocean Optics, Inc.). We calibrated the three fluorescence-quenched, optical fiber probes initially and then inserted them into different regions of the tumors to monitor the changes of oxygen tension (ΔpO_2) in response to the respiratory challenges as described previously.[24] The probes were placed in such a way that at least one was in a relatively poorly oxygenated region (low baseline pO_2) and at least one was in a relatively well-oxygenated region (high baseline pO_2). A mean ΔpO_2 of the tumor was obtained by averaging the three readings. A typical FOXY channel used a pulsed blue light-emitting diode (LED) at $\sim$475 nm and was coupled into one branch of a bifurcated optical fiber bundle. Then, the fluorescent light was propagated to the FOXY probe tip. Each probe had two 300-μm-diameter optical fibers with an aluminum jacket. The distal end of the probes was coated with a thin layer of a hydrophobic sol-gel material, so that an oxygen-sensing ruthenium complex was effectively trapped and protected from water. The ruthenium complex at the probe tips was excited by the blue LED and emitted fluorescence at $\sim$600 nm. If the excited ruthenium complex encountered an oxygen molecule, the excess energy would be transferred to the oxygen molecule in a nonradiative transition, decreasing or quenching the fluorescence signal.

The fluorescence response of the ruthenium complex was highly temperature dependent, so the probe calibration was accomplished by streaming gases of known oxygen concentrations (100, 20.9, 10, 2, and 0%) through a cylindrical water jacket heated to 37°. Calibration curves were calculated by the vendor-supplied software, using the second-order, polynomial calibration:

$$\frac{I_0}{I} = 1 + K_1[O] + K_2[O]^2 \tag{12}$$

where I_0 is the fluorescence intensity at zero concentration (nitrogen), I is the measured intensity of fluorescence at a pressure p of oxygen, $[O]$ represents the oxygen concentration (related to pO_2), and K_1 and K_2 are the first- and second-order coefficients and are automatically supplied by the curve-fitting routine from the calibration measurements. After the system calibration, the oxygen concentration in tumor–tissue sample measurements was deduced using Eq. (12).

NIR Measurements of Breast–Prostate Tumors Under Interventions

We conducted all animal experiments in a darkened room, and the measurements were initiated while the rats breathed air for 10 min to get a stable baseline. The inhaled gas was switched to carbogen (5% CO_2 and 95% O_2) or pure oxygen for 15–20 min, and then back to air for 15 min. Sometimes, repeated cycles were taken either to check the reproducibility or to minimize preconditioning effects. Raw amplitude data from either the one-channel or three-channel detectors were recorded simultaneously during the experiments, processed, and displayed through the computer to obtain tumor $\Delta[HbO]$ and $\Delta[Hb]_{total}$. The dynamic data were fitted with a single-exponential and/or double-exponential expression, using Kaleidagraph software (Synergy Software, Reading, PA), to obtain time constants and amplitudes. Those fitted parameters allowed us to further develop a two-region tumor model, which will be discussed in a later section.

In many cases, we performed simultaneous measurements of changes in tumor HbO and in tumor pO_2, using either the needle electrode or the fiber-optic FOXY system along with the NIRS readings, as shown in Fig. 1, for both breast and prostate tumors.

Results and Model Development

HbO and pO_2 Changes in Prostate Tumors Under Carbogen Intervention

We have measured relative changes of [HbO], [Hb]$_{total}$, and tumor tissue pO_2 from several Dunning prostate R3327-HI tumors, and Fig. 7 shows two representative data sets. Figure 7A shows the temporal profiles of Δ[HbO] and pO_2 in a Dunning prostate R3327-HI tumor (3.6 cm^3) measured simultaneously with the NIRS and pO_2 needle electrode during carbogen respiratory challenge. After the breathing gas was switched from air to carbogen, Δ[HbO] increased rapidly, whereas Δ[Hb]$_{total}$ seemed to have much smaller responses. In this case, tumor tissue pO_2 increased at a much slower rate, from a baseline of 60 mmHg to ~80 mmHg during the entire carbogen intervention. Figure 7B is obtained from another prostate tumor (3.1 cm^3); the electrode readings showed a slower pO_2 response from ~15 to 40 mmHg, whereas the NIRS response was biphasic, a sharp rise in Δ[HbO] followed by a further slow, gradual significant increase over the next ~15 min. In this case, Δ[Hb]$_{total}$ showed little change, ~2% of the maximum in Δ[HbO]. The biphasic feature of Δ[HbO] has been a commonly observed dynamic characteristic under tumor vascular oxygenation, as reported previously.[19,20,24]

HbO and pO_2 Changes in Breast Tumors Under Carbogen and
*   Oxygen Intervention*

For a representative 13762NF breast tumor (3.2 cm^3), typical time profiles of the normalized Δ[HbO] and mean ΔpO_2 in response to carbogen and then oxygen intervention are shown in Fig. 8. When the inspired gas was switched from air to carbogen, the normalized Δ[HbO] showed a sharp initial rise in the first minute ($p < 0.0001$), followed by a slower, gradual, but further significant increase over the next 19 min ($p < 0.001$). The mean ΔpO_2 profile was averaged over the three individual ΔpO_2 readings (FOXY), and it increased rapidly by about 50 torr ($=$ mmHg) within 8 min ($p < 0.0005$) and also continued a slower and gradual increase over the next 12 min ($p < 0.005$). Return to breathing air produced a significant

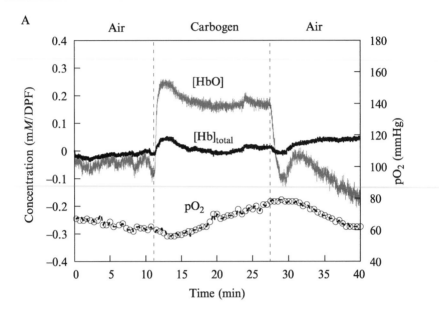

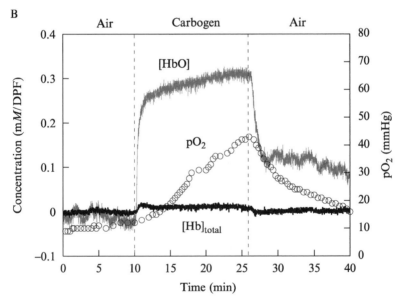

Fig. 7. Simultaneous dynamic changes of $\Delta[HbO]$ and pO_2 in R3327-HI rat prostate tumors using NIRS and pO_2 needle electrode. (A) A representative tumor (3.6 cm³) showed a rapid NIR response, whereas (B) another tumor (3.1 cm³) showed a clear biphasic feature in $\Delta[HbO]$ with a slow pO_2 response. The unit of $\Delta[HbO]$ is mM/DPF, where DPF is equal to the optical pathlength divided by the source–detector separation. Dotted vertical line marks the time when the gas was changed.

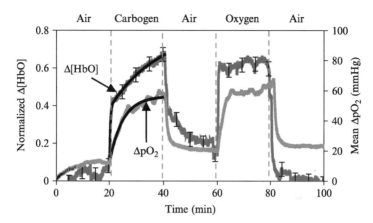

FIG. 8. Dynamic responses of $\Delta[HbO]$ and mean ΔpO_2 to hyperoxic gas intervention in a rat breast tumor (3.2 cm^3). In response to carbogen breathing, single-exponential curve fitting yielded $\Delta pO_2 = 42.68 \{1 - \exp[-(t - 21.01)/4.56]\} + 16.66$ ($r = 0.98$), and biexponential fitting resulted in $\Delta[HbO] = 0.373 \{1 - \exp[-(t-20.36)/0.61]\} + 0.648 \{1-\exp[-(t-20.36)/21]\}$ ($r = 0.97$). The smooth black curves plotted along with the data are obtained according to the previous exponential expressions, respectively.

decline for both signals ($p < 0.0001$). When the hyperoxic intervention was repeated with pure oxygen, the dynamic features of $\Delta[HbO]$ and ΔpO_2 remained consistent, except that the biphasic behavior of $\Delta[HbO]$ was less apparent. A detailed investigation on $\Delta[HbO]$ differences in response to carbogen and oxygen interventions has been given in our earlier study.[24]

In response to carbogen intervention, the pO_2 profile displayed a single-phase dynamic behavior, whereas $\Delta[HbO]$ showed an apparent biphasic response. These dynamics may be characterized by time constants of single- and double-exponential responses, respectively. A single exponential fitting gives rise to a slow ΔpO_2 response of $\tau(\Delta pO_2) = 4.56 \pm 0.06$ min ($r = 0.98$). A double-exponential expression with two time constants, τ_1 and τ_2, was used to fit the normalized $\Delta[HbO]$, yielding fast and slow time constants of 0.61 min and 21 min, respectively. Figure 8 plots the fitted curves along with the experimental data. During the course of our study, we have often observed biphasic characteristic of $\Delta[HbO]$, which motivated us to develop a model to interpret the experimental finding, to be described in a later section.

The advantage of using FOXY pO_2 probes is to detect distinct heterogeneity in tumor pO_2. The individual responses at different locations within the tumor to the hyperoxic gas were diverse: those probes that indicated

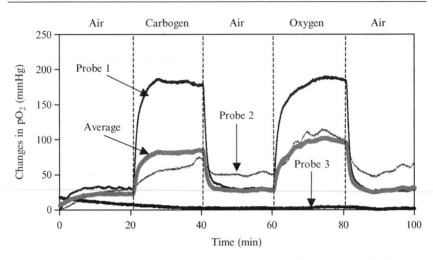

FIG. 9. Time profiles of tumor oxygen tension changes, ΔpO_2, measured with three channels of a FOXY fiber-optic, oxygen-sensing system with respect to different gas inhalations for a breast tumor (4.6 cm^3). The mean signal for the three channels was calculated and plotted by the thicker trace. Modified from Gu *et al.*[24]

apparently well-oxygenated regions usually showed a large and rapid response, whereas those with lower baseline pO_2 often showed little change (Fig. 9). By studying a group of five rats with both carbogen and oxygen interventions, we obtained a distinct correlation between the maximal values of global $\Delta[HbO]$ and the mean ΔpO_2, as shown in Fig. 10. Because of heterogeneity in regional pO_2, the standard deviations of the mean pO_2 values were quite large.

Tumor Heterogeneity Observed by a Multichannel NIRS System

To study tumor heterogeneity, the multichannel NIRS system also has been used to detect changes in [HbO] with different source-detector separations. Fig. 11A shows three temporal profiles of $\Delta[HbO]$ obtained from three detectors in a breast tumor (16.6 cm^3), with a source–detector separation of 1.5 cm for detector 1, 2.5 cm for detector 2, and 2.8 cm for detector 3, respectively (Fig. 3). The measurement uncertainties are plotted at discrete times in Fig. 11A. After 10 min of air breathing measurement as the baseline, the inhaled gas was switched from air to carbogen, causing a sharp increase in $\Delta[HbO]$ ($p < 0.0001$ after 1 min from gas switch), followed by a further gradual, but significant, increase over the next 15 min ($p < 0.0001$).

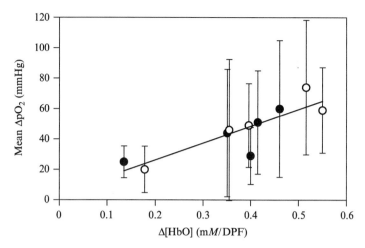

Fig. 10. Correlation between mean ΔpO_2 and ΔHbO for five breast tumors ($r > 0.86$) with a linear relationship of $\Delta pO_2 = 110.65\ \Delta HbO + 4.1285$. Solid circles are for transition air to carbogen, whereas open circles are for transition air to oxygen.[24]

As with the single-channel NIRS measurement, the biphasic features appear clearly in the multichannel results. The rising parts of $\Delta[HbO]$ from detectors 1, 2, and 3 after gas switch to carbogen are shown in Fig. 11B–11D, respectively, along with the fitted curves. The two equations used for the curve fitting are

$$\Delta[HbO]_{single} = A\{1 - \exp[-(t - t_0)/\tau]\} \tag{13}$$

$$\Delta[HbO]_{double} = A_1\{1 - \exp[-(t - t_0)/\tau_1]\} + A_2\{1 - \exp[-(t - t_0)/\tau_2]\} \tag{14}$$

respectively, for the single-exponential and double-exponential expressions. It is clear that the double-exponential expression gives a much better fit, as confirmed by the respective chi squared (χ^2) and R^2 values. Table I summarizes all the fitted parameters for the respective detectors.

Mathematical Model Development to Interpret Data

As shown in the last section, the temporal profiles of tumor $\Delta[HbO]$ caused by respiratory challenge can be well fitted with a double-exponential equation, represented by two time constants (fast and slow). To understand these time constants and to interpret the experimental findings, we further developed a hemoperfusion model,[19] briefly described as follows.

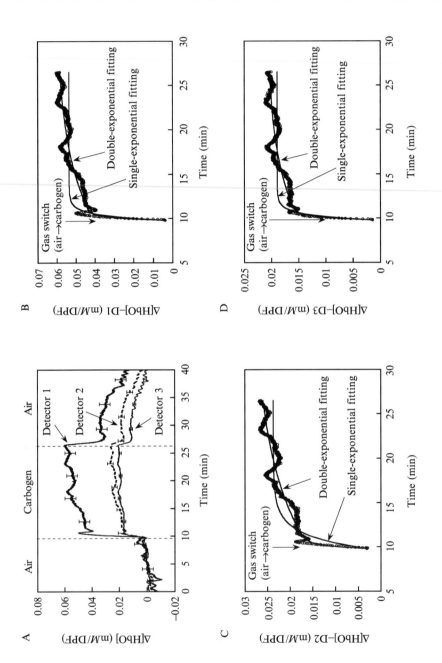

TABLE I
SUMMARY OF VASCULAR OXYGEN DYNAMICS FROM THREE DETECTORS FROM FIG. 11

Double-exponential fitting $\Delta HbO_2 = A_1[1 - \exp(-t/\tau_1)] + A_2[1 - \exp(-t/\tau_2)]$

Parameters	Detector 1	Detector 2	Detector 3
Separation: d (cm)	1.5	2.5	2.8
A_1 (mM/DPF)	0.037 ± 0.001	0.0125 ± 0.0002	0.0134 ± 0.0002
τ_1 (min)	0.24 ± 0.01	0.30 ± 0.02	0.27 ± 0.01
A_2 (mM/DPF)	0.020 ± 0.001	0.0130 ± 0.0004	0.0060 ± 0.0002
τ_2 (min)	8.27 ± 0.72	9.87 ± 0.84	7.00 ± 0.58
χ^2	0.005	0.001	0.0005
R^2	0.95	0.96	0.95
$\gamma_1/\gamma_2 = A_1/A_2$	1.85 ± 0.07	0.96 ± 0.05	2.23 ± 0.08
τ_1/τ_2	0.029 ± 0.004	0.030 ± 0.004	0.039 ± 0.005
$f_1/f_2 = (A_1/A_2)/(\tau_1/\tau_2)$	64 ± 9	32 ± 5	58 ± 9

Following an approach used to measure regional cerebral blood flow (rCBF) with diffusible radiotracers,[57–59] we made an analogy to evaluate tumor hemodynamics, using the respiratory intervention gas as a tracer.

In response to respiratory intervention, a sudden small change introduced in arterial O_2 saturation (S_aO_2) results in an increase in arterial HbO concentration (ΔHbO^{artery}). This increase in HbO^{artery} can be considered as an intravascular tracer.[60] Following Kety's method and assuming that changes in dissolved O_2 are negligible,[60] we have

$$\frac{d}{dt}\left(\Delta HbO^{vasculature}\right) = f\left(\Delta HbO^{artery} - \frac{\Delta HbO^{vasculature}}{\gamma}\right) \qquad (15)$$

[57] S. S. Kety, *Pharmacol. Rev.* **3,** 1 (1951).
[58] S. S. Kety, *Israel J. Med. Sci.* **23,** 3 (1987).
[59] H. Watabe, M. Itoh, V. Cunningham, A. A. Lammertsma, P. Bloomfield, M. Mejia, T. Fujiwara, A. K. P. Johes, T. Johes, and T. Nakamura, *J. Cerebr. Blood Flow Metab.* **16,** 311 (1996).
[60] A. D. Edwards, C. Richardson, P. van der Zee, C. Elwell, J. S. Wyatt, M. Cope, D. T. Delpy, and E. O. R. Reynolds, *J. Appl. Physiol.* **75,** 1884 (1993).

FIG. 11. (A) Dynamic changes of [HbO] measured at three detectors from a rat breast tumor (volume: 16.6 cm^3). Dotted vertical lines mark the points when the gas was changed. The rising parts of Δ[HbO] obtained from three detectors were fitted using both single-exponential and double-exponential expressions, and (B) to (D) are from detectors 1 to 3, respectively.

where f represents blood perfusion rate, and γ is defined as a vasculature coefficient of the tumor. The coefficient, γ, is the ratio of HbO concentration change in the vascular bed to that in veins and equals $(\Delta HbO^{vasculature})/(\Delta HbO^{vein})$.

In Eq. (15), whereas f and γ are constants, $\Delta HbO^{vasculature}$ is a time-dependent variable. In principle, $\Delta HbO^{vasculature}$ can be solved rigorously given a constant input, H_0, for ΔHbO^{artery} after time 0. Our previous study demonstrates that changes in arterial HbO (S_aO_2) are much faster than that in the vascular bed.[19] Then we arrive at Eq. (16):

$$\Delta HbO^{vasculature}(t) = \gamma \times H_0 \times \left(1 - e^{-ft/\gamma}\right) \tag{16}$$

Equation (16) indicates that the change in oxygenated hemoglobin concentration in tumor vasculature, $\Delta HbO^{vasculature}(t)$, depends on the blood perfusion rate, f, the arterial oxygenation input, H_0, and the vasculature coefficient of the tumor, γ.

Our NIR instrument indeed measures changes in vascular HbO concentration (i.e., $\Delta[HbO]^{vasculature}$). Then, Eq. (16) can be used to interpret the NIR measurements, since it gives the same exponential form as Eqs. (13) and (14), used to fit the experimental data. The comparison between Eqs. (13) and (16) reveals that the measured time constant is closely associated with the blood perfusion rate, f, and the vasculature coefficient of the tumor, γ in the measured volume. Furthermore, if the measured volume involves two distinct regions, then the detected signal will involve two different blood perfusion rates, f_1 and f_2, and/or two different vasculature coefficients, γ_1 and γ_2. In this case, it is reasonable to assume that the measured signal results from both of the regions, as illustrated in Fig. 12A. Consequently, Eq. (16) can be modified with a two-exponential expression and two time constants:

$$
\begin{aligned}
\Delta HbO^{vasculature}(t) &= \gamma_1 \times H_0 \times \left(1 - e^{-f_1 t/\gamma_1}\right) + \gamma_2 \times H_0 \times \left(1 - e^{-f_2 t/\gamma_2}\right) \\
&= A_1 \times \left(1 - e^{-f_1 t/\gamma_1}\right) + A_2 \times \left(1 - e^{-f_2 t/\gamma_2}\right)
\end{aligned}
\tag{17}
$$

where f_1 and γ_1 are the blood perfusion rate and vasculature coefficient from region 1, respectively, f_2 and γ_2 are from region 2, and $A_1 = \gamma_1 \times H_0$ and $A_2 = \gamma_2 \times H_0$. The two time constants are equal to $\tau_1 = \gamma_1/f_1$ and $\tau_1 = \gamma_2/f_2$, respectively. When A_1, A_2, and the two time constants are determined from our NIR measurements, we arrive at the ratios for the two vasculature coefficients and the two blood perfusion rates:

$$\frac{\gamma_1}{\gamma_2} = \frac{A_1}{A_2}; \quad \frac{f_1}{f_2} = \frac{A_1/A_2}{\tau_1/\tau_2} \tag{18}$$

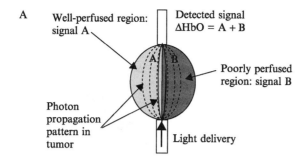

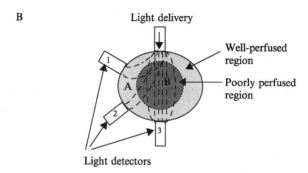

Fig. 12. (A) Schematic diagram to show a tumor model with two vascular perfusion regions, along with the light patterns propagating within the tumor tissue. "A" represents a portion of the detected signal interrogating the well-perfused region, and "B" represents another portion of the detected signal, mainly passing through the poorly perfused region. Our assumption is that the total detected signal is a sum of A and B. (B) A more realistic tumor tissue model with the center poorly perfused (leading to signal A) and the periphery well perfused (leading to signal B). It demonstrates possible tumor volumes interrogated by the multichannel detectors. The total detected signal is still a sum of A plus B.

In this way, by quantifying these two ratios, we are able to obtain insight into the tumor vasculature and blood perfusion. For example, a rather uniform vascular structure will result in a ratio of γ_1/γ_2 near 1, whereas the coexistence of two time constants possibly reveals two mechanisms of regional blood perfusion in the tumor. A large time constant implies a slow perfusion through a poorly perfused area, whereas a small time constant indicates a fast perfusion through a well-perfused area. In the meantime, we can determine the ratio of the perfusion rates in these two areas after obtaining the ratios of amplitudes and time constants from the two-exponential curve fitting. In this way, we have developed a hemodynamic model for the NIR data interpretation, which allows us to associate tumor blood oxygenation dynamics with regional blood perfusion and vascular structures of the tumor within the measured volume.

Discussions

NIRS is noninvasive and provides a real-time assessment of changes in tumor vascular hemoglobin oxygenation. In this chapter, we basically provide demonstration of the ease and utility of NIRS studies of tumors. By switching the inhaled gas from air to carbogen, the NIRS measurement produces a rapid biphasic elevation in $\Delta[\text{HbO}]$. The rapid time constant is in the range of seconds to a minute, whereas the slow component (10 to 50 times slower) continues for many minutes (see Figs. 7B and 8). The elevated oxygenation process is completely reversible upon returning air breathing, but still present 10 to 20 min after the baseline inhalation in many cases (see Figs. 7B and 8). The high reproducibility of results suggests that one can apply repeated interventions to explore the efficacy of interventions designed to alter tumor vascular oxygenation (e.g., vascular targeting agents). Our recent study has also shown that tumor response to oxygen was much more rapid and fit well to a monoexponential curve. For almost every tumor in a group of seven breast tumors, the time to reach 80% of maximum elevation in $\Delta[\text{HbO}]$ was longer with carbogen intervention than with oxygen breathing.[24]

With single-channel and multichannel NIRS, we have measured relative changes in $[\text{Hb}]_{\text{total}}$ and $[\text{HbO}]$ in breast and prostate rat tumors in response to hyperoxic respiratory intervention. We observed that respiratory challenge caused $\Delta[\text{HbO}]$ to rise promptly and significantly in both breast and prostate tumors. However, the total concentration of hemoglobin did not always behave consistently. The dynamic changes of tumor oxygenation can be modeled by a two-exponential expression with a fast and slow time constant. Based on the model, we suggest that the NIRS measurement can "see" two vascular mechanisms during tumor oxygenation under hyperoxic gas intervention. By fitting the two-exponential model with the NIR experimental data, we can determine the two time constants and their corresponding amplitudes, leading to the relationships between the two perfusion rates and between the vascular structures, as expressed in Eq. (18).

With this tumor model, we are able to obtain information on the blood perfusion of tumor: a large time constant usually represents a slow blood perfusion, whereas a small time constant indicates a fast blood perfusion. A combination of well-perfused and poorly perfused mechanisms in tumor vasculature will result in coexistence of two time constants. Indeed, some tumor lines have been reported with only 20–85% of vessels perfused,[61] and it is known that tumor structures and oxygen distribution are highly heterogeneous.[46,62] Therefore it is highly possible that our NIRS readings often detect a mixture of both well-perfused and poorly perfused regions in the tumor, depending on the locations of sources and detectors of the NIR system.

In the process of developing the model, we introduced a vasculature coefficient, γ. We expect that γ depends on (1) oxygen consumption and (2) capillary density of the tumor. Further studies are underway to quantify more about this coefficient and to confirm our speculations.

As an example to demonstrate the direct use of the newly developed model, the multichannel NIR results are fitted with the model, and the corresponding fitted parameters are listed in Table I. It is reasonable to expect that the central region of a solid tumor is close to necrosis (i.e., a poorly perfused region), whereas the tumor peripheral region is highly vascularized. Then, Fig. 12B can be used to schematically illustrate possible tumor volumes interrogated by the multichannel detectors. Close inspection of Table I reveals that the tumor structure ratios of γ_1/γ_2 from the three respective detectors are significantly different, whereas the perfusion ratios of f_1/f_2 from detectors 1 and 3 are almost doubled with respect to that from detector 2. Besides the results presented here, we have observed similar ΔHbO profiles and fitted parameters from other breast tumors. In such a way, we can obtain more details of tumor heterogeneity by having more source–detector pairs in the measurements. Indeed, we are now in the process of developing a multichannel NIR imaging system, along with the imaging reconstruction algorithm, so that tumor heterogeneity can be studied with better spatial resolution.

The time constants are not source–detector separation-sensitive. Equations (6), (7), and (11) demonstrate that ΔHbO and ΔHb are proportional to $1/d$, where d is the source–detector separation. This indicates that a different d value will only stretch or compress an entire temporal profile of ΔHbO, but it does not change the transient behavior of time response. The same argument can apply to DPF. Currently we have grouped DPF into the unit of [HbO] for simplicity. If the DPF value is larger than 1, the values of $\Delta[HbO]$ and $\Delta[Hb]_{total}$ will decrease by a factor of DPF. However, this decrease does not affect the time constants, τ_1 and τ_2 the dynamic responses of tumor $\Delta[HbO]$ to respiratory intervention.

[61] H. J. J. A. Bernsen, P. F. J. W. Rijken, T. Oostendorp, and A. J. van der Kogel, *Br. J. Cancer* **71,** 721 (1995).

[62] R. P. Mason, P. P. Antich, E. E. Babcock, A. Constantinescu, P. Peschke, and E. W. Hahn, *Int. J. Radiat. Oncol. Biol. Phys.* **29,** 95 (1994).

The simultaneous measurements of $\Delta[\text{HbO}]$ and ΔpO_2, shown in this chapter, demonstrate the compatibility of the NIRS system with the needle electrode and the FOXY fiber-optic oxygen-sensing system, without interference. All three systems are relatively inexpensive and provide real-time measurements, but the needle electrode and the three-channel FOXY system monitor pO_2 (viz., ΔpO_2) in specific locations, whereas the NIRS system provides global measurements. There are a few advantages of the FOXY system over the needle electrode: (1) it allows multiple locations to be interrogated simultaneously and (2) it is much easier to use than electrodes, particularly in calibration and stability. However, the FOXY probes may not provide accurate absolute pO_2 readings.

It is well known that measurements of tumor pO_2 have prognostic value in the clinic.[28,63] In this chapter, we have shown a linear correlation between tumor ΔHbO and mean ΔpO_2 (Fig. 10) when a tumor undergoes a hyperoxic respiratory intervention, and thus the noninvasive NIRS monitoring could have a potential value for clinical practice. The major deficiency in our current approach is lack of spatial resolution, so implementation of an NIR imaging system for tumor monitoring is our current effort.

NIRS provides a global assessment, in contrast to blood oxygenation level dependent (BOLD)-MRI, which can provide high-resolution images.[64,65] Although the latter approach can show heterogeneity in both temporal and spatial response, the results are often summarized to show mean values only. As such, an a priori global measurement can provide similar insight into dynamic tumor physiology or drug pharmacodynamics, while being more cost-effective, portable, and easier to implement and operate. For example, several groups of researchers have previously used NIR studies of tumors to observe the influence of chemotherapy,[52] pentobarbital overdose,[52] ischemic clamping,[66] and infusion of perfluorocarbon blood substitute.[67] All of these investigations demonstrate potential versatility of the NIRS application for diverse future studies.

Furthermore, the setup in our NIRS uses transmission mode between the source and detector, which we believe probes large and deep portions of the tumor, including the periphery on each side, as well as the center (see Fig. 3). This is in contrast to the setups used by most other investigators,

[63] M. Höckel and P. Vaupel, *J. Natl. Cancer Inst.* **93,** 266 (2001).

[64] S. P. Robinson, D. R. Collingridge, F. A. Howe et al., *NMR Biomed.* **12,** 98 (1999).

[65] X. Fan, J. N. River, M. Zamora et al., *Int. J. Radiat. Oncol. Biol. Phys.* **54,** 1202 (2002).

[66] F. Steinberg, H. J. Röhrborn, K. M. Scheufler, T. Otto, and C. Streffer, *Adv. Exp. Med. Biol.* **428,** 69 (1996).

[67] H. D. Sostman, S. Rockwell, A. L. Sylva, D. Madwed, G. Cofer, H. C. Charles, R. Negro-Villar, and D. Moore, *Magn. Reson. Med.* **20,** 253 (1991).

who apply reflectance mode,[21,66,68] which predominately detects the behavior of the peripheral vasculature. Indeed, the dynamic results measured from breast tumors reported by Hull et al.[21] are consistent with the fast component that we have observed and attributed to the well-perfused regions of our tumors.

Given the evidence for intratumoral heterogeneity from MRI[46,69] and histology,[70] it will be important to advance the NIRS system to an NIR imaging system so as to study not only dynamic, but also spatial aspects of blood oxygenation in tumor vasculature. In the meantime, we believe the preliminary results reported here are proof of principle for NIR imaging of tumor vascular oxygenation, laying a foundation for more extensive studies to correlate NIR imaging measurements with tumor heterogeneity and heterogeneous responses to various tumor therapeutic interventions and treatments.

Conclusions

In conclusion, we believe that NIRS presents a new potential imaging modality to examine tumor vasculature rapidly, noninvasively, and cost-effectively. Ease of implementation and operation permit rapid application to accessible tumors in cancer patients. The inherent compatibility of fiber-optics technology and light with other modalities, such as electrodes[20,24] and MRI,[71] will facilitate multiparametric multimodality investigations of tumor heterogeneity and vasculature in the near future.

In summary, we have demonstrated in this chapter that the NIR technology can provide an efficient, real-time, noninvasive means for monitoring vascular oxygenation dynamics in tumors during hyperoxic respiratory challenge. Concentration changes in HbO measured from both breast and prostate tumors often display a very prompt rise, followed by a gradual persistence throughout the intervention. By developing a hemoperfusion model with two exponential terms and fitting the model to the ΔHbO data, we have recognized two perfusion mechanisms in tumor vasculature and quantified the ratios of the two perfusion rates.

Furthermore, we have also obtained tumor pO_2 measurements using a needle electrode or multichannel, fiber-optic, FOXY probe in simultaneous conjunction with the noninvasive NIRS measurement. The comparative

[68] M. Kragh, B. Quistorff, and P. E. Kristjansen, Eur. J. Cancer 37, 924 (2001).

[69] S. P. Robinson, F. A. Howe, L. M. Rodrigues, M. Stubbs, and J. R. Griffiths, Semin. Radiat. Oncol. 8, 198 (1998).

[70] B. M. Fenton, Radiother. Oncol. 44, 183 (1997).

[71] G. Gulsen, H. Yu, J. Wang, O. Nalcioglu, S. Merritt, F. Bevilacqua, A. J. Durkin, D. J. Cuccia, R. Lanning, and B. J. Tromberg, Technol. Cancer Res. Treatment 1, 497 (2002).

results exhibit a linear correlation between ΔHbO and ΔpO_2 of the tumors under hyperoxic gas intervention, suggesting that the NIRS approach could have a good potential value in the clinic. Finally, the newly developed tumor hemodynamic model allows us to reveal tumor heterogeneities at different tumor locations based on the multichannel NIRS results. Through this chapter, we lay a foundation for an NIR imaging technique to be further developed to facilitate investigations of tumor heterogeneity and vascular perfusion. Such a noninvasive imaging approach can enhance our understanding of the dynamics of tumor oxygenation and the mechanism of tumor physiology under baseline and perturbed conditions.

Acknowledgments

This work was supported in part by the Department of Defense Breast Cancer Research grants BC990287 (HL) and BC000833 (YG), and NIH R01 CA79515 (NCI)/EB002762 (NIBIB) (RPM). We are grateful to Vincent Bourke for his collaborative work on multichannel pO_2 measurements and Dr. Anca Constantinescu for her assistance with all the tumor investigations. We also gratefully acknowledge Dr. Britton Chance for his technical support on the multichannel NIR system.

[18] Measuring Changes in Tumor Oxygenation

By DAWEN ZHAO, LAN JIANG, and RALPH P. MASON

Introduction

Significance of pO$_2$ in Oncology

It has long been appreciated that hypoxic tumor cells are more resistant to radiotherapy.[1] Indeed, a 3-fold increase in radio resistance may occur when cells are irradiated under hypoxic conditions compared with oxygen pressure $pO_2 > 15$ torr for a given single radiation dose. However, recent modeling has indicated that the proportion of cells in the range 0–20 torr may be most significant in terms of surviving a course of fractionated radiotherapy.[2] Certain chemotherapeutic drugs also present differential efficacy, depending on hypoxia.[3,4] Increasingly, there is evidence that hypoxia also

[1] L. Gray, A. Conger *et al.*, *Br. J. Radiol.* **26,** 638 (1953).
[2] B. G. Wouters and J. M. Brown, *Radiat. Res.* **147,** 514 (1997).
[3] B. Teicher, J. Lazo *et al.*, *Cancer Res.* **41,** 73 (1981).
[4] A. C. Sartorelli, *Cancer Res.* **48,** 775 (1988).

Copyright 2004, Elsevier Inc.
All rights reserved.
0076-6879/04 $35.00

influences such critical characteristics as angiogenesis, tumor invasion, and metastasis.[5–8] Moreover, repeated bouts of intermittent hypoxic stress may be important in stimulating tumor progression.[9] Thus the ability to measure pO_2 noninvasively and repeatedly, with respect to acute or chronic interventions, becomes increasingly important.

Early work examined cells *in vitro*, where ambient oxygen concentrations are readily controlled. *In vivo*, hypoxia may be achieved by clamping the blood supply to a tumor,[10] but other levels of oxygenation reflect the interplay of supply and consumption.[11,12] Robust fine-needle polarographic electrodes opened the possibility of measuring pO_2 in tumors *in situ* and *in vivo* to define local pO_2 under baseline conditions or with respect to interventions. In early work, Cater and Silver[13] showed the ability to monitor pO_2 at individual locations in patients' tumors with respect to breathing oxygen. Later, Gatenby *et al.*[14] showed that pO_2 in a tumor was correlated with clinical outcome. Tumor oximetry received its greatest boost with the development of the Eppendorf Histograph polarographic needle electrode system.[15] This computer-controlled device equipped with a stepper motor can reveal distributions of tumor oxygenation and has been applied extensively to clinical trials. Many reports have now shown that tumors are highly heterogeneous and have extensive hypoxia; furthermore, strong correlations have been shown in cervix and head and neck tumors between median pO_2 or hypoxic fraction and survival or disease-free survival.[5,16–20] Extensive hypoxia also has been found in tumors of the prostate and breast.[21–23] Thus tumor oxygenation is now recognized as a strong

[5] E. K. Rofstad, K. Sundfor *et al.*, *Br. J. Cancer* **83,** 354 (2000).

[6] K. De Jaeger, M. C. Kavanagh *et al.*, *Br. J. Cancer* **84,** 1280 (2001).

[7] M. Höckel and P. Vaupel, *J. Natl. Cancer Inst.* **93,** 266 (2001).

[8] H. J. Knowles and A. L. Harris, *Breast Cancer Res.* **3,** 318 (2001).

[9] R. A. Cairns, T. Kalliomaki *et al.*, *Cancer Res.* **61,** 8903 (2001).

[10] J. Moulder and S. Rockwell, *Int. J. Radiat. Oncol. Biol. Phys.* **10,** 695 (1984).

[11] T. W. Secomb, R. Hsu *et al.*, *Adv. Exp. Med. Biol.* **454,** 629 (1998).

[12] M. W. Dewhirst, B. Klitzman *et al.*, *Int. J. Cancer* **90,** 237 (2000).

[13] D. Cater and I. Silver, *Acta Radiol.* **53,** 233 (1960).

[14] R. A. Gatenby, H. B. Kessler *et al.*, *Int. J. Radiat. Oncol. Biol. Phys.* **14,** 831 (1988).

[15] M. Nozue, I. Lee *et al.*, *J. Surg. Oncol.* **66,** 30 (1997).

[16] D. M. Brizel, S. P. Scully *et al.*, *Cancer Res.* **56,** 941 (1996).

[17] M. Höckel, K. Schlenger *et al.*, *Cancer Res.* **56,** 4509 (1996).

[18] M. Nordsmark, M. Overgaard *et al.*, *Radiother. Oncol.* **41,** 31 (1996).

[19] A. W. Fyles, M. Milosevic *et al.*, *Radiother. Oncol.* **48,** 149 (1998).

[20] T. H. Knocke, H. D. Weitmann *et al.*, *Radiother. Oncol.* **53,** 99 (1999).

[21] P. W. Vaupel, K. Schlenger *et al.*, *Cancer Res.* **51,** 3316 (1991).

[22] P. Hohenberger, C. Felger *et al.*, *Breast Cancer Res. Treat.* **48,** 97 (1998).

[23] B. Movsas, J. D. Chapman *et al.*, *Urology* **53,** 11 (1999).

prognostic indicator, and this device has laid a convincing foundation for the value of measuring pO_2 in patients. However, the Histograph is highly invasive, and it is not possible to make repeated measurements at individual locations, precluding dynamic studies to assess the influence of interventions on tumor pO_2.

Given that hypoxic tumors are more resistant to certain therapies, it becomes important to assess tumor oxygenation as part of therapeutic planning. Patients could be stratified according to baseline hypoxia to receive adjuvant interventions designed to modulate pO_2, or more intense therapy as facilitated by intensity modulated radiation therapy (IMRT). Tumors, which do not respond to interventions, may be ideal candidates for hypoxia-selective cytotoxins (e.g., tirapazamine[24]). Noting that any therapy and intervention may have side effects or simply add to clinical costs, it is vital that efficacy be established and therapy be optimized for an individual patient. Whether initially hypoxic regions of a tumor can be modified to become better oxygenated has long been considered a key to improving outcome of irradiation. However, many attempts to improve therapeutic outcome by manipulation of tumor oxygenation have shown only modest success in the clinic,[25] and it is thought that lack of success may have resulted from inability to identify those patients who would benefit from adjuvant interventions.

Although pO_2 determinations could be of great clinical value, they are also vital to many laboratory investigations of new drugs and studies of tumor development. Given the potential importance of measuring pO_2, many diverse techniques have been developed, as reviewed by others previously,[26–29] and here, in the next section.

Methods of Measuring Tumor Oxygenation

Table I lists various techniques that have been reported to provide quantitative estimates of pO_2. Historically, polarographic needle oxygen electrodes have been considered a "gold standard," and they have been applied in the clinic since the 1950s. One or more electrodes may be placed in a tumor, facilitating measurement of baseline pO_2 and dynamic response to

[24] J. M. Brown, *Mol. Med. Today* **6**, 157 (2000).

[25] J. Overgaard and M. R. Horsman, *Semin. Radiat. Oncol.* **6**, 10 (1996).

[26] H. B. Stone, J. M. Brown et al., *Radiat. Res.* **136**, 422 (1993).

[27] R. P. Mason, S. Ran et al., *J. Cell. Biochem.* **87S**, (2002).

[28] H. M. Swartz, *Biochem. Soc. Trans.* **30**, 248 (2002).

[29] H. M. Swartz and J. F. Dunn, *in* "Oxygen Transport to Tissue XXIV" (J. F. Dunn and H. M. Swartz, eds.), Vol. 530, p. 1. Kluwer Academic, New York, 2003.

interventions.[13,30–32] Initially, the focus was on generating finer needles, which would be less invasive, and tips as fine as a few microns have been applied to animal tissues.[33] However, such needles are progressively brittle and generate such small current that stray electromagnetic fields can interfere. Stationary electrodes sample limited volumes, and recognizing tumor heterogeneity, the Eppendorf Histograph was developed to generate multiple measurements along tracks in tumors.[15,34,35] Following extensive studies in animals, the Histograph has found widespread application in the clinical setting and has unequivocally revealed hypoxia in many tumor types, for example, head and neck,[36,37] cervix,[19,38] breast,[21,22] and prostate.[23] Moreover, pO_2 distributions have been found to have prognostic value. Disease-free survival is significantly worse for patients with hypoxic tumors, though the optimal prognostic parameter has variously been median pO_2 or percent measurements <5 torr (HF_5).

Although the Eppendorf Histograph uses a large invasive needle (size = 26 G or about 0.35 mm), it has provided great impetus for further investigations. One aspect is the application of less invasive probes. Fiber-optic probes are typically finer and do not consume oxygen during measurement. Typically, only two or four locations are sampled simultaneously, but as with the earlier electrodes, these optical probes facilitate observation of dynamic changes in pO_2 in response to interventions.[39–43] Both the current commercial systems, the OxyLite (http://www.oxford-optronix.com/tissmon/oxylite/oxylite.htm) and FOXY (http://www.oceanoptics.com/Products/foxyfaqs.asp), exploit the fluorescent quenching by oxygen of a ruthenium complex coating. OxyLite measures fluorescent lifetime, whereas FOXY uses a simple intensity integration and is correspondingly much cheaper. Fibers are fragile, and coatings have a limited lifetime.

[30] N. Evans and P. Naylor, Br. J. Radiol. 36, 418 (1963).

[31] C. Song, I. Lee et al., Cancer Res. 47, 442 (1987).

[32] D. Zhao, A. Constantinescu et al., Int. J. Radiat. Oncol. Biol. Phys. 53, 744 (2002).

[33] D. W. Crawford and M. A. Cole, J. Appl. Physiol. 58, 1400 (1985).

[34] F. Kallinowski, R. Zander et al., Int. J. Radiat. Oncol. Biol. Phys. 19, 953 (1990).

[35] P. W. Vaupel, D. K. Kelleher, and M. Günderoth, "Tumor Oxygenation." Gustav Fischer Verlag, Stuttgart, Germany, 1995.

[36] D. M. Brizel, G. S. Sibly et al., Int. J. Radiat. Oncol. Biol. Phys. 38, 285 (1997).

[37] V. Rudat, B. Vanselow et al., Radiother. Oncol. 57, 31 (2000).

[38] C. Aquino-Parsons, A. Green et al., Radiother. Oncol. 57, 45 (2000).

[39] J. R. Griffiths, Br. J. Radiol. 72, 627 (1999).

[40] J. Bussink, J. H. A. M. Kaanders et al., Radiat. Res. 154, 547 (2000).

[41] R. D. Braun, J. L. Lanzen et al., Am. J. Physiol. Heart Circ. Physiol. 280, H2533 (2001).

[42] D. Zhao, A. Constantinescu et al., Am. J. Clin. Oncol. 24, 462 (2001).

[43] Y. Gu, V. Bourke et al., Appl. Opt. 42, 1 (2003).

TABLE I
TUMOR OXIMETRY METHODS

Technique	Reporter	Parameter measured	Invasiveness	Spatial	Temporal	References
				Characteristic resolution		
FREDOM	HFB	R_1	Minimal 32-G needle	Map multiple locations each 8 mm^3	6.5 min	Hunjan, Zhao et al.[159]; Zhao, Constantinescu et al.[134]; Song, Constantinescu et al.[135]; Zhao, Constantinescu et al.[32]; Kim, Zhao et al.[106]; Zhao, Constantinescu et al.[136]; Zhao, Ran et al.[137]
^{19}F MRI	PFC	R_1	IV	Perfused regions	min	Hees and Sotak[153]; Dardzinski and Sotak[117]; McIntyre, McCoy et al.[158]; Fan, River et al.[95]; Wang, Su et al.[88]
^{19}F MRS	PFC	R_1	IV	Perfused regions	s to min	Hees and Sotak[153]; Mason, Antich et al.[154]; Baldwin and Ng[156]; McIntyre, McCoy et al.[158]; van der Sanden, Heerschap et al.[138]
DCE MRI	Gd-DTPA	Contrast kinetics	IV	Maps	min	Cooper, Carrington et al.[87]; Lyng, Vorren et al.[a]; Wang, Su et al.[88]
ESR/EPR	Charcoal, phthalocyanine	Linewidth	Needle	Single location 23 G IT	seconds	O'Hara, Goda et al.[b]; Goda, Bacic et al.[c]; O'Hara, Goda et al.[59]; Gallez, Jordan et al.[d];

ESR/EPR	Nitroxides	Linewidth	IV	Global or map mm^3 maps	s to min	Jordan, Misson et al.[48]; He, Beghein et al.[49]; Jiang, Beghein et al.[50]; O'Hara, Blumenthal et al.[e]; Baudelet and Gallez[94]; Dunn, O'Hara et al.[f]; Jordan, Gregoire et al.[52]; Mahy, De Bast et al.[55]
OMRI	Free radical	Overhauser enhancement	IV		10 min	Elas, Williams et al.[53]; Krishna, English et al.[60]
Needle electrode	Oxygen	Current	Needle 26 G IT	Single location	~1 s	Cater and Silver[13]; Evans and Naylor[30]; Hasegawa, Rhee et al.[g]; Gatenby, Kessler et al.[14]; Song, Shakil et al.[h]; Zhao, Constantinescu et al.[32]
(Histograph)	Oxygen	Current	Needle 26 G IT	Multiple tracks	1 s per location	Eble, Wenz et al.[i]; Falk, Laurence et al.[j]; Vaupel, Kelleher et al.[35]; Brizel, Scully et al.[16]; Höckel, Schlenger et al.[k]; Nozue, Lee et al.[15]; Fyles, Milosevic et al[19]; Siemann, Johansen et al.[l]; Mason, Constantinescu et al.[175]; Aquino-Parsons, Green et al.[38]; Jenkins, Evans et al.[75]; Höckel and Vaupel[7]

(continued)

TABLE I (continued)

Technique	Reporter	Parameter measured	Invasiveness	Characteristic resolution		References
				Spatial	Temporal	
Optical probe (OxyLite, FOXY)	Rh complex	Fluorescent lifetime	Needle (26 G)	2–4 locations	Real time	Griffiths[39], Bussink, Kaanders et al.[40]; Braun, Lanzen et al.[41]; Zhao, Constantinescu et al.[134]; Gu, Bourke et al.[43]; Jordan, Beghein et al.[54]
Phosphorescence	PD complex	Lifetime	IV	Maps	<1 min	Wilson[m]; Vinogradov, Lo et al.[n]; Dewhirst, Ong et al.[o]; Wilson, Vinogradov et al.[63]; Erickson, Braun et al.[p]
Fluorescence	EF5	Fluorescent intensity	IV + biopsy	Maps microscopic	Once	Koch[79]
Mass spectrometry	Oxygen	Atoms	Needle	Single location		Potapov, Sirovskii et al.[q]

[a] H. Lyng, A. O. Vorren et al., J. Magn. Reson. Imaging **14**, 750 (2001).

[b] J. A. O'Hara, F. Goda et al., Radiat. Res. **144**, 222 (1995).

[c] F. Goda, G. Bacic et al., Cancer Res. **56**, 3344 (1996).

[d] B. Gallez, B. F. Jordan et al., Magn. Reson. Med. **42**, 627 (1999).

[e] J. A. O'Hara, R. D. Blumenthal et al., Radiat. Res. **155**, 466 (2001).

[f] J. F. Dunn, J. A. O'Hara et al., J. Magn. Reson. Imaging **16**, 511 (2002).

[g] T. Hasegawa, J. G. Rhee et al., Int. J. Radiat. Oncol. Biol. Phys. **13**, 569 (1987).

[h] C. W. Song, A. Shakil et al., Int. J. Hyperthermia **12**, 367 (1996).

[i] M. J. Eble, F. Wenz et al., in "Tumor Oxygenation" (P. W. Vaupel, D. K. Kelleher, and M. Günderoth, eds.), p. 95. Gustav Fischer, Stuttgart, Germany, 1995.

[j] S. Falk, V. Laurence et al., in "Tumor Oxygenation" (P. W. Vaupel, D. K. Kelleher, and M. Günderoth, eds.), p. 281. Gustav Fischer, Stuttgart, Germany, 1995.

[k] M. Höckel, K. Schlenger et al., Semin. Radiat. Oncol. **6**, 3 (1996).

[l] D. W. Siemann, I. M. Johansen et al., Int. J. Radiat. Oncol. Biol. Phys. **40**, 1171 (1998).

[m] D. F. Wilson, in "Oxygen Transport to Tissue XIV" (W. Erdmann and D. F. Bruley, eds.), p. 195. Plenum Press, New York, 1992.

[n] S. A. Vinogradov, L.-W. Lo et al., Biophys. J. **70**, 1609 (1996).

[o] M. W. Dewhirst, E. T. Ong et al., Br. J. Cancer **79**, 1717 (1999).

[p] K. Erickson, R. D. Braun et al., Cancer Res. **63**, 4705 (2003).

[q] A. A. Potapov, E. B. Sirovskii et al., Vopre. Neirokhir. **1**, 20 (1979).

Reporter molecules have been developed for use with electron spin resonance (ESR or EPR), where the line width is highly sensitive to oxygen.[28,44–55] Two primary approaches are used: (1) direct intratumoral (IT) injection of char crystals,[49,56] phthalocyanine,[57] or India ink[58] into a tissue or (2) intravenous (IV) infusion of water-soluble agents which disseminate throughout the tumor vasculature.[47,53] Direct IT injection is invasive and has generally been applied as a spectroscopic approach to report pO$_2$ at single locations only. Nonetheless, significant data have been achieved demonstrating hypoxiation and reoxygenation with respect to irradiation, and the importance of timing successive radiation doses to coincide with reoxygenation.[59] Char particles may be stable in tissue for weeks to years, allowing measurements of chronic changes in tissues (e.g., accompanying tumor growth).[28] The IV approach is noninvasive, but reporter molecules may predominately distribute in the well-perfused vasculature, potentially biasing measurements toward the well-oxygenated tumor regions. Progressive uptake and clearance of agents produces variable concentrations, and some agents degrade in tissue requiring appropriate correction factors.[47] Nonetheless, images of tumor oxygen distribution have been reported, including three-dimensional representations.[53] Spin radicals also may be applied to a combined ESR–NMR (nuclear magnetic resonance) approach, Overhauser-enhanced magnetic resonance imaging (OMRI), exploiting the Overhauser enhancement in the tissue water proton MRI signal that occurs by polarization transfer from free radicals upon electromagnetic irradiation.[60]

[44] H. M. Swartz, S. Boyer et al., Magn. Reson. Med. 20, 333 (1991).

[45] H. M. Swartz, S. Boyer et al., in "Oxygen Transport to Tissue XIV" (W. Erdmann and D. F. Bruley, eds.), p. 221. Plenum Press, New York, 1992.

[46] H. M. Swartz, K. J. Liu et al., Magn. Reson. Med. 31, 229 (1994).

[47] P. Kuppusamy, R. Afeworki et al., Cancer Res. 58, 1562 (1998).

[48] B. F. Jordan, P. Misson et al., Int. J. Radiat. Oncol. Biol. Phys. 48, 565 (2000).

[49] J. He, N. Beghein et al., Magn. Reson. Med. 46, 610 (2001).

[50] H. Jiang, N. Beghei et al., Phys. Med. Biol. 46, 3323 (2001).

[51] P. E. James and H. M. Swartz, Methods Enzymol. 350, 52 (2002).

[52] B. F. Jordan, V. Gregoire et al., Cancer Res. 62, 3555 (2002).

[53] M. Elas, B. B. Williams et al., Magn. Reson. Med. 49, 682 (2003).

[54] B. F. Jordan, N. Beghein et al., Int. J. Cancer 103, 138 (2003).

[55] P. Mahy, M. De Bast et al., Radiother. Oncol. 67, 53 (2003).

[56] N. Vahidi, R. B. Clarkson et al., Magn. Reson. Med. 31, 139 (1994).

[57] J. F. Glockner and H. M. Swartz, in "Oxygen Transport to Tissue XIV" (W. Erdmann and D. F. Bruley, eds.), p. 229. Plenum Press, New York, 1992.

[58] F. Goda, K. Jian Lu et al., Magn. Reson. Med. 33, 237 (1995).

[59] J. A. O'Hara, F. Goda et al., Radiat. Res. 150, 549 (1998).

[60] M. C. Krishna, S. English et al., Proc. Natl. Acad. Sci. USA 99, 2216 (2002).

Vascular oxygenation has been probed by fluorescence or phosphorescence imaging based on reporter complexes delivered IV.[61,62] Historically, the approach was limited to superficial tissues due to limited light penetration. The latest molecules are active in the near-infrared, permitting greater depth of signal penetration.[63]

NMR facilitates interrogation of deep tissues noninvasively, and ^{19}F NMR approaches will be reviewed in detail in the following section. The methods discussed earlier provide direct quantitative measurements of pO_2 based on various physiochemical parameters, such as electric current, fluorescent lifetime, magnetic resonance linewidth, or relaxation. Other approaches are less direct, but can reveal hypoxia or correlates of pO_2.

Specific classes of reporter molecules have been developed to reveal hypoxia[26,64] (e.g., pimonidazole,[65,66] EF5,[67,68] CCl-103F,[69] Cu-ATSM[70,71] galactopyranoside IAZA[72]). Following IV infusion, these agents become reduced in tissues and are trapped. However, in the presence of oxygen they are reoxidized and ultimately clear from the body. Histologic assessment of the distribution of these agents provides microscopic indications of local hypoxia. EF5, pimonidazole, and Cu-ATSM are currently being tested in clinical trials, and correlations have been reported with clinical outcome.[66,67,70] Many variants have been proposed over the past 20 years, and incorporation of radionuclides has facilitated noninvasive investigations using positron emission tomography (PET) or single photon emission computed tomography (SPECT), while ^{19}F labels permitted NMR spectroscopy.[72–74] Generally, only a single time point is investigated, but dynamic variations in hypoxia may be assessed, even in biopsy specimens, by applying pairs of hypoxia reporters in a pulse-chase fashion with respect to an intervention, as shown by Ljungkvist et al.[69]

[61] D. Wilson and G. Cerniglia, *Cancer Res.* **52,** 3988 (1992).

[62] G. Helminger, F. Yuan et al., *Nature Med.* **3,** 177 (1997).

[63] D. F. Wilson, S. A. Vinogradov et al., *Comp. Biochem. Physiol., Part A Mol. Integr. Physiol.* **132,** 153 (2002).

[64] J. R. Ballinger, *Semin. Nucl. Med.* **31,** 321 (2001).

[65] J. A. Raleigh, S. C. Chou et al., *Radiat. Res.* **151,** 580 (1999).

[66] J. H. A. M. Kaanders, K. I. E. M. Wijffels et al., *Cancer Res.* **62,** 7066 (2002).

[67] S. M. Evans, S. Hahn et al., *Cancer Res.* **60,** 2018 (2000).

[68] W. R. Dolbier, Jr., A. R. Li et al., *Appl. Radiat. Isotop.* **54,** 73 (2001).

[69] A. S. E. Ljungkvist, J. Bussink et al., *Int. J. Radiat. Oncol. Biol. Phys.* **48,** 1529 (2000).

[70] F. Dehdashti, P. W. Grigsby et al., *Int. J. Radiat. Oncol. Biol. Phys.* **55,** 1233 (2003).

[71] F. Dehdashti, M. A. Mintun et al., *Eur. J. Nucl. Med. Molec. Imaging* **30,** 844 (2003).

[72] J. D. Chapman, E. L. Engelhardt et al., *Radiother. Oncol.* **46,** 229 (1998).

[73] R. J. Maxwell, P. Workman et al., *Int. J. Radiat. Oncol. Biol. Phys.* **16,** 925 (1989).

[74] R. J. Hodgkiss, *Anticancer Drug Des.* **13,** 687 (1998).

Several studies have shown a lack of correlation between hypoxic marker binding and pO_2 assessed using the Eppendorf Histograph, which may be related to chronic versus acute hypoxia, or the extent of necrosis.[75–77] The ultimate value of the techniques is evidenced by correlations between uptake and outcome.[66,70,71,78] Recent data also indicate that EF5 fluorescence may be correlated with pO_2.[79]

The techniques discussed so far all depend on exogenous reporter molecules or probes. Ideally, oxygenation could be related to endogenous characteristics. Because many biochemical pathways are under oxygen regulation, they can provide an elegant window on hypoxia, for example, induction of hypoxia-inducible factor 1 (HIF-1) and glucose transporter 1 (Glut-1) together with secondary responses, such as increased production of vascular endothelial growth factor (VEGF), NIP3 and tumor-associated macrophage activity.[8] Such molecules indicate hypoxia, though they may be induced by other factors. Intrinsic radiation sensitivity also may be assessed using the Comet assay.[80] These assays each require biopsy. Other markers potentially associated with hypoxia may be found in the plasma or urine and have been correlated with clinical outcome.[81] An attractive alternative is the introduction of transgenes with hypoxic response elements (HREs) as promoter sequences coupled to reporter genes such as GFP (green fluorescent protein)[82,83] or luciferase.[84,85] GFP synthesis is an energetic process, which could be hindered under hypoxia conditions. Likewise, bioluminescence accompanying action of luciferase on luciferin requires adenosine triphosphate (ATP) and O_2, but reports suggest that even under exceedingly low pO_2, sufficient oxygen remains to reveal hypoxia.

Many practical considerations govern clinical application of oximetry methods. Proton MRI is routinely applied for anatomic evaluation of tumors and would provide an ideal conduit for prognostic investigations. Application of contrast agents may reveal tumor boundaries to enhance detectability, and the dynamic contrast enhancement (DCE) changes provide insight into vascular perfusion and surface permeability area.[86]

[75] W. T. Jenkins, S. M. Evans et al., Int. J. Radiat. Oncol. Biol. Phys. **46**, 1005 (2000).
[76] P. L. Olive, J. P. Banath et al., Acta Oncol. **40**, 917 (2001).
[77] M. Nordsmark, J. Loncaster et al., Radiother. Oncol. **67**, 35 (2003).
[78] C. J. Koch, S. M. Hahn et al., Cancer Chemother. Pharmacol. **48**, 177 (2001).
[79] C. J. Koch, Methods Enzymol. **352**, 3 (2002).
[80] P. L. Olive, P. J. Johnston et al., Nature Med. **4**, 103 (1998).
[81] Q. T. Le, P. D. Sutphin et al., Clin. Cancer Res. **9**, 59 (2003).
[82] Y. Cao, C. Li et al., "A Study of Hypoxia-Induced Gene Expression in Human Tumors." 48th Annual Meeting of Radiation Research, San Juan, Puerto Rico, 2001.
[83] D. Vordermark, T. Shibata et al., Neoplasia **3**, 527 (2001).
[84] T. Shibata, A. J. Giaccia et al., Gene Ther. **7**, 493 (2000).
[85] E. Payen, M. Bettan et al., J. Gene Med. **3**, 498 (2001).
[86] A. R. Padhani, J. Magn. Reson. Imaging **16**, 407 (2002).

Specific studies have shown a correlation between DCE and pO_2,[87] and indeed, a theoretical underpinning has been provided based on the Krogh cylinder model.[88] However, the correlation is unlikely to be widely applicable, since DCE is sensitive to vascular flow, perfusion, and permeability, where pO_2 depends on oxygen consumption as well as delivery.

Blood oxygen level dependent (BOLD) contrast proton NMR facilitates rapid interrogation of vascular oxygenation and is particularly appropriate for examining dynamic responses to interventions.[48,89–91] Deoxyhemoglobin is paramagnetic and induces signal loss in T_2^*-weighted images. However, BOLD does not provide absolute pO_2 values and is confounded by the influence of blood flow, as investigated extensively by Howe et al.,[92] who termed the expression *FLOOD* (flow and oxygen level dependent) contrast. In addition, variation in vascular volume can introduce signal perturbation.[93] Nonetheless, some studies have indicated a correlation with relative pO_2, but poor indication of absolute pO_2.[53,94,95]

Near-infrared spectroscopy (NIRS) offers an alternative approach based on the differential light absorption of the strong chromophores oxyhemoglobin and deoxyhemoglobin. NIRS provides a noninvasive means to monitor global tumor vascular oxygenation in real time based on endogenous molecules. Although many NIRS investigations have been conducted in the brain and breast in both laboratory and clinical settings over the past decade, there have been relatively few reports regarding solid tumors.[43,96–106] Most studies to date have used reflectance mode. By contrast, we have favored transmission mode, so as to interrogate deep tumor regions, and we have presented preliminary studies in rat breast

[87] R. A. Cooper, B. M. Carrington *et al.*, *Radiother. Oncol.* **57**, 53 (2000).
[88] Z. Wang, M.-Y. Su *et al.*, *Technol. Cancer Res. Treat.* **1**, 29 (2002).
[89] H. A. Al-Hallaq, J. N. River *et al.*, *Int. J. Radiat. Oncol. Biol. Phys.* **41**, 151 (1998).
[90] S. P. Robinson, F. A. Howe *et al.*, *Semin. Radiat. Oncol.* **8**, 198 (1998).
[91] R. Mazurchuk, R. Zhou *et al.*, *Magn. Reson. Imaging* **17**, 537 (1999).
[92] F. A. Howe, S. P. Robinson *et al.*, *Magn. Reson. Imaging* **17**, 1307 (1999).
[93] F. A. Howe, S. P. Robinson *et al.*, *NMR Biomed.* **14**, 497 (2001).
[94] C. Baudelet and G. Gallez, *Magn. Reson. Med.* **48**, 980 (2002).
[95] X. Fan, J. N. River *et al.*, *Int. J. Radiat. Oncol. Biol. Phys.* **54**, 1202 (2002).
[96] H. D. Sostman, S. Rockwell *et al.*, *Magn. Reson. Med.* **20**, 253 (1991).
[97] R. G. Steen, K. Kitagishi *et al.*, *J. Neurooncol.* **22**, 209 (1994).
[98] F. Steinberg, H. J. Rohrborn *et al.*, *Adv. Exp. Med. Biol.* **428**, 553 (1997).
[99] E. L. Hull, D. L. Conover *et al.*, *Br. J. Cancer* **79**, 1709 (1998).
[100] B. J. P. van der Sanden, A. Heerschap *et al.*, *Magn. Reson. Med.* **42**, 490 (1999).
[101] H. Liu, Y. Song *et al.*, *Appl. Op.* **39**, 5231 (2000).
[102] M. Kragh, B. Quistorff *et al.*, *Eur. J. Cancer* **37**, 924 (2001).
[103] G. Gulsen, H. Yu *et al.*, *Technol. Cancer Res. Treat.* **1**, 497 (2002).
[104] E. L. Heffer and S. Fantini, *Appl. Opt.* **41**, 3827 (2002).
[105] T. O. McBride, B. W. Pogue *et al.*, *J. Biomed. Opt.* **7**, 72 (2002).
[106] J. G. Kim, D. Zhao *et al.*, *J. Biomed. Optics* **8**, 53 (2003).

and prostate tumors with respect to various interventions.[43,101,106] NIR approaches are presented in detail in Chapter 17 of this volume.

Each technique has specific virtues and drawbacks, which must be considered for any given application, particularly the degree of invasiveness, the ability to generate maps of heterogeneity, and the ability to assess dynamic changes. In addition, the location of a measurement (e.g., vascular versus tissue compartments, the precision of measurements, and spatial and temporal resolution) must be considered. For further details of the techniques described earlier, the reader is referred to the references. In the next section, we present ^{19}F NMR approaches in greater detail.

^{19}F NMR Approaches to Measuring pO_2

Nuclear magnetic resonance (NMR) is attractive because it is inherently noninvasive. Liquid-state NMR is characterized by several parameters, including signal amplitude, chemical shift (δ), spin–spin relaxation (T_2), and spin–lattice relaxation (T_1). Oxygen could be quantified using ^{17}O NMR, but this is rather esoteric.[107] Alternatively, it has long been recognized that the oxygen molecule (O_2) is paramagnetic, causing increased spin–lattice relaxation rates ($R_1 = 1/T_1$). Indeed, physical and theoretical chemists must go to great lengths to rigorously remove oxygen from solutions (using freeze–thaw procedures) to achieve inherent relaxation rates for studying nuclear interactions.[108] Proton NMR studies have reported changes in the water relaxation rate as a result of tissue oxygenation,[109] but many other processes (metal ions, cellularity, pH, ionic strength) also cause relaxation, and thus it is not suitable for detecting pO_2, except under rare circumstances, such as with the eye.[110] There is also a substantial temperature response, whereas the relaxivity due to oxygen is only 0.0002 s^{-1}/torr.[110]

However, several investigators showed that the ^{19}F NMR spin–lattice relaxation rates for fluorocarbons are much more sensitive to pO_2.[111,112] Thomas et al.[113] pioneered the application of ^{19}F NMR relaxometry to measure pO_2 in tissues, in vivo, including lung, liver, and spleen; several other investigators demonstrated feasibility and applications,[114–120] as

[107] X. H. Zhu, H. Merkle et al., Magn. Reson. Med. **45,** 543 (2001).

[108] M. A. Hamza, G. Serratice et al., J. Magn. Reson. **42,** 227 (1981).

[109] E. Tadamura, H. Hatabu et al., J. Magn. Reson. Imaging **7,** 220 (1997).

[110] B. A. Berkowitz, C. McDonald et al., Magn. Reson. Med. **46,** 412 (2001).

[111] J.-J. Delpuech, M. A. Hamza et al., J. Chem. Phys. **70,** 2680 (1979).

[112] M. A. Hamza, G. Serratice et al., J. Am. Chem. Soc. **103,** 3733 (1981).

[113] S. R. Thomas, in "Magnetic Resonance Imaging" (C. L. Partain, R. R. Price, J. A. Patton, M. V. Kulkarni, and A. E. J. James, eds.), Vol. 2, p. 1536. W. B. Saunders, London, 1988.

[114] J. E. Fishman, P. M. Joseph et al., Magn. Reson. Imaging **5,** 279 (1987).

reviewed some years ago by Mason.[121] The ^{19}F NMR R_1 of perfluorocarbons (PFCs) varies linearly with pO_2,[113,122] and each resonance is sensitive to pO_2, temperature, and magnetic field, but importantly, is essentially unresponsive to pH, CO_2, charged paramagnetic ions, mixing with blood, or emulsification.[123–127]

A particular PFC molecule may have multiple resonances, and each resonance has a characteristic R_1 response to pO_2. This is attributed to steric effects of O_2, as it approaches the molecule,[112] which implies that perfluorinated groups, which are both geometrically and magnetically comparable, should have similar R_1 responses to oxygen tension. At a fixed temperature and magnetic-field strength, the R_1 response to pO_2 of any single resonance obeys the simple formula

$$R_1 = R_{1a} + (R_{1p}X) \tag{1}$$

where X is the mole fraction of O_2 dissolved in the PFC, R_{1a} is the anoxic relaxation rate, and R_{1p} is the relaxation rate due to the paramagnetic contribution of oxygen. According to Henry's law, the dissolved mole fraction is related directly to the partial pressure of oxygen,

$$pO_2 = KX \tag{2}$$

where K represents Henry's constant for a given solution of gas at a specified temperature. By substitution,

$$R_1 = R_{1a} + (R_{1p}/K)pO_2 \tag{3}$$

The slope (R_{1p}/K) indicates the response of a particular resonance to pO_2.

PFCs essentially act as molecular amplifiers, since the solubility of oxygen is greater than in water, but thermodynamics require that the pO_2 in the PFC will rapidly equilibrate with the surrounding medium, and estimates of diffusion suggest the equilibration can occur within seconds.

[115] S. K. Holland, R. P. Kennan et al., Magn. Reson. Med. 29, 446 (1993).

[116] B. R. Barker, R. P. Mason et al., J. Magn. Reson. Imaging 4, 595 (1994).

[117] B. J. Dardzinski and C. H. Sotak, Magn. Reson. Med. 32, 88 (1994).

[118] U. Noth, S. P. Morrissey et al., Magn. Reson. Med. 34, 738 (1995).

[119] H. T. Tran, Q. Guo et al., Acad. Radiol. 2, 756 (1995).

[120] S. Laukemper-Ostendorf, A. Scholz et al., Magn. Reson. Med. 47, 82 (2002).

[121] R. P. Mason, Artif. Cells Blood Substit. Immobil. Biotechnol. 22, 1141 (1994).

[122] P. Parhami and B. N. Fung, J. Phys. Chem. 87, 1928 (1983).

[123] L. C. Clark, Jr., J. Ackerman et al., in "Oxygen Transport to Tissue VI" (D. Bruley, H. I. Bicher, and D. Reneau, eds.), p. 835. Plenum Press, New York,.

[124] C. F. Kong, G. M. Holloway et al., J. Phys. Chem. 88, 6308 (1984).

[125] C.-S. Lai, S. Stair et al., J. Magn. Reson. 57, 447 (1984).

[126] D. Eidelberg, G. Johnson et al., Magn. Reson. Med. 6, 344 (1988).

[127] S. R. Thomas, R. G. Pratt et al., Radiology 18, 159 (1991).

Because relaxation is proportional to oxygen concentration, the effect will be greater at a given pO_2 than for water. Importantly, ions do not enter the hydrophobic PFC phase, and thus do not affect the bulk relaxation. Indeed, PFCs are typically exceedingly hydrophobic and do not mix with the aqueous phases, but rather form droplets or emulsions. Based on these principles, PFCs have been applied to *in vivo* pO_2 measurements. Characteristics of many diverse PFCs are summarized in Table II.

At any given magnetic field (Bo) and temperature (T), sensitivity to changes in pO_2 is given by $R_1 = a + bpO_2$. Thus a greater slope is important, and the ratio $\eta = b/a$ has been proposed as a sensitivity index.[128] Generally, a small "a" value (intercept) represents greater sensitivity, but it also generates longer T_1 values under hypoxic conditions, potentially increasing data acquisition times. Indeed, the T_1 of hexafluorobenzene (HFB) at 4.7 T may reach 12 s, potentially creating long imaging cycles, but this is readily overcome by applying single-shot (echo planar) imaging techniques, as presented in a later section.

Many PFCs, such as perfluorotributylamine (PFTB), perflubron (formerly referred to as perfluorooctyl bromide; PFOB), and Therox (F44-E), have several ^{19}F NMR resonances, which can be exploited to provide additional information in spectroscopic studies, but seriously hamper effective imaging. Multiple resonances can lead to chemical shift artifacts in images, which compromise the integrity of relaxation time measurements, though they can be avoided by selective excitation, or detection, chemical shift imaging, deconvolution, or sophisticated tricks of NMR spin physics.[116,119,129–133] These approaches add to experimental complexity and are generally associated with lost signal to noise ratio (SNR). Thus we strongly favor PFCs with a single resonance, and we will describe the use of HFB,[27,32,42,106,134–137] though some research groups favor 5-crown-5-ether (15C5).[88,117,138,139]

[128] S. R. Thomas, R. G. Pratt *et al.*, *Magn. Reson. Imaging* **14,** 103 (1996).

[129] L. J. Busse, R. G. Pratt *et al.*, *J. Comp. Ast. Tomagr.* **12,** 824 (1988).

[130] R. P. Mason, N. Bansal *et al.*, *Magn. Reson. Imaging* **8,** 729 (1990).

[131] H. K. Lee and O. Nalcioglu, *J. Magn. Reson. Imaging* **2,** 53 (1992).

[132] U. Nöth, R. Deichmann *et al.*, *J. Magn. Reson. B* **105,** 233 (1994).

[133] R. G. Pratt, J. Zheng *et al.*, *Magn. Reson. Med.* **37,** 307 (1997).

[134] D. Zhao, A. Constantinescu *et al.*, *Radiat. Res.* **156,** 510 (2001).

[135] Y. Song, A. Constantinescu *et al.*, *Technol. Cancer Res. Treat.* **1,** 471 (2002).

[136] D. Zhao, A. Constantinescu *et al.*, *Radiat. Res.* **159,** 621 (2003).

[137] D. Zhao, S. Ran *et al.*, *Neoplasia* **5,** 308 (2003).

[138] B. P. J. van der Sanden, A. Heerschap *et al.*, *Int. J. Radiat. Oncol. Biol. Phys.* **44,** 649 (1999).

[139] T. Q. Duong, C. Ladecola *et al.*, *Magn. Reson. Med.* **45,** 61 (2001).

TABLE II

^{19}F NMR Characteristics and Applications of PFCs for Tissue Oximetry

PFC	Sensitivity to pO_2[a]	Temp. sensitivity (torr/°)	Magnetic field B_0(T)	Application/ comments	References
Hexafluorobenzene (HFB)	A = 0.0835 B = 0.001876	0.13	4.7	Rat breast tumor, prostate tumor, human lymphoma xenograft	Hunjan, Zhao et al.[159] Zhao, Constantinescu et al.[32,134,136]; Mason, Ran et al.[27]; Song, Constantinescu et al.[135]; Zhao, Ran et al.[137]
HFB	A = 0.074 B = 0.00158		4.7	Rat prostate tumor	Hunjan, Mason et al.[161]; Mason, Constantinescu et al.[175]
HFB	A = 0.093 B = 0.103	1.40	7	Phantom	Mason, Rodbumrung et al.[143]
Perfluoro-15-Crown-5-ether (15C5)	A = 0.345 B = 0.0034	2.94	2.0	Tumor cells	Helmer, Han et al.[157]
15C5	A = 0.333 B = 0.0033	3.98	2.0	Mouse tumor, spleen, liver	Dardzinski and Sotak[117]
15C5	A = 0.44 B = 0.0028		4.3	Human glioma tumor in mice	van der Sanden, Heerschap et al.[100]; van der Sanden, Heerschap et al.[138]
15C5	A = 0.375 B = 0.00198		4.7	Rat breast tumor	Fan, River et al.[95]
15C5	A = 0.362 B = 0.1239		4.7	Rat brain	Duong, Ladecola et al.[139]
Perfluorotributyl-amine (FC-43) PFTB	A = 1.09 B = 0.00623	4.43	0.14	Pig liver, spleen, lung	Thomas, Pratt et al.[128]
PFTB			0.14	Rat liver, spleen, lung	Pratt, Zheng et al.[133]; Thomas, Gradon et al.[163]

(continued)

TABLE II (continued)

PFC	Sensitivity to pO_2[a]	Temp. sensitivity (torr/°)	Magnetic field B_0(T)	Application/ comments	References
PFTB	A = 0.684 B = 0.00305		4.7	Rabbit eye	Berkowitz, Wilson et al.[166]; Wilson, Berkowitz et al.[b]
PFTB	A = 0.9072 B = 0.004486		1.5	Human eye	Wilson, Berkowitz et al.[167]
PFTB	A = 0.8848 B = 0.1307	8.17	7	Mouse Meth-A tumor and heart	Mason, Shukla et al.[141]
Perfluorotripropylamine (FTPA)	A = 0.314 B = 0.002760⁻³		1.9	Rat subcutaneous tumor	Fishman, Joseph et al.[c]
FTPA	A = 0.301 B = 0.00312		1.4	Rat spleen, lung, liver	Fishman, Joseph et al.[114]
FTPA	A = 0.4052 B = 0.0023		2.0	Rat liver, spleen	Holland, Kennan et al.[115]
FTPA			4.4	Cells	Taylor and Deutsch[d]
Bis-perfluoro-butylethylene (F-44E)	A = 0.3421 B = 0.11172		7.05	Rat, alginate capsules	Noth, Grohn et al.[164]
F-44E	A = 0.342 B = 0.1201		7.05	Rat spleen, liver, abdominal aorta, vena cava	Noth, Morrissey et al.[118]
F-44E	A = 0.2525 B = 0.16527	0.59	2.0	Mouse tumor	Hees and Sotak[153]
Perfluorooctyl-bromide (perflubron) (PFOB)	A = 0.517 B = 0.0038		9.4	Rat heart	Shukla, Mason et al.[e]
PFOB	A = 0.2677 B = 0.12259	1.26	4.7	Prostate tumor in rat	Antich et al.[162]

Compound	Coefficients		Sample	Reference
PFOB	A = 0.328 B = 0.12137	2.85	Phantom	Mason, Shukla et al.[142]
PFOB	A = 0.085 B = 0.0033	2.0	Rat tumor	Sostman, Rockwell et al.[96]
PFOB		1.5	Rabbit liver	Tran, Guo et al.[119]
PFOB		8.5	Pig liver, lung, spleen	Millard and McGoron[f]
PFOB		1.45	Rat lung, Mouse lung	Thomas, Clark, Jr. et al.[g]
PFOB		1.5	Pig lung	Laukemper-Ostendorf, Scholz et al.[120]
Perfluoro-2,2,2',2'-tetramethyl-4,4'-bis(1,3-fioxolane) (PTBD)	A = 0.50104 B = 0.1672	2.0	Phantom	Sotak, Hees et al.[h]

Some original papers presented calibration curves in other forms (e.g., including coefficients for temperature dependence). In those cases, equations have been derived assuming 37°. Where a PFC has more than one resonance, the equation presented is either for the most sensitive signal, or the equation used where the signal may have been unresolved.

[a] R_1 $(s^{-1}) = A + B \times pO_2$ (torr).
[b] C. A. Wilson, B. A. Berkowitz et al., Exp. Eye Res. **55**, 119 (1992).
[c] J. E. Fishman, P. M. Joseph et al., Invest. Radiol. **24**, 65 (1989).
[d] J. Taylor and C. J. Deutsch, Biophys. J. **53**, 227 (1988).
[e] H. P. Shukla, R. P. Mason et al., Magn. Reson. Med. **35**, 827 (1996).
[f] R. W. Millard and A. J. McGoron, Artif. Cells Blood Subst. Immobil. Biotechnol. **22**, 1251 (1994).
[g] S. R. Thomas, L. C. Clark, Jr. et al., J. Comput. Assist. Tomogr. **10**, 1 (1986).
[h] C. H. Sotak, P. S. Hees et al., Magn. Reson. Med. **29**, 188 (1993).

R_1 is sensitive to temperature, although the response varies greatly between PFCs and between individual resonances of each individual PFC. Over small temperature ranges, a linear correction to calibration curves is appropriate, but over larger temperature ranges, the response can be complex, as investigated extensively by Shukla et al.[140] for several PFCs. Differential sensitivity of pairs of resonances to pO_2 and temperature allowed Mason et al.[141] to simultaneously determine both parameters by solving simultaneous equations. However, generally it is preferable for a pO_2 sensor to exhibit minimal response to temperature, since this is not always known precisely in vivo and temperature gradients may occur across tumors. As shown in Table II, even a relatively small error in temperature estimate can introduce a sizable discrepancy into the apparent pO_2; for example, the relative error introduced into a pO_2 determination by a $1°$ error in temperature estimate ranges from 8 torr/$°$ for PFTB[141] to 3 torr/$°$ for PFOB (perflubron)[142] or 15C5[117] and 0.1 torr/$°$ for HFB,[143] when pO_2 is actually 5 torr. It must be noted that error depends on actual pO_2 and the error varies with magnetic field and temperature. R_1 response does depend on magnetic field, necessitating calibration curves for each type of magnet system (e.g., 1.5, 4.7, or 7 T). Thus comparison of PFC utility for pO_2 measurements is complicated by the field used for specific published investigations, and in Table II, we consider sensitivity as presented.

Choice of PFC may be governed by practical considerations, such as cost and availability, since several products, particularly proprietary emulsions, may be difficult to obtain. HFB and 15C5 offer the immediate advantage of a high symmetry and a single ^{19}F NMR resonance. This offers maximum SNR and simplifies imaging, which may otherwise require frequency selective excitation, deconvolution, or other NMR tricks to avoid chemical shift artifacts.

Route of Administration. The most popular route for the delivery of PFCs is as emulsions injected intravenously. Given the extremely hydrophobic nature of PFCs, they do not dissolve in blood directly, but may be formulated as biocompatible emulsions. Much effort has been applied to formulate stable homogenous emulsions, as reviewed elsewhere.[144] Following IV infusion, the emulsion circulates in the vasculature with a

[140] H. P. Shukla, R. P. Mason et al., J. Magn. Reson. B **106**, 131 (1995).
[141] R. P. Mason, H. P. Shukla et al., Magn. Reson. Med. **29**, 296 (1993).
[142] R. P. Mason, H. P. Shukla et al., Biomater. Artif. Cells Immobilization Biotechnol. **20**, 929 (1992).
[143] R. P. Mason, W. Rodbumrung et al., NMR Biomed. **9**, 125 (1996).
[144] J. G. Riess, Biomater. Artif. Cells Immobilization Biotechnol. **20**, 183 (1992).

typical half-life of 12 h, (depending on the nature of the emulsion) providing substantial clearance within 2 days.[145,146] Primary clearance is by macrophage activity, leading to extensive accumulation in the liver, spleen, and bone marrow.[113,147] Indeed, this is a major shortcoming of IV delivery, since animals may exhibit extensive hepatomegaly or splenomegaly.[148] The emulsions are not toxic, and other than causing swelling, appear not to cause health problems. PFC clearance occurs from the liver with a typical half-life of 60 days for perfluorotripropylamine and 3 days for perflubron, with primary clearance by migration to the lungs and exhalation.[149]

Some investigators have examined pO_2 of tissues, while PFC remained in the blood, providing a vascular pO_2.[114,118,150] Flow can generate artifacts, and correction algorithms have been proposed.[151] Many investigators have measured pO_2 in liver, spleen, and tumors following clearance from the blood, thus providing measurement of tissue pO_2.[95,100,115,117–119,152–157]

Both spectroscopic and imaging approaches have been applied to tissue pO_2 measurements depending on the available SNR. It appears that uptake and distribution efficiency vary with tumor type, but in general, maximum signal is detected from the tumor periphery corresponding with regions of greater perfusion.[100,117,138,154,158,159] Several reports have examined changes in tumor pO_2 in response to acute interventions such as vasoactive drugs and hyperoxic gases.[32,95,100,135–137,153,160] Spectroscopic time resolution has ranged from seconds to minutes,[161,162] whereas imaging often takes longer.[160]

Long tissue retention facilitates chronic studies during tumor development, and progressive tumor hypoxiation has been observed over extended time periods of many days.[154,156] Correlated ^{19}F and proton MRI suggest

[145] M. C. Malet-Martino, D. Betbeder et al., J. Pharm. Pharmacol. **36**, 556 (1984).

[146] T. F. Zuck and J. G. Riess, Crit. Rev. Clin. Lab. Sci. **31**, 295 (1994).

[147] R. P. Mason, P. P. Antich et al., Magn. Reson. Imaging **7**, 475 (1989).

[148] W. I. Rosenblum, M. G. Hadfield et al., Arch. Pathol. Lab. Med. **100**, 213 (1976).

[149] R. F. Mattrey and D. C. Long, Invest. Radiol. **23**, s298 (1988).

[150] D. Eidelberg, G. Johnson et al., J. Cereb. Blood Flow P. Metab. **8**, 276 (1988).

[151] T. Higuchi, S. Naruse et al., in "7th SMRM," p. 435, 1988.

[152] R. P. Mason, R. L. Nunnally et al., Magn. Reson. Med. **18**, 71 (1991).

[153] P. S. Hees and C. H. Sotak, Magn. Reson. Med. **29**, 303 and erratum **29**, 716 (1993).

[154] R. P. Mason, P. P. Antich et al., Int. J. Radiat. Oncol. Biol. Phys. **29**, 95 (1994).

[155] S. R. Thomas, R. W. Millard et al., Artif. Cells Blood Subst. Immobil. Biotechnol. **22**, 1029 (1994).

[156] N. J. Baldwin and T. C. Ng, Magn. Reson. Imaging **14**, 541 (1996).

[157] K. G. Helmer, S. Han et al., NMR Biomed. **11**, 120 (1998).

[158] P. P. Antich, R. P. Mason, A. Constantinescu et al., Proc. Soc. Nucl. Med. **35**, 216P (1994).

[159] D. J. O. McIntyre, C. L. McCoy et al., Curr. Sci. **76**, 753 (1999).

[160] S. Hunjan, D. Zhao et al., Int. J. Radiat. Oncol. Biol. Phys. **49**, 1097 (2001).

[161] R. P. Mason, F. M. H. Jeffrey et al., Magn. Reson. Med. **27**, 310 (1992).

[162] S. Hunjan, R. P. Mason et al., Int. J. Radiat. Oncol. Biol. Phys. **40**, 161 (1998).

that PFC does not redistribute, but remains associated with specific tissues, analogous to tree rings.[154,156] Thus in principle, a whole tumor can be investigated by administering successive doses of PFC emulsion during growth.

PFC emulsions may also be administered intraperitoneally (IP), resulting in similar distribution to IV administration (unpublished observations). Given the volatile nature of many PFCs, they could be inhaled, but although this is a popular route for delivery of anesthetics and blood flow tracers, it does not appear to have been widely exploited for oximetry. Nonetheless, aerosols have been delivered to the lungs by inhalation to facilitate pO_2 measurements.[163]

Two approaches have been applied to circumvent reticuloendothelial uptake. PFC has been incorporated in polyalginate beads for direct implantation at a site of interest.[164,165] We favor direct IT injection of neat PFC, allowing any region of interest in a tumor to be interrogated immediately. Use of a fine needle ensures minimal tissue damage, as described in detail in a later section. Others have used direct injection of emulsions into tumors, but this increases the volume considerably, making it more invasive.[88] Investigators have suggested that emulsification improves retention at the site of injection. Direct injection of neat PFC also has been used to investigate retinal oxygenation[166–168] and cerebral oxygenation in the interstitial and ventricular spaces.[139]

As described in the following section we favor direct intratumoral injection of neat HFB followed by echo planar imaging to generate pO_2 maps in tumors.

FREDOM (Fluorocarbon Relaxometry using Echo Planar Imaging for Dynamic Oxygen Mapping)

Recognizing that tumors are heterogeneous and that pO_2 may fluctuate, we developed a procedure, which allows repeated quantitative maps of regional pO_2 to be achieved with multiple individual locations simultaneously in 6.5 min with a precision of 1–3 torr, when pO_2 is in the

[163] S. R. Thomas, L. Gradon et al., Invest. Radiol. **32,** 29 (1997).
[164] U. Nöth, P. Grohn et al., Magn. Reson. Med. **42,** 1039 (1999).
[165] U. Zimmermann, U. Noth et al., Artif. Cells Blood Subst. Immobil. Biotechnol. **28,** 129 (2000).
[166] B. A. Berkowitz, C. A. Wilson et al., Invest. Ophthalmol. Vis. Sci. **32,** 2382 (1991).
[167] C. Wilson, B. Berkowitz et al., Arch. Ophthalmol. **110,** 1098 (1992).
[168] W. Zhang, Y. Ito et al., Invest. Ophthalmol. Visual Sci. **44,** 3119 (2003).

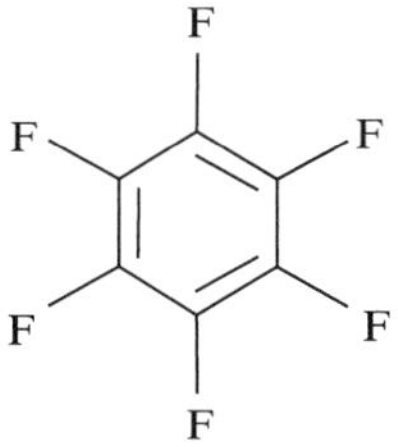

Fig. 1. Hexafluorobenzene (HFB) is a perfluorocarbon (PFC) exhibiting extensive symmetry.

range 0–10 torr.[160] We have applied FREDOM to diverse tumor types and interventions, as reviewed in a later section.

MRI is attractive because it is readily available at many institutions. For small animal work, ^{19}F NMR is widely available at 4.7, 7, and 9.4 T by minor adaptation of routine instrumentation, [e.g., retuning proton radio-frequency (RF) coils]. Within the recent past ^{19}F MRI is also becoming available on clinical systems, facilitating translation of these techniques to patients. ^{19}F NMR is particularly facile because there is essentially no background signal in tissues to interfere with measurements, yet the resonance frequency and sensitivity approach that of proton NMR. The pioneering work of Thomas[113] showed that tissue pO_2 could be imaged in various organs based on the ^{19}F NMR spin–lattice relaxation rate (R_1) of PFC reporter molecules following IV infusion. Prompted by these studies, we surveyed a number of PFCs and identified that HFB (Fig. 1) has many virtues as a pO_2 reporter.[143] Symmetry provides a single narrow ^{19}F NMR signal, and the spin–lattice relaxation rate is highly sensitive to changes in pO_2, yet minimally responsive to temperature (Fig. 2).[143] HFB also has a long spin–spin relaxation time (T_2), which is particularly important for imaging investigations. From a practical perspective, HFB is cheap (<\$2/g) and readily available commercially in high purity (≥99%). We obtain supplies from Lancaster Synthesis (Windham, NH), though many other fine chemical supply houses also offer HFB. We do favor bottles over sealed ampoules, since they are easier to handle. HFB is well characterized in terms of lack of toxicity,[169,170] exhibiting no mutagenicity,[171] teratogenicity, or fetotoxicity,[172] and the manufacturer's material

[169] Y. S. Gorsman and T. A. Kapitonenko, *Izv. Estestvennonauchu. Inst. Pevinsk.* **15,** 155 (1973).

[170] I. M. C. M. Rietjens, A. Steensma *et al., Eur. J. Pharmacol.* **293,** 292 (1995).

[171] K. E. Mortelmans and V. F. Simmon, *Gov. Rep. Announce. Index (US)* **81,** 2555 (1981).

[172] K. D. Courtney and J. E. Andrews, *J. Environ. Sci. Health B* **19,** 83 (1984).

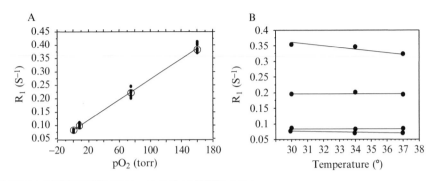

Fig. 2. (A) The ^{19}F NMR spin–lattice relaxation rate (R_1) of HFB shows a linear response to pO_2: at $37°$ and 4.7 T, $R_1 = 0.0835 + 0.001876 \times pO_2$. (B) The ^{19}F NMR R_1 of HFB shows minimal response to temperature in the physiologic range; separate curves are shown at four different pO_2s: 0, 1, 9.8, and 21% O_2.

data safety sheet indicates $LD_{50} > 25$ g/kg (oral—rat) and LC_{50} 95 g/m^3/2 h (inhalation—mouse).

HFB had been proposed as a veterinary anesthetic and has been used in many species, including ponies, sheep, cats, dogs, rats, and mice, but was abandoned because of its high volatility (boiling point $81°$) and low flash point $(10°)$.[173] This presented unacceptable dangers in veterinary suites for inhalation anesthetics. It is not a problem in our studies, where small quantities of liquid (typically, 50 μl) are injected directly into the tumor. HFB requires no special storage, other than a sealed bottle to prevent evaporation. Melting point is 4–6$°$, and density is 1.62 g/ml.

Methodology

Tumor Preparation. We have applied this technique to tumors in rats and mice, but our methodology will focus on our standard application to rats. As for electrode techniques, the tumor must be accessible (e.g., subcutaneous in flank of thigh), though we favor the pedicle model, which provides a tumor remote from the body, analogous to an additional limb.[174] This model is optimal for selective therapy such as local hyperthermia, irradiation, or excision. Noting the increasing interest in orthotopic tumors, the approach is also facile in breast tumors in the mammary fat pad and applicable in the prostate with a little practice.

[173] L. W. Hall, S. R. K. Jackson *et al.*, *in* "Recent Progress in Anaesthesiology and Resuscitation" (A. Arias, R. Llaurado, M. A. Nalda, and J. N. Lunn, eds.), p. 201. Excerpta Medica, Oxford, UK, 1975.
[174] E. W. Hahn, P. Peschke *et al.*, *Magn. Reson. Imaging* **11**, 1007 (1993).

Rats are preanesthetized with ketamine hydrochloride (100 mg/ml) as a relaxant and maintained under general gaseous anesthesia with air (1 dm^3/min) and 1.2% isoflurane [we do note that the appropriate concentration of anesthesia may depend on strain; for example, Copenhagen rats require (and tolerate) higher concentrations than Fisher rats]. In our latest refinement of the procedure, isoflurane may be stopped for a period of minutes during HFB administration, since rats occasionally exhibit respiratory distress, which may be caused by the anesthetic properties of HFB interacting with isoflurane. Within a few minutes, rats are stable and routine isoflurane anesthesia is maintained. In our earlier work we used 0.5% methoxyflurane in 33% oxygen with 66% N_2O,[160,175] but isoflurane appears to be a less stressful anesthetic.

HFB is deoxygenated by bubbling nitrogen for 5 min before use. We have previously shown in hypoxic tumor biopsies that use of aerated HFB may introduce a systematic apparent elevation in pO_2 ($\sim$2–3 torr). HFB is injected directly into tumors using a gas tight syringe with a custom-made fine sharp needle (32G #7803–04; Hamilton, Reno, NV). Generally, HFB is administered along three to five tracks in the form of a fan in a single central plane of the tumor coronal to the rat's body (Fig. 3A). The needle is inserted manually to penetrate across the whole tumor and withdrawn $\sim$1 mm to reduce pressure, and 3 μl HFB is deposited. The needle is repeatedly withdrawn a further 2–3 mm, and additional HFB is deposited. Typically, HFB is deliberately deposited at about 16 individual locations per tumor, in both the central and peripheral regions of the tumors, to ensure that the interrogated regions are representative of the whole tumor.

The animal is placed on its side in a cradle with a thermal blanket to maintain body temperature. A fiber-optic probe is inserted rectally to monitor core temperature. Temperature measurement is optional, since the R_1 is essentially invariant with temperature and does not need correction for pO_2 estimates. Of course, temperature regulation and measurement is important to ensure stable tumor physiology. An MR-compatible pulse oximeter equipped with "rat" software (8600V; Nonin, Inc., Plymouth, MN) may be applied to a hind foot to monitor arterial oxygenation (S_aO_2) and heart rate (optional, but provides additional useful data regarding animal health and physiology).

Most of our MR experiments were performed using an Omega CSI 4.7 horizontal bore magnet system with actively shielded gradients (GE systems, acquired by Bruker Instrument, Inc., Fremont, CA). Recently, our MR system has been upgraded to a Varian Unity INOVA, providing

[175] R. P. Mason, A. Constantinescu *et al.*, *Radiat. Res.* **152,** 239 (1999).

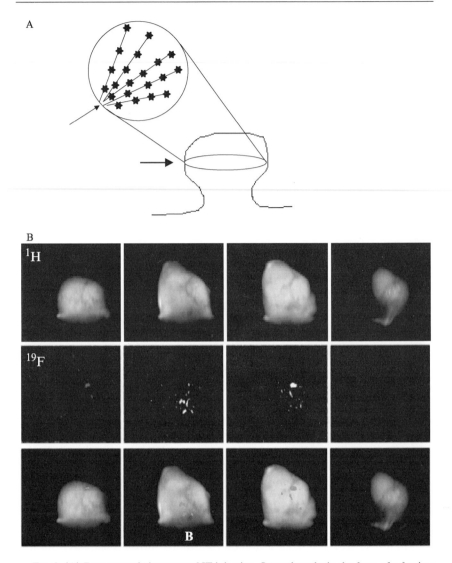

FIG. 3. (A) Recommended pattern of IT injection. Several tracks in the form of a fan in a single plane. (B) Four contiguous MR images (5 mm thick) of a representative Dunning prostate R3327-AT1 tumor (volume = 1.6 cm^3). *Top:* ^{1}H MRI; *middle:* corresponding ^{19}F; *bottom:* overlay to show interrogated regions. *B* marks point of attachment of the tumor to back of rat. (See color insert.)

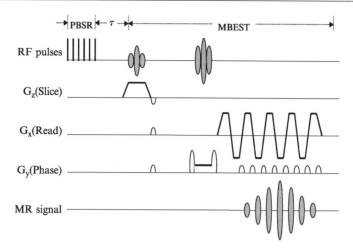

FIG. 4. MRI sequence used to achieve relaxation curves.

enhanced data acquisition and processing software and stronger imaging gradients. A tunable (^{1}H/^{19}F) MR coil, 2 or 3 cm in diameter matched to the tumor size (constructed from a cylindrical copper tube about 2 cm deep and acting as a single-turn solenoid), is placed around the tumor-bearing pedicle. Shimming is performed on the proton water signal to a typical linewidth of 60–100 Hz. Proton images are obtained for anatomic reference using a three-dimensional (3D) spin–echo sequence. The coil is then retuned in place to 188.27 MHz, and corresponding ^{19}F MR images are obtained. Overlaying the ^{19}F MR images on the corresponding proton images reveals the distribution of HFB (Fig. 3B). Typically, ^{19}F NMR signal is obtained from 6–10% of the total tumor voxels.[136,137,175]

pO$_2$ maps are obtained using our standard MR oximetry protocol.[116,160,176] This applies pulse burst saturation recovery (PBSR) echo planar imaging (EPI) relaxometry using the ARDVARC (*a*lternated *r*elaxation *d*elays with *v*ariable *a*cquisitions to *r*educe *c*learance effects) acquisition protocol to map the tumors (Fig. 4).[175,177] EPI uses a single spin–echo with "blipped" phase encoding (modulus-blipped echo-planar single-pulse technique [MBEST]), although other EPI sequences should be equally applicable. We chose the PBSR approach to T$_1$ relaxation measurements for historical reasons; our earliest work used ^{19}F NMR spectroscopy of tumors with excitation and detection based on surface coils.[152] The PBSR approach is suitable for use with the nonuniform excitation typical of surface

[176] B. R. Barker, R. P. Mason *et al.*, *Magn. Reson. Imaging* **11,** 1165 (1993).
[177] D. Le, R. P. Mason *et al.*, *Magn. Reson. Imaging* **15,** 971 (1997).

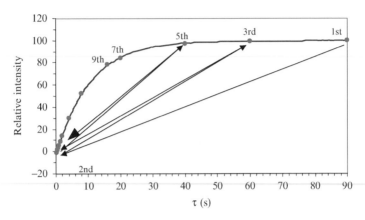

Fig. 5. Typical relaxation curve showing order of data acquisition. $y = A [1 - (1 + W) \exp(-R_1 \times \tau)]$ and $T_1 = 10.34 \pm 0.16$ s, A = 99.5, W = 0.04, and hence $pO_2 = 7.0 \pm 0.7$ torr.

coils as presented by Evelhoch and Ackerman.[178] For EPI, we do apply carefully calibrated $\pi/2$ pulses to ensure accurate refocusing of images, but we continue to favor the PBSR approach, since it provides particularly rapid estimation of T_1 by avoiding the need for extended relaxation recovery times between acquisitions. Saturation is achieved by a series of 20 non-spatially selective $\pi/2$ pulses with 50-ms spacing, which is sometimes called a "Comb" format.

Typical FREDOM parameters use 32×32 data points across a field of view of 40×40 mm, providing 1.25 mm in-plane resolution. Recently, we have applied a slice selection gradient, providing 5-mm-thick slices, and hence, 8-μl voxels. However, the deposition of HFB is designed to occur in a plane, and thus the discrete distribution of the reporter molecule itself can be used to define the slice. We apply a PBSR preparation sequence because it is ideally suited for measuring diverse, long T_1 values. Unlike more traditional inversion recovery sequences, there is no need to wait $>5 T_1$ between successive images. As shown in Fig. 5, we use 14 delays in the order 90 s, 200 ms, 60 s, 400 ms, 40 s, 600 ms, 20 s, 800 ms, 16 s, 1 s, 8 s, 1.5 s, 4 s, and 2 s, selected to cover the whole range of T_1s (viz pO_2 values). Many papers have been published on optimizing relaxation curves and choosing parameters to enhance precision.[179–181] Our experience suggests that our

[178] J. L. Evelhoch and J. J. H. Ackerman, *J. Magn. Reson.* **53,** 52 (1983).
[179] A. P. Crawley and R. M. Henkelman, *Magn. Reson. Med.* **7,** 23 (1988).
[180] S. J. Doran, J. J. Attard *et al.*, *J. Magn. Reson.* **100,** 101 (1992).
[181] F. Franconi, F. Seguin *et al.*, *Medical Phys.* **22,** 1763 (1995).

parameters are appropriate for interrogating a broad range of pO_2 values, though they have not been rigorously or theoretically optimized. More delays could improve curve fitting, but would increase experiment time. We believe fewer delays would degrade quality of the curve fitting.

Traditional T_1 measurement sequences acquire data with delays in monotonic order, whereas we alternate longer and shorter delays to minimize any systematic errors, which would be introduced if the signal amplitude varies during the measurement. We have found that HFB clears from tissue with a typical half-life of 600 min,[162] which would introduce errors into the amplitude. Although the total acquisition time for a T_1 map is 6.5 min, we reduce the time between first and last acquisitions further by applying the longest delay first. Our experience shows that it is important to measure at least two data points on the plateau of the curve (i.e., $>5 \times T_1$, and we choose 60 and 90 s, respectively. The data points with longer recovery times have greater SNR, and we find that a minimum SNR = 10 is required to produce satisfactory T_1 curves and pO_2 estimates. The variation in amplitude between longest and shortest delays approaches 100-fold. The poorest SNR data points (short delay; τ) could compromise the quality of the T_1 curves, and thus we obtain multiple acquisitions at these times to provide enhanced SNR by signal averaging. Because the τ values are short, it adds little to the overall experimental time. We use NA = 12 for $\tau < 1$ s; NA = 8 for $\tau = 1$ s; NA = 4 for $\tau = 1.5, 2,$ and 4 s; NA = 2 for $\tau = 8$ s; and 16 s and NA = 1 for the longer delays. Signal amplitudes are corrected for the additional acquisitions, and a 2-d Fourier transform (FT) is applied to each image. Curve fitting is then applied on a voxel-by-voxel basis to each image set. Data quality could probably be enhanced by application of apodization and filtering functions, such as Fermi and Hanning filters, though we have not yet implemented these approaches. The acquisition protocol is not part of the standard software supplied by the manufacturer, but we can assist interested investigators.

Data are transferred to a personal computer (PC) for further analysis using a program created in our laboratory (T_1map [Fig. 6], PASCAL by Dennis Le). T_1map recognizes issues in data analysis and provides filters, clustering algorithms, and temporospatial correlation to assist in effective data reduction. At top right of Fig. 6 is a map, which may be toggled to display signal amplitude, T_1, T_1 error ($T_{1\ err}$), or pO_2 (Fig. 7). Blank pixels indicate that no curve fit could be achieved. Gray pixels indicate a successful curve fit and potential pO_2 value, but with errors beyond the specified range. In this example, thresholds were set as $T_{1\ err} < 2.5$ s and the ratio $T_{1\ err}/T_1 < 50\%$. Tightening these thresholds can produce higher-quality data, but eliminates more data. Colored pixels report measurements satisfying the threshold criteria, and these are tabulated in the form of a

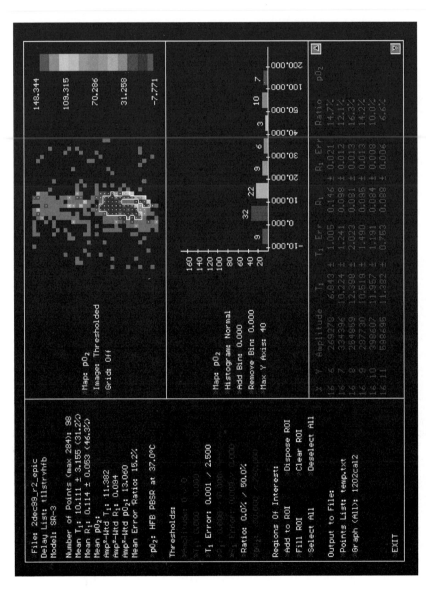

FIG. 6. Example T_1map used for data filtering, clustering, and analysis. (See color insert.)

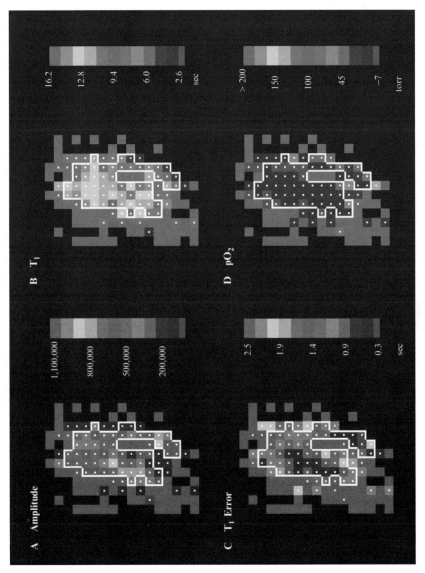

Fig. 7. (*continued*)

histogram (center), which can be plotted with selected data bins and ranges of interest. At bottom right of Fig. 6 are data for each pixel, including co-ordinate and fitted parameters. In the box on the left are shown the number of fitted data points, number of accepted data points, and statistics such as mean T_1, pO_2, and corresponding amplitude-weighted values.

pO_2 is calculated on the basis of a curve fit to exponential data. Any curve fitting is associated with uncertainty, and indeed our procedure deter-mines $T_{1 err}$ values. Although the concept of negative pO_2 values may appear impossible, it is perfectly reasonable, provided that error bars and uncer-tainties are considered. Thus a value of -2 torr $\pm$ 3 torr legitimately indi-cates an actual pO_2 very close to zero. Some oximetry approaches ignore negative values or bin them all as zero. We accept any value providing that the error estimated for T_1 is within a specified range.

The most powerful aspect of FREDOM is the ability to follow the fate of individual voxels, and thus we usually acquire at least three baseline pO_2 maps, followed by further maps accompanying interventions, such as hy-peroxic gas breathing. Even under baseline conditions, fluctuations in T_1 are apparent. These may arise from uncertainty in T_1, which may be reflected in $T_{1 err}$ or transient fluctuations in pO_2.

T_1 map allows us to investigate sequential maps and select only those pixels that consistently show small $T_{1 err}$s, for example, white regions of interest (ROI). In subsequent maps we may include only these regions, which satisfy the inclusion criteria for every map. This allows us to follow population dynamics based on specific tumor regions, avoiding potential anomalies due to varying numbers of "good quality" pixels in sequential images. Typically, 50–150 pixels provide high-quality data in any given map, but generally 20–80 pixels may be followed for a series of 23 maps associated with interventions.

The ROI tool also facilitates clustering, for example, selection of only those tumor regions, which consistently have $pO_2 < 10$ torr throughout the baseline period (these may be considered as chronically hypoxic, as opposed to regions, which fluctuate and may only be <10 torr in some

---

FIG. 7. Maps showing (A) ^{19}F NMR signal intensity, (B) spin–lattice relaxation times (T_1), (C) estimated errors in T_1, and (D) pO_2 values. Black squares show regions where signal to noise or data quality was so poor as to not provide a T_1 curve fit. Gray voxels show regions that provided a curve fit, but where the uncertainty in the data exceeded the threshold criteria (i.e., $T_{1 err} > 2.5$ s or $T_{1 err}/T_1 > 50\%$). All colored voxels ($n = 100$) provided a curve fit within the acceptance criteria in this map, but only those 56 voxels within the white regions of interest (ROI) provided consistently high-quality data throughout the sequence of eight maps acquired during baseline and with oxygen breathing. Data were obtained using the FREDOM approach from a representative Dunning prostate R3327-AT1 tumor (2 cm³). (See color insert.)

baseline maps). The ability to follow groups of pixels with particular baseline characteristics has revealed heterogeneity in response to many interventions—often those regions initially well oxygenated show rapid and large response to hyperoxic gas breathing, whereas those that are initially poorly oxygenated show little and sluggish response (e.g., Fig. 8). This approach can also reveal phenomena such as the "steal" effect, whereby initially well-oxygenated regions decrease in pO_2, while others increase, which might appear as "no change," when histogram-based population statistics are used.

Figure 9 shows that the data quality is strongly related to signal amplitude. Although these data represent 56 voxels from a single tumor, we have previously shown similar data for the alternate PFC perflubron.[116] Given the high solubility of O_2 in PFCs, there could be a concern that PFCs act as reservoirs, perturbing local pO_2. Fig. 9B shows that there is no correlation ($r^2 < 0.05$) between signal amplitude (viz. HFB concentration) and pO_2.

As with any measurement, sampling is a critical issue. FREDOM is analogous to the Eppendorf Histograph in that it samples multiple locations, which appear to reflect interstitial pO_2. Histograph data suggest that a minimum of 100–140 data points[182] along five tracks[183] are required to accurately represent the pO_2 distribution of a tumor, though such criteria do depend on tumor size and heterogeneity. We applied Monte Carlo simulation to assess the data requirements for FREDOM. A data set was selected, which provided 120 high-quality data points, and these data points were accessed in random order with continuous calculation of mean pO_2 to assess the asymptotic trend lines toward the actual mean pO_2 (Fig. 10). It appears that about 50 data points are required to well represent the pO_2 distribution of the tumor, which is generally achievable in any individual pO_2 map. However, the unique ability to observe dynamic changes in pO_2 at multiple locations simultaneously is the greatest strength of FREDOM. Detection of heterogeneous responses is useful even if fewer data points are examined, since each location serves as its own control.

Validation of Measurements. The spin–lattice relaxation rate (R_1) is determined for each voxel using a three-parameter fit of signal intensities to

$$y_i = A[1 - (1 + W) \exp(-R_1 \times \tau)] \quad (4)$$

using the Levenberg–Marquardt least squares fitting protocol.[116] Typically, ~100–300 voxels provide an R_1 fit and potential pO_2 value. Because noise

[182] K. A. Yeh, S. Biade *et al.*, *Int. J. Radiat. Oncol. Biol. Phys.* **33,** 111 (1995).
[183] O. Thews, D. K. Kelleher *et al.*, *in* "Tumor Oxygenation" (P. W. Vaupel, D. K. Kelleher, and Günderoth, eds.), Vol. 24, p. 39. Gustav Fischer Verlag, Stuttgart, Germany, 1995.

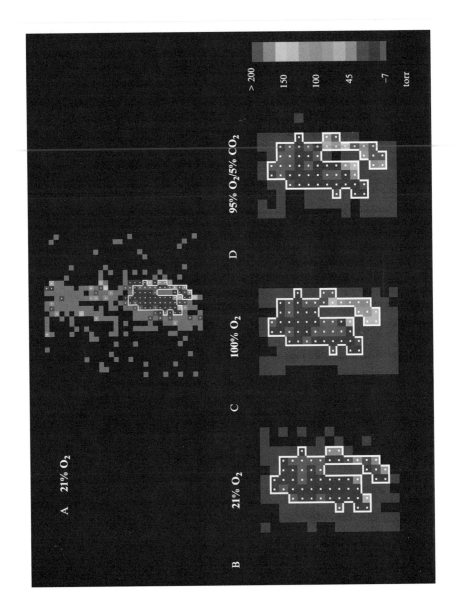

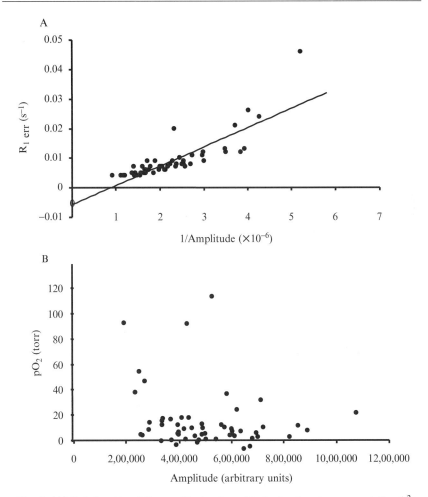

FIG. 9. (A) Relationship of R_1 err with signal amplitude showing strong correlation ($r^2 >$ 0.69). (B) Relationship of pO_2 and signal amplitude showing lack of correlation ($r^2 < 0.05$).

FIG. 8. PO_2 maps obtained using the FREDOM approach from the tumor shown in Fig. 3. (A) Under baseline conditions, those voxels ($n = 56$) within the white region provided consistently reliable data during repeated measurements. (B) Baseline map (breathing air: $FO_2 = 21\%$): mean $pO_2 = 7.2 \pm 2.6$ (SE) torr, median $pO_2 = 1.3$ torr (range -7–88 torr). (C) Breathing oxygen ($FO_2 = 100\%$): fourth map obtained 24–32 min after switching from air: mean $pO_2 = 47.2 \pm 10.7$ torr ($p < 0.0001$ compared with baseline), median $pO_2 = 8.3$ torr (range -3–204 torr). (D) Breathing carbogen ($FO_2 = 95\%$): fifth map after switching to carbogen: mean $pO_2 = 43.1 \pm 9.2$ torr ($p < 0.0001$ compared with baseline), median $pO_2 = 5.6$ torr (range -10–216 torr). (See color insert.)

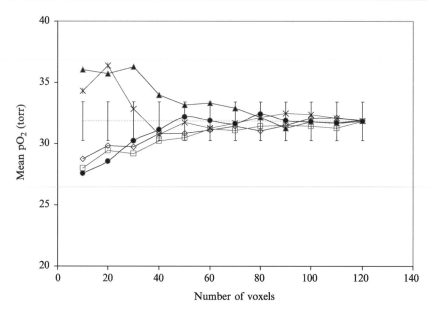

FIG. 10. Monte Carlo simulation of tumor oxygenation. Asymptotic behavior shows that tumor may be characterized by about 50 data points. For this large AT1 tumor in a rat breathing oxygen, all estimates converge on the mean $pO_2 = 31.7 + 1.0$ (SE) torr.

itself may give an apparent relaxation curve (R_1) fit, data are selected within a region of interest, having $T_{1\ err} < 2.5$ s and the ratio $T_{1\ err}/T_1 < 50\%$. With respect to respiratory interventions, only those voxels that provided consistently reliable data throughout the measurements are included for further analysis. At 37° and 4.7 T

$$pO_2(\text{torr}) = [R_1(s^{-1}) - 0.0835]/0.001876 \qquad (5)$$

Equation (5) provides pO_2 in units of torr. The literature can be complicated by use of various units, and it may be instructive to provide conversion factors. We favor torr (1 torr = 1 mmHg), since radiobiologic hypoxia develops in the range 0–15 torr and this is the traditional unit favored by radiation biologists and oncologists. Further, 760 torr = 1 standard atmosphere (atm.), and gases are often quoted in %atm. For SI units, 101, 325 Pa (or N/m^2) = 1.01325×10^6 dynes/cm^2 = 1 atm. Some investigators quote oxygen concentrations in μM, often assuming that the solubility of oxygen in water at 37° is 1.35 μM/torr.[29] Quoting concentrations can be confusing because the solubility of oxygen is highly variable in solutions. FREDOM calibration is based on pO_2 values, and as shown by Eqs. 1–3, these are directly related to the solubility of oxygen in HFB. However, partitioning of oxygen between the aqueous and PFC phases depends on pO_2, not concentrations. Thus the pO_2 determined by FREDOM will accurately reflect the ambient tissue pO_2 and if the solubility of oxygen can be estimated in the particular milieu, then $[O_2]$ could be calculated.

Both systematic and random errors may interfere. Random errors may be diminished by performing multiple repeat measurements (provided that the biologic system is stable). Systematic errors are more complex and could arise *inter alia* from erroneous calibrations, inappropriate curve fitting, and temperature changes. Appropriate curve fitting may be the greatest problem in relaxation analysis. Provided signal/noise > 10 for the most intense signals, we generally obtain excellent curve fits. Each voxel (or ROI) will comprise HFB at a range of pO_2 values, creating a multiexponential curve. However, modeling shows that the fit provides an "average" value. For population data, we sometimes determine error-weighted means $[\Sigma(x/\sigma^2)]/[\Sigma(1/\sigma^2)]$ in order to exploit as much available data as possible.

In terms of absolute pO_2 values, the most critical aspect is effective calibration curves. Over the years we have achieved pO_2-dependent ^{19}F NMR relaxation curves for many PFCs[116,141–143,160] and encountered potential pitfalls. The calibration curve we recommend at 4.7 T and 37° is given by Eq. (5) which was originally presented by Hunjan *et al.*[160] Briefly, 125 μl HFB was added to each of four gas-tight NMR tubes together with 0.5 ml water and saturated at 37° by bubbling with carbon dioxide, 1% O_2 (balance N_2), 9.8% O_2 (balance N_2), or air, respectively. Tubes were sealed, and the phantom was maintained at 37° in a water bath within a coil in the magnet. FREDOM was applied using the parameters described earlier, and the spin–lattice relaxation rates were estimated on a voxel-by-voxel basis using three-parameter fit. Equation (5) was established using linear regression analysis of amplitude squared weighted mean values for each gas.[159]

Although the relationship $R_1 = f(pO_2)$ is theoretically expected to be linear and empirically found to be so, we believe that it is important to use calibration gases in the range of physiologic pO_2. We recommend purchase of rigorously calibrated gases. We bubble gases for 30 min and use gas-tight Wilmad NMR tubes, which may be sealed with ground glass joints. Samples are used within hours of saturation. As desired, multiple tubes may be prepared at given pO_2, but since a linear relationship is expected, it may be equally appropriate to use additional gases. T_1 may be determined multiple times for each sample, and we recommend using the pulse sequence to be applied for *in vivo* investigations. In 1997, we published a calibration curve

$$R_1(s^{-1}) = 0.074(\pm0.003) + 0.012(\pm0.0002)pO_2(\%atm.) \qquad (6)$$

for HFB at 37° and 4.7 T.[177] Considering temperature sensitivity gave

$$R_1(s^{-1}) = 0.77(\pm0.03) - (0.00009 \pm 0.001)T(°C)$$
$$+ (0.018 \pm 0.003)P(\%atm.) - (0.00017 \pm 0.00008)TP(°C\% \ atm.)$$
$$(7)$$

More recently, we achieved Eq. (5) as a new calibration curve at 37° and 4.7 T using the ARDVARC approach.[160] This calibration curve is based on higher-quality data and provides superior pO_2 estimates. In particular, when tissue is expected to be hypoxic (excised tissue), we find fewer apparently negative pO_2 values.

A major strength of the FREDOM approach is that calibration curves remain valid between samples and across experimental platforms. Calibration curves obtained *in vitro* are valid *in vivo*. Thus we believe other investigators can apply Eq. (5) without the need to establish their own calibrations, provided that studies are undertaken at about 4.7 T and in the range 30–45°. As discussed in an earlier section and Table II, a 1° error in temperature estimates will only introduce a 0.13-torr error in pO_2 estimate, when pO_2 is about 5 torr. We also note that the calibration curves for HFB are relatively insensitive to magnetic field.[111,143]

We have previously investigated microbiodistribution of HFB based on Oil Red O stain, which indicated that HFB occurs as microscopic droplets (1–20 μm) widely distributed across tumor tissue.[175] Significantly, there was no evidence for formation of films, which could act as conduction conduits, causing oxygen equilibration. Occasionally, an animal (viz. tumor) will move slightly during a long time series of measurements. In this case the requirement of consistently high-quality curve fits throughout the data set for individual voxels fails because the tissues have moved relative to the voxel grid. Such motion is immediately apparent when examining the images. There is a choice of eliminating such a data set or relaxing the acceptance criteria and examining pO_2 population distribution without spatial continuity. It is important to recognize that all methods include a degree of sampling. In some approaches, this involves the selective placement and tracking of electrodes; in others, the choice of locations for biopsy and the number of microscopic fields of view. Sampling can be avoided by obtaining global measurements, as commonly acquired using near-infrared approaches, but this may itself mask the fundamental tumor heterogeneity[101] (see Chapter 17 in this volume).

We have previously shown that HFB shows little macroscopic redistribution over a period of hours.[177] It does clear from the tumors with a typical half-life of HFB of about 600 min, though some tumors show essentially no detectable clearance over a period of 6 h.[143,162] Clearance of HFB precludes long-term studies of chronic oxygenation, unless further doses of HFB are administered.

Comparison of pO_2 distributions using FREDOM or the Eppendorf Histograph has shown close similarity in both small and large tumors.[175] Dynamic studies in several tumor types have shown equivalent behavior when assessed using polarographic oxygen electrodes or OxyLite or FOXY optical probes.[32,43,134] Relative hypoxia has been compared with the histologic reporter pimonidazole, revealing similar trends across tumor types.[137]

Applications of FREDOM

FREDOM has been applied to investigations of diverse tumor types (syngeneic rat prostate and breast tumors and xenograft human lymphomas) with respect to growth and acute interventions.[27,32,134–137,160] Perhaps the greatest strength is the ability to investigate regional dynamic changes in pO_2 accompanying acute interventions. Figure 8 shows changes in regional pO_2 in an AT1 tumor accompanying hyperoxic gas breathing. As published previously,[137,160,162] the areas that were initially better oxygenated responded with elevated pO_2, whereas those relatively hypoxic regions showed little response. Data also may be presented as histograms (Fig. 11), revealing significant differences between mean and median pO_2 and hypoxic fractions between small and large tumors, and between the slow- and fast-growing sublines H and AT1 of the Dunning prostate R3327. Figure 12 shows variation in global mean pO_2 accompanying respiratory challenge with oxygen or carbogen and return to baseline following the intervention. Because individual tumor regions may be observed, local response may be compared at identical locations, efficiently comparing the efficacy of interventions (Fig. 13). Figure 14 shows the differential behavior of regions in undifferentiated Dunning prostate R3327-AT1 tumors versus highly differentiated H tumors. In each tumor type there are both well and poorly oxygenated regions under baseline conditions (see Fig. 11). In response to hyperoxic gas breathing, well-oxygenated regions in both tumor types show a rapid and significant elevation in pO_2, which is reversible when inhaled gas is returned to air (Fig. 14). Poorly oxygenated regions in H tumors respond slowly, but ultimately rise above the range of radiobiologic hypoxia, whereas corresponding regions in AT1 tumors do not. We also have investigated vasoactive drugs[42] and vascular targeting agents,[27] and the short-term changes in pO_2 following irradiation also have been examined.[184] Perhaps the most significant results to date

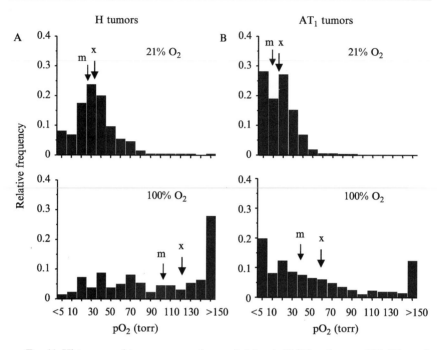

Fig. 11. Histograms of tumor oxygenation pooled for six H (A) and seven AT1 (B) small Dunning prostate R3327 rat tumors. The H tumors (219 voxels) showed significantly higher baseline pO_2 (mean = 31 ± 2 torr, median = 27 torr) than the size-matched AT1 tumors (338 voxels; mean = 14 ± 1 torr, median = 11 torr; $p < 0.0001$). With respect to oxygen challenge *(bottom)*, mean pO_2 in both groups increased significantly (mean = 122 ± 7 in the H versus, mean = 58 ± 4 in the AT1; $p < 0.0001$). Arrows indicate mean (x) and median (m) pO_2, respectively.

show that pO_2 measurements and detection of changes in pO_2 accompanying interventions correlate with the efficacy of tumor irradiation.[136]

Future

Ultimately, the value of a technique depends on its robustness, ease of use, and widespread implementation. To date, few laboratories had adopted the FREDOM approach because efficient investigation of HFB requires an unusual NMR pulse sequence. With the recent upgrade of our own instrumentation to the Varian Unity INOVA, the software is

[184] R. P. Mason, S. Hunjan *et al.*, *Int. J. Radiat. Oncol. Biol. Phys.* **42,** 747 (1998).

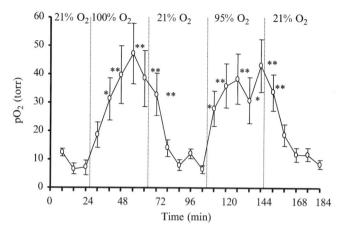

FIG. 12. Dynamic pO$_2$ (mean ± SE) obtained from sequential maps of the AT1 tumor shown in Fig. 8 with respect to respiratory challenge. *, $p < 0.001$; **, $p < 0.0001$ versus baseline.

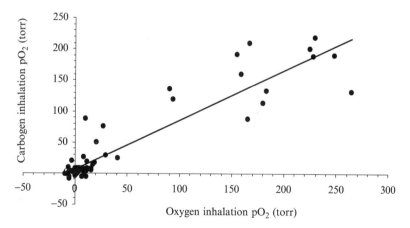

FIG. 13. Correlation between maximum pO$_2$ detected in each of 56 voxels from the AT1 tumor shown in Fig. 8, when the rat breathed oxygen versus carbogen ($r^2 > 0.85$).

now available on this popular platform, facilitating ready implementation elsewhere. In terms of research applications, it is known that tumor tissue pO$_2$ varies rapidly in response to many acute interventions, ranging from irradiation to photodynamic therapy, and various chemotherapies. We foresee FREDOM as a valuable tool for assessing the dynamic time course

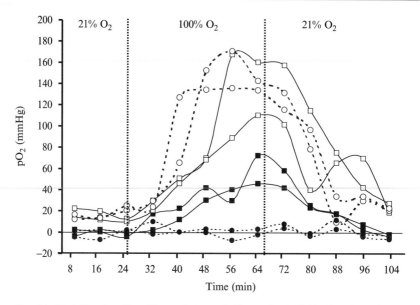

Fig. 14. Variation in pO_2 with respiratory challenge for individual regions chosen as initially well oxygenated ($pO_2 > 10$ torr) or hypoxic ($pO_2 < 5$ torr) from an AT1 (dotted lines) and H (solid lines) tumor, respectively. All the well-oxygenated regions in the two sublines increased significantly in pO_2 in response to oxygen breathing ($p < 0.01$). The hypoxic regions from the H tumor increased, whereas those in the AT1 tumor did not.

of such interventions to provide clear insight into the mode of action of therapeutic approaches and aid in the high-throughput screening of new drugs, such as vascular targeting and antiangiogenic agents.

Acknowledgments

This work was supported in part by NIH R01 CA79515 (NCl)/EB002762 (NIBIB), DOD Breast Cancer Initiative IDEA Award (DAMD 17–03–1–0363) (DZ) and predoctoral scholarship (DAMD 17–02–1–0592) (LJ) in conjunction with Cancer Imaging Program P20 CA 86354 and NIH BRTP Facility P41-RR02584. We are grateful to Dr. Anca Constantinescu for facilitating all the tumor investigations, Ms. Soon-Hee Sul for undertaking the Monte Carlo simulations, and Professor Eric Hahn for mentoring us in tumor biology.

Author Index

Numbers in parentheses are footnote reference numbers and indicate that an author's work is referred to although the name is not cited in the text.

A

Abdallah, B., 303
Abenhaim, J., 37
Aboagye, E. O., 30, 41
Abramovitch, R., 6
Abramovitz, R., 308
Abrams, T., 193
Abramsky, O., 287
Achilefu, S., 257
Ackerman, E. J., 31, 39(87), 299
Ackerman, J. J., 50, 171, 391, 404
Ackerman, J. L., 12, 13, 13(17)
Ackerstaff, E., 3, 21, 24(41), 41, 49, 52
Acred, P., 181
Adams, G. P., 253(66), 254
Adams, H. P., Jr., 69
Adler, S., 89
Adriaens, P., 245
Afeworki, R., 386
Agarwal, S., 60
Agema, W. R., 127
Aguayo, J. B., 31, 35(91)
Aguinaldo, G., 173
Ahier, R. G., 167, 169
Ahlstrom, H., 99
Ahn, C. B., 31
Ahn, S. J., 42
Ahrens, E. T., 25, 282, 308,
    310(47), 314
Aicardi, G., 305
Aiken, N., 306(27), 307
Akita, Y., 24
Akiyama, Y., 277
Akpalu, Y. A., 285
Akutsu, T., 99
Albensi, B. C., 201
Albers, G. W., 62
Albert, M. S., 179

Aldrich, E. F., 348
Aletras, A. H., 51
Alexander, S., 73
Alfano, R. R., 350
Alger, J. R., 50, 164(33), 166
Al-Hallaq, H. A., 389
Allamand, V., 308
Allegrini, P. R., 178
Allen, B. J., 24
Allen, M., 23, 26(46), 283
Allen, P. M., 254, 259(70), 261(70)
Allison, J. D., 127, 130(7)
Almenar, L., 99
Alperin, N., 323, 327, 331, 334, 335, 338(18),
    339(15), 340(15), 343, 344
Al-Saffar, N. M., 39
Alsop, D. C., 161
Altes, T. A., 34
Althaus, U., 99
Altieri, M., 60
Alvarez, J., 50
Alvord, E.-C. J., 175, 176
Amado, L. C., 292, 293(70; 71)
Amann, F. W., 144
Amedio, J. C., Jr., 22(45), 23
Amemiya, H., 99
Ametamey, S., 253
Amis, E. J., 285
Amor, S., 174
Amrein, C., 99
Anday, E., 350, 354(6)
Andersen, A., 173
Anderson, A.-L. J., 254, 259(78)
Anderson, C. J., 237, 240, 243(15), 244,
    244(15), 245, 246, 247, 249, 251,
    251(36; 55), 252(36), 253, 253(36; 58),
    254, 254(42; 59; 60), 256(72), 257,
    257(42; 59; 71; 73), 258,
    258(59; 71–73; 93), 259(77), 261(93)

419

W

Subject Index

A

Adipose tissue, magnetic resonance imaging
  brown adipose tissue, 196–197
  ob–ob mice, 197–199
  proton imaging, 196
  rationale, 195–196
Alzheimer's disease
  brain gene expression voxelation
    studies, 315
  magnetic resonance imaging, 173
Aminoguanidine
  nitric oxide synthase inhibition, 202
  traumatic brain injury therapy studies, 211
Angiogenesis
  animal window models
    angiogenesis inhibitor
      studies, 112
    arthritis angiogenesis studies, 115
    dorsal skin window chamber model
      intravital microscopy, 107
      preparation, 107
    multiphoton laser-scanning microscopy,
      112, 114
    overview, 106–107
    tumor angiogenesis and development
      imaging parameters
      blood flow, 109
      hypoxia, 109–110
      tumor size, 110
      tumor vascular length density,
        108–109
      vessel dilation, 109
      vessel permeability, 109
      vessel tortuosity, 109
  green fluorescent protein-labeled tumor
    cell studies of host interaction and
    angiogenesis initiation, 110–111
  inhibition rationale in cancer therapy, 105
  magnetic resonance imaging of integrins,
    230–232
  polymerized vesicle nanoparticle
    angiogenesis inhibitors, 232–234, 236

SU6668 inhibitor study of colon carcinoma
  with magnetic resonance imaging,
  190–195
Tie-2–green fluorescent protein transgenic
  mouse studies
  blood flow velocity measurement,
    120–122
  capillary perfusion measurement,
    118–119
  confocal microscopy, 117
  corneal angiogenesis assay, 117–118
Apoptosis
  superparamagnetic iron
    oxide-labeled cells, 297
  traumatic brain injury assay, 210–211
Arrythmogenic right ventricular dysplasia,
  magnetic resonance imaging, 130–131
Arterial spin labeling
  cerebral blood flow measurement with
    magnetic resonance imaging
    continuous arterial spin labeling,
      158–159
    overview, 158
    pulsed arterial spin labeling, 159, 161
  renal perfusion imaging with magnetic
    resonance imaging, 89–90
Arthritis, angiogenesis studies, 115
ARVD, see Arrythmogenic right
  ventricular dysplasia
ASL, see Arterial spin labeling

B

Blood oxygenation level dependent contrast
  functional studies of plasticity in stroke
    recovery, 168–169
  theory, 6–7
  tumor oximetry, 389
BOLD contrast, see Blood oxygenation level
  dependent contrast
Boyden Chamber, Metabolic Boyden
  Chamber assay of cancer cell invasion
  with magnetic resonance imaging, 51–54

457

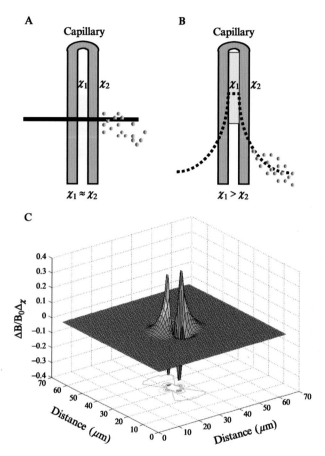

PATHAK *ET AL.*, CHAPTER 1, FIG. 1. Schematic illustrating the premise of the BOLD effect. (A) In the absence of a susceptibility difference between (oxygenated) blood (χ_1) and the surrounding tissue (χ_2), no microscopic field gradient is set up and diffusing water protons "see" the same local magnetic field. (B) When there is a susceptibility difference between (deoxygenated) blood (χ_1) and the surrounding tissue (χ_2), a microscopic field gradient (—) is set up and diffusing water protons "see" different local magnetic fields, leading to loss of phase coherence, reduction in T_2^*, and MR signal attenuation. (C) Surface plot illustrating the three-dimensional aspects of mathematically simulated microscopic field gradients induced around a microvessel.

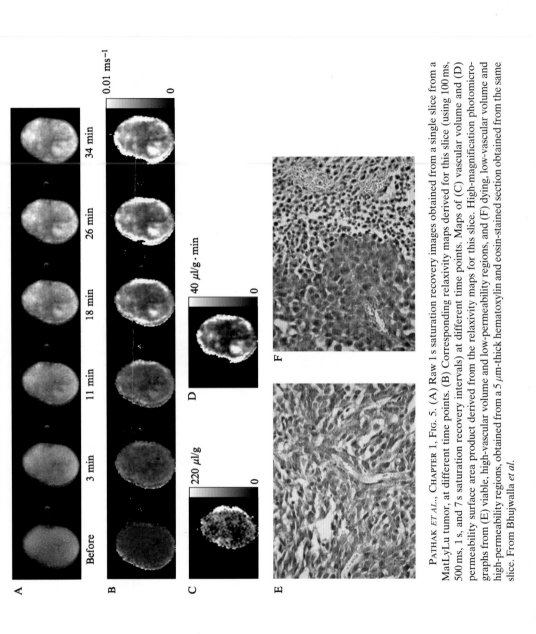

PATHAK *ET AL.*, CHAPTER 1, FIG. 5. (A) Raw 1 s saturation recovery images obtained from a single slice from a MatLyLu tumor, at different time points. (B) Corresponding relaxivity maps derived for this slice (using 100 ms, 500 ms, 1 s, and 7 s saturation recovery intervals) at different time points. Maps of (C) vascular volume and (D) permeability surface area product derived from the relaxivity maps for this slice. High-magnification photomicrographs from (E) viable, high-vascular volume and low-permeability regions, and (F) dying, low-vascular volume and high-permeability regions, obtained from a 5 μm-thick hematoxylin and eosin-stained section obtained from the same slice. From Bhujwalla *et al.*

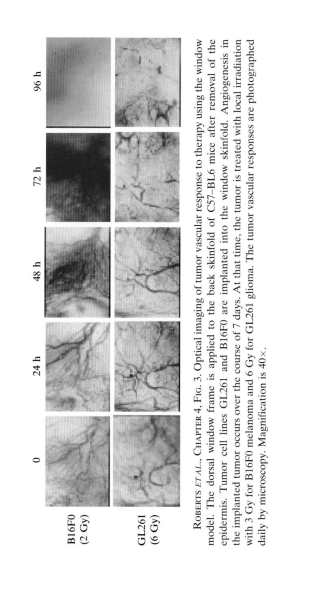

ROBERTS *ET AL.*, CHAPTER 4, FIG. 3. Optical imaging of tumor vascular response to therapy using the window model. The dorsal window frame is applied to the back skinfold of C57–BL6 mice after removal of the epidermis. Tumor cell lines GL261 and B16F0 are implanted into the window skinfold. Angiogenesis in the implanted tumor occurs over the course of 7 days. At that time, the tumor is treated with local irradiation with 3 Gy for B16F0 melanoma and 6 Gy for GL261 glioma. The tumor vascular responses are photographed daily by microscopy. Magnification is 40×.

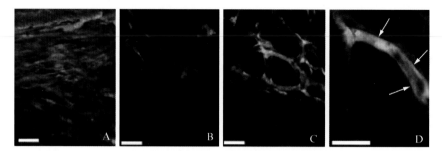

ROBERTS *ET AL.*, CHAPTER 4, FIG. 4. Multiphoton laser-scanning microscopy imaging of tumor angiogenesis. This figure is adapted from Rekesh Jain's publication, with permission. A MCaIV tumor is grown in transgenic mice that express EGFP under the control of the VEGF promoter (green). The tumor vasculature is highlighted by TMR-BSA injected systemically (red). A–B. Two optical sections of the same region within the tumor, at different depths, showing the three-dimensional resolution and depth of penetration of MPLSM. (A) A highly fluorescent (GFP-expressing) layer is seen within the first 35–50 m of the tumor–host interface. (B) 200 m deeper inside the tumor, EGFP-expressing host cells have successfully migrated into the tumor and tend to colocalize with angiogenic blood vessels. (C) A maximum intensity projection of 27 optical sections, spaced 5 m apart and beginning 45 m from the tumor surface. Colocalization of VEGF-expressing host cells with angiogenic vessels is readily apparent. (D) A single optical section of a vessel from (C) taken at twice the magnification. A thin, nonfluorescent layer is apparent at many locations (arrows). Scale bars are 50 μm.

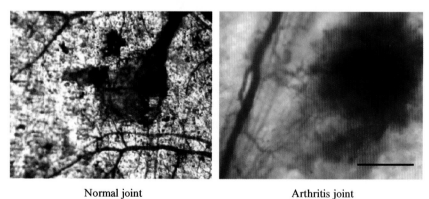

Normal joint   Arthritis joint

ROBERTS *ET AL.*, CHAPTER 4, FIG. 5. Arthritis synovium window chamber model. A novel vascular window model is established on the back skinfold of a DBA mouse to study arthritis-induced angiogenesis. Synovium samples were isolated from a donor mouse paw with collagen-induced arthritis (CIA). A small piece of CIA-synovium (right panel) or a normal joint tissue (left panel) was implanted in the vascular window. The photos were live pictures taken from mouse synovium windows. Magnification is 40×. Bar represents 1 mm.

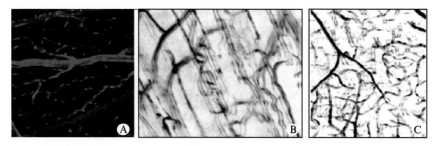

Roberts *et al.*, Chapter 4, Fig. 6. (A) Conjunctiva capillaries in a Tie2–GFP mouse. Often all elements of the microcirculatory system show uniformly strong fluorescence. This image was obtained noninvasively in a lightly anesthetized Tie2–GFP mouse using a Zeiss 510 confocal microscope. (B) Anaglyph showing 3D structure of microcirculatory network in skeletal muscle of Tie2–GFP mouse. The Z-series stack of images was obtained from the medial quadriceps muscle of an anesthetized mouse through the incised skin. Arterioles (A) and venules (V) can be distinguished from capillaries. (C) Anaglyph showing 3D structure of subpial microvessel network in fixed mouse brain.

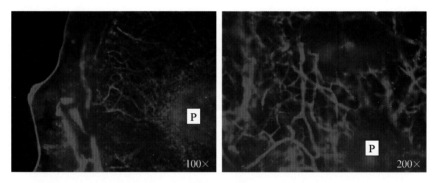

Roberts *et al.*, Chapter 4, Fig. 7. Corneal micropocket assay in Tie2–GFP mice. A small surgical micropocket is created in the Tie2–GFP mouse cornea 1 mm from the limbus, and a small pellet containing bFGF (25 ng/pellet) is implanted in the pocket afterwards. Mouse cornea is harvested 5 days after pellet implantation and flatly mounted on a glass cover slip. Images are taken from fluorescent microscopy. P, Pellet.

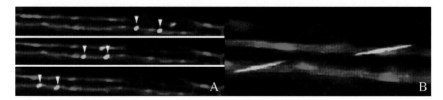

Roberts *et al.*, Chapter 4, Fig. 8. Blood velocity measurements by LSCM. (A) Skeletal muscle capillaries from Tie2–GFP mice are depicted and show three successive frames from a time-lapse sequence, and it is possible to track a pair of fluorescent RBCs in each of the three frames. (B) The same capillaries imaged with slow scan speed show streaked RBCs. In *A*, the direct calculation of $\Delta d/\Delta t$ (comparing top to bottom) measurement of the distance traveled reveals the particle velocity at 0.26 pixel per millisecond. Velocity calculated from $\Delta d/\Delta t$ in streaks in *B* is 0.27 pixel per millisecond.

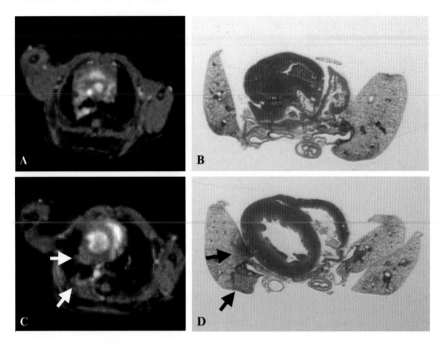

Marzola and Sbarbati, Chapter 7, Fig. 4. Pneumonia. Comparison between (A, C) MRI and (B, D) histologic slices obtained at a similar anatomic position in the same animal. Arrows indicate regions of infection.

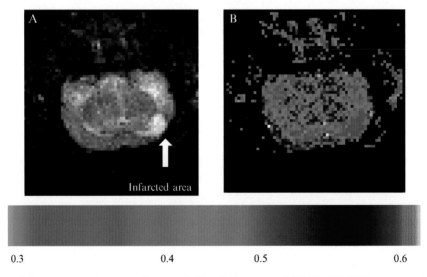

Marzola and Sbarbati, Chapter 7, Fig. 5. Permanent MCAO. (A) T_2W transversal image acquired 30 days after MCAO and (B) rCBV map obtained for the same slice. A region with reduced blood volume is observed in the cerebral cortex. This region corresponds to an hyperintensity area in T_2W.

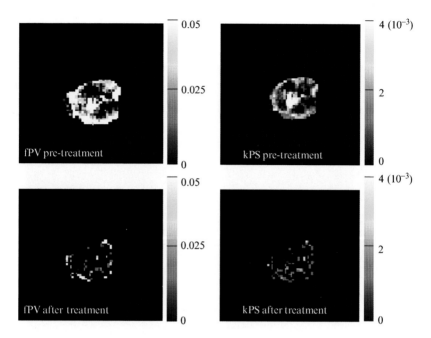

Marzola and Sbarbati, Chapter 7, Fig. 7. Colon carcinoma. *Upper line:* fPV and kPS maps obtained for the slice shown in Fig. 6 (i.e., before treatment). *Lower line:* fPV and kPS maps obtained for the same animal but 24 h after treatment. Adopted from Morzola *et al.*, with permission.

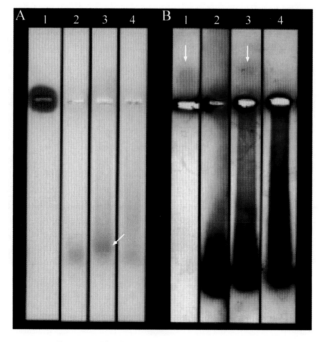

Guccione *et al.*, Chapter 10, Fig. 1. Demonstration of anti-VCAM antibody–avidin conjugation to biotinylated vesicles by immunodetection.

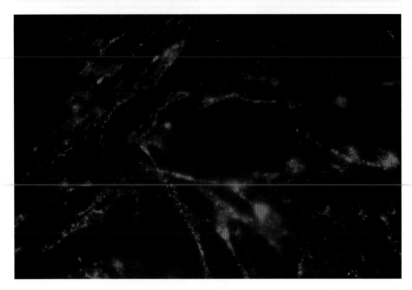

GUCCIONE *ET AL.*, CHAPTER 10, FIG. 2. Fluorescent ACPV labeling of stimulated endothelial cells expressing ICAM-1 cell adhesion molecules.

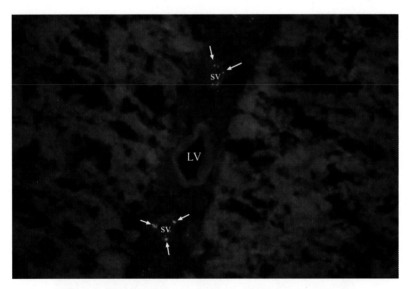

GUCCIONE *ET AL.*, CHAPTER 10, FIG. 3. Fluorescent anti-ICAM-1 ACPVs bind *in vivo* to cerebellar vasculature of a mouse with grade 2 EAE. Anti-ICAM-1 ACPVs (arrows) are shown bound to small vessels (sv), but not to the central large vessel (LV).

Sᴜɴ ᴀɴᴅ Aɴᴅᴇʀsᴏɴ, Cʜᴀᴘᴛᴇʀ 11, Fɪɢ. 1. Gold disk *(top)* and gold disk electroplated with ⁶⁴Ni *(bottom)* for use as a target to prepare ⁶⁴Cu on a biomedical cyclotron via the ⁶⁴Ni(p,n)⁶⁴Cu reaction. Courtesy of Lucie Tang and Michael J. Welch.

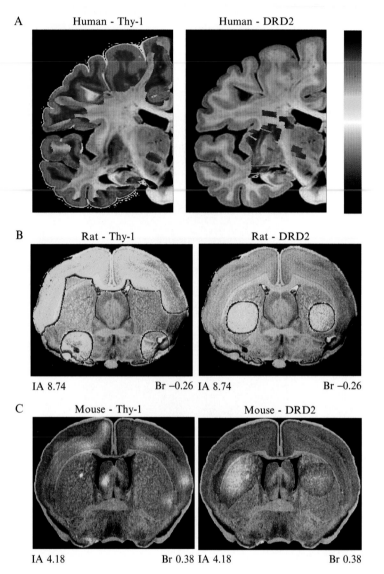

SFORZA AND SMITH, CHAPTER 15, FIG. 1. Human and rodent brain voxelation gene expression images for Thy-1 and DRD2 (dopamine D2 receptor) genes. Images are the result of the methods explained in the text. Gene expression patterns are shown in pseudocolor and smoothed across voxels. Thy-1 is expressed in the cortex, DRD2 in the striatum (caudate/putamen). (A) Human brain, atlas section from Virtual Hospital: The Human Brain (http://www.vh.org). The voxel size was 3.3 mm. (B) Rat brain, coronal atlas section from Paxinos and Watson. The voxel size was 1 mm. Coordinates in mm: IA, interacural; Br, bregma. (C) Mouse brain, coronal atlas section from the Mouse Brain Library (http://www.mbl.org). The voxel size was 1 mm.

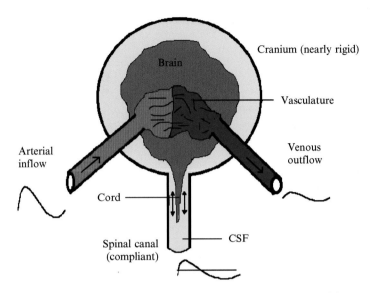

ALPERIN, CHAPTER 16, FIG. 3. The craniospinal flow–volume–pressure model proposed by Alperin *et al.* This model shows the arterial inflow, the venous outflow, and the CSF flow that oscillates between the cranium and spinal canal. During systole, arterial blood inflow is greater than the venous outflow, which is accomodated by the CSF movement to the spinal canal. The CSF moves to the cranium during diastole as the venous outflow becomes greater.

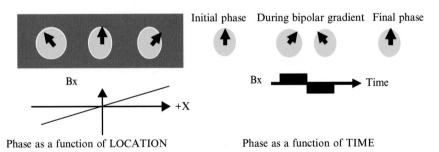

ALPERIN, CHAPTER 16, FIG. 4. A schematic representation of the effect a magnetic field gradient has on stationary protons. *Left:* Protons at different locations along the x-axis accumulate a different phase when a magnetic field gradient is applied. *Right:* Time evolution of the phase of a stationary proton located at $+X$ that is experiencing a bipolar field gradient. The positive phase shift caused by the positive lobe is canceled after the application of the negative lobe, resulting with no phase shift for a stationary proton. A moving proton will accumulate a phase shift that is proportional to its velocity.

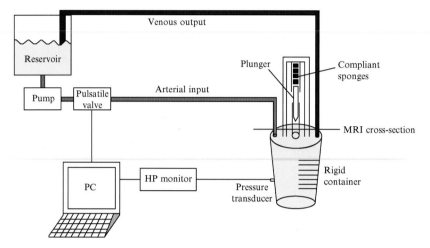

ALPERIN, CHAPTER 16, FIG. 11. The experimental setup of the flow phantom study to quantify the inherent accuracy and reproducibility of the method for ICVC measurement. The rigid container represents the cranial compartment, and the glass syringe attached to the container simulates the CSF channel and spinal canal. A pulsatile flow pump was used to drive the arterial input (red) through the tube to the cranial compartment. The venous outflow (blue) was drained to a reservoir.

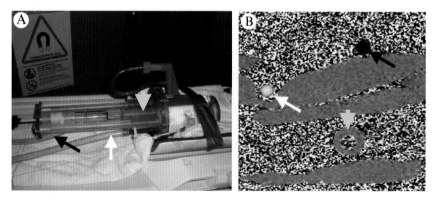

ALPERIN, CHAPTER 16, FIG. 12. (A) A picture of the flow phantom described in the experimental setup shown in Fig. 11. (B) The MRI velocity image for the phantom obtained at the cross section indicated in Fig. 11. The arterial inflow (black arrow), the venous outflow (white arrow), and CSF flow channel (short arrow) are indicated on the physical phantom, as well as on the MRI velocity image.

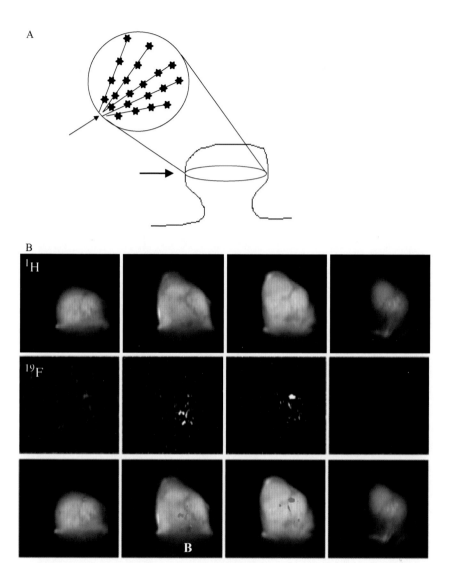

Zhao *ET AL.*, Chapter 18, Fig. 3. (A) Recommended pattern of IT injection. Several tracks in the form of a fan in a single plane. (B) Four contiguous MR images (5 mm thick) of a representative Dunning prostate R3327-AT1 tumor (volume $= 1.6$ cm^3). *Top:* ^{1}H MRI; *middle:* corresponding ^{19}F; *bottom:* overlay to show interrogated regions. *B* marks point of attachment of the tumor to back of rat.

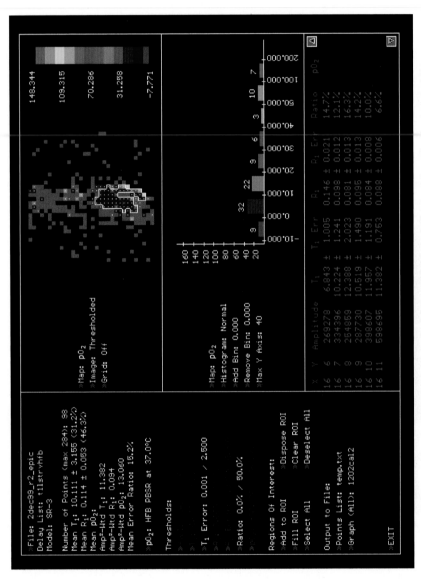

ZHAO *ET AL.*, CHAPTER 18, Fig. 6. Example T₁map used for data filtering, clustering, and analysis.

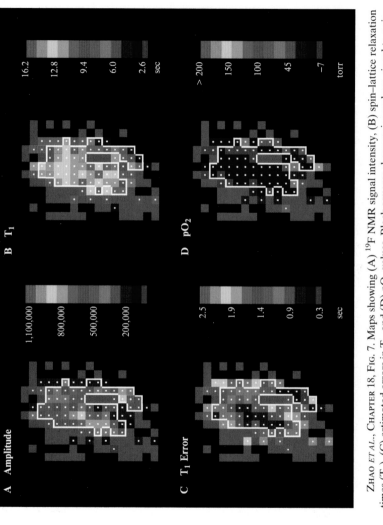

ZHAO *ET AL.*, CHAPTER 18, FIG. 7. Maps showing (A) ^{19}F NMR signal intensity, (B) spin–lattice relaxation times (T_1), (C) estimated errors in T_1, and (D) pO_2 values. Black squares show regions where signal to noise or data quality was so poor as to not provide a T_1 curve fit. Gray voxels show regions that provided a curve fit, but where the uncertainty in the data exceeded the threshold criteria (i.e., $T_{1\ err} > 2.5$ s or $T_{1\ err}/T_1 > 50\%$). All colored voxels ($n = 100$) provided a curve fit within the acceptance criteria in this map, but only those 56 voxels within the white regions of interest (ROI) provided consistently high-quality data throughout the sequence of eight maps acquired during baseline and with oxygen breathing. Data were obtained using the FREDOM approach from a representative Dunning prostate R3327-AT1 tumor (2 cm^3).

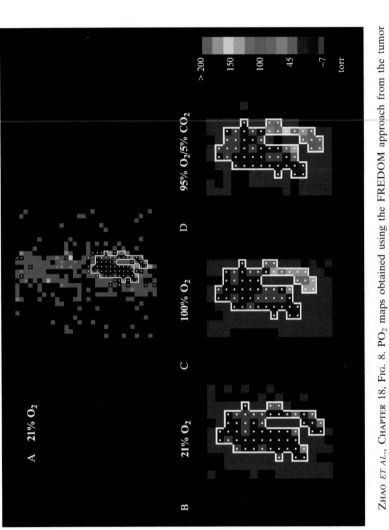

ZHAO *ET AL.*, CHAPTER 18, FIG. 8. PO$_2$ maps obtained using the FREDOM approach from the tumor shown in Fig. 3. (A) Under baseline conditions, those voxels ($n = 56$) within the white region provided consistently reliable data during repeated measurements. (B) Baseline map (breathing air: FO$_2$ = 21%): mean pO$_2$ = 7.2 ± 2.6 (SE) torr, median pO$_2$ = 1.3 torr (range −7–88 torr). (C) Breathing oxygen (FO$_2$ = 100%): fourth map obtained 24–32 min after switching from air: mean pO$_2$ = 47.2 ± 10.7 torr ($p < 0.0001$ compared with baseline), median pO$_2$ = 8.3 torr (range −3–204 torr). (D) Breathing carbogen (FO$_2$ = 95%): fifth map after switching to carbogen: mean pO$_2$ = 43.1 ± 9.2 torr ($p < 0.0001$ compared with baseline), median pO$_2$ = 5.6 torr (range −10–216 torr).